Geology

Geology

Volume 2
Marine Terraces –
Weathering and Erosion
Index
389-752

edited by
James A. Woodhead

SALEM PRESS, INC.
Pasadena, California   Hackensack, New Jersey

Essays originally appeared in *Magill's Survey of Science: Earth Science Series* (1990), consulting editor James A. Woodhead, and *Magill's Survey of Science: Earth Science Series, Supplement*, edited by Roger Smith. New material has been added.

∞ The paper used in these volumes conforms to the American National Standard for Permanence of Paper for Printed Library Materials, Z39.48-1992.

Library of Congress Cataloging-in-Publication Data
Geology / edited by James A. Woodhead
    p.   cm. — (Magill's choice)
Includes bibliographical references and index.
ISBN 0-89356-522-9 (set : alk. paper). — ISBN 0-89356-523-7 (v. 1 : alk. paper). — ISBN 0-89356-524-5 (v. 2 : alk. paper)
    1. Geology—Encyclopedias. I. Woodhead, James A. II. Series.
QE5.G465   1999
550'.3—dc21   98-53009
                                                                              CIP

Second Printing

Contents

Marine Terraces 389
Metamorphic Rocks 398
Meteorite and Comet Impacts 407
Minerals: Physical Properties 415
Minerals: Structure 428
Mountain Belts 438

Normal Faults 448

Ocean Basins 456
The Ocean Ridge System 464
The Oceanic Crust 471
Oil and Gas: Distribution 481
Oil and Gas: Origins 489

Permafrost 498
Plate Tectonics 505
Pyroclastic Rocks 512

Reefs 518
River Flow 526
River Valleys 534
Rock Magnetism 542
Rocks: Physical Properties 550

Sand Dunes 559
Sedimentary Rocks 568
Shield Volcanoes 577
Siliciclastic Rocks 584
Stratigraphic Correlation 593
Stratovolcanoes 599
Stress and Strain 607
Subduction and Orogeny 615
Supercontinent Cycle 623

Thrust Faults 631
Transform Faults 638
Transgression and Regression 646

Ultramafic Rocks 655
Uniformitarianism 665

Volcanoes: Recent Eruptions 675
Volcanoes: Types of Eruption 685

Water-Rock Interactions 692
Weathering and Erosion 701

Glossary 709
Categorized List of Essays 751
Subject Index 755

Geology

Marine Terraces

Marine terraces are ancient (fossil) coastlines that result from several different modes of origin. They are a common coastal feature worldwide and are at elevations well above or well below present-day sea level. Global cycles of sea-level change and tectonic uplift and subsidence of coasts are responsible for creating multiple sets of parallel marine terraces. Global sea-level history and tectonic history of coastal areas can be determined by the study of the age and elevation of marine terraces.

Field of study: Geomorphology

Principal terms

BARRIER ISLAND: a long, low sand island parallel to the coast and separated from the mainland by a salt marsh and lagoon; a common coastal feature on depositional coasts worldwide

BIOABRASION: physical and chemical erosion or removal of rock as a result of the activities of marine organisms

GUYOT: an oceanic volcano, presently submerged far below sea level, with a top that has been beveled flat by wave erosion; the tops of some guyots are capped by coral reefs at or near their summits

INNER EDGE POINT: the landward edge of a wave-cut terrace at the base of the sea cliff; the elevation of this point is the position of highest sea level during formation of the terrace

ISOSTATIC READJUSTMENT: rapid tectonic uplift or subsidence of continental areas in response to the addition or removal of the weight of overlying deposits of glacial ice or seawater

LITHOSPHERE: the outer shell of the earth, where the rocks are less dense but more brittle and coherent than those in the underlying layer (asthenosphere)

NOTCH OR NIP: an erosional feature found at the base of a sea cliff, the result of undercutting by wave erosion, biobrasion from marine organisms, and dissolution of rock by groundwater seepage

PLEISTOCENE EPOCH: the time of earth history, from about 2 million to 10,000 years ago, during which the earth experienced cycles of warming and cooling, resulting in cycles of glaciation and sea level change

STRANDLINE: the position or elevation of the portion of the shoreline between high and low tide (at sea level); usually synonymous with "beach" and "shoreline"

SUBSIDENCE: the sinking of a block of the earth's lithosphere because of a force pushing it down; coastal areas undergoing rapid subsidence tend to be submerged below sea level

TECTONIC: pertaining to large-scale movements of the earth's lithosphere

UPLIFT: the rising of a block of the earth's lithosphere because of a force pushing it up; coastal areas undergoing uplift tend to emerge above sea level

WAVE-CUT BENCH: a gently seaward-sloping platform cut into the bedrock of a coast by wave erosion and landsliding; wave-cut benches are proof of sea-level variations and tectonic uplift and subsidence of coastal areas

Summary

The coastline of continents and islands represents a fundamental boundary between the earth's solid landmass and the constructional and destructional energy of the sea. The landforms in coastal areas are the result of continuous dynamic interaction between these competing geological agents. Marine processes of erosion and sedimentation construct a shoreline profile on the edge of the landmass that defines the strandline, a narrow zone of wave- and tide-washed coast. The coastal strandline usually exhibits a gently sloping platform with its top at an elevation between high and low tides; it is often bounded by adjacent steeper slopes and is a reliable indicator of sea level. Familiar strandline features are beaches, coral reefs, and wave-cut platforms, strandline platforms cut into the bedrock by wave erosion.

If erosion and sedimentation were the only active geological conditions, most coasts would have a single type of strandline landform whose position remained constant through time. The volume of seawater in the world's oceans, however, fluctuates directly with changes in the volume of ice in continental glaciers, causing sea level to rise and fall in response. As the average temperature of the earth's atmosphere rises or falls, glacial ice on the continents melts (sea-level rise) or accumulates (sea-level fall) in response to the temperature change. The temperature of the surface ocean reacts in a similar manner: Warming causes expansion of the seawater (sea-level rise); cooling results in contraction (sea-level fall). It has been estimated from oxygen isotope evidence in deep-sea cores that worldwide sea level has been 430-530 feet lower than today several times during the last 2-3 million years of earth history. There is also abundant physical evidence for large-scale sea-level excursions. On the Atlantic coast of the United States, elephant teeth 25,000 years old have been dredged up by fishermen from more than forty locations on the continental shelf, some as far as 75 miles offshore. Three-thousand-year-old oak tree stumps rooted in life position and many cultural artifacts used by coastal American Indians have also been found below sea level in this region.

The land masses are also surprisingly dynamic, rising or falling in elevation in response to a variety of tectonic factors. On some coasts, isostatic readjustment results in uplift and subsidence. The weight of increasing thicknesses of overlying glacial ice or seawater will cause the earth's litho-sphere to be depressed downward; when the weight is removed, the land-mass will rebound upward. Volcanic activity will cause expansion (uplift) when the earth's lithosphere is heated; contraction (subsidence) occurs as the lithosphere cools. An extreme example of this effect is the guyot, an inactive volcanic seamount that has subsided far below sea level as the hot lithosphere on which it formed cooled and contracted. Guyots have been beveled flat by wave erosion and rimmed with banks of dead coral reef during the period when their tops were in shallow water. Coastal landmasses also undergo uplift and subsidence in response to the forces of plate tectonic movement. Although usually a slow, incremental process, a sudden, rapid coastal uplift occurred in Prince William Sound, Alaska, during the 1964 "Good Friday Earthquake," which resulted in elevation of a wave-cut platform one-quarter mile wide.

During the Pleistocene epoch (about 2 million to 10,000 years ago), strandline features formed on coasts worldwide that were strongly influ-enced by these complex and rapidly changing geologic conditions. Most coasts are not tectonically stable (located at one elevation) for a very long period of geologic time, and processes of landscape erosion and decay often act much more slowly than the rates of sea-level and land-level change. Abandoned strandline features, or "fossil coasts," are common to many areas of the world. First recognized in nineteenth century Europe, parallel sets of these distinctive landforms are present on cliffs and coastal hills at elevations high above present-day sea level and on the ocean floor far below present-day sea level. Many contain marine fossils (discovered by Leonardo da Vinci about A.D. 1500), demonstrating their marine origin, and create a steplike topographic profile of the terrain, leading up the seaward-facing hills. These features are referred to as marine terraces and are found on many coasts worldwide both above and below sea level.

Because their formation requires time, strandline landforms usually form when coastal uplift/subsidence and sea-level change are in balance and the position of the strandline is relatively stable for a period of thousands of years, a situation referred to as a sea-level stillstand. In many areas, sea level tends to vary in cycles separated by stillstand events, during which marine terraces can develop. Occasionally, the processes of terrace formation will destroy or obscure an older terrace, so most coasts exhibit only one or two terraces. An unbroken flight of marine terraces climbing inland indicates a coast that is being continuously uplifted. One of the most striking and well-developed sequences of marine terraces in the world is on the seaward slopes of the Palos Verdes Peninsula of Southern California. Here, thirteen distinct terrace levels, from 50 to 1,480 feet in elevation, form a complete record of coastal uplift and sea-level change spanning several hundred thousand years.

Submarine terraces have been reported from many parts of the world at depths several hundred feet below present-day sea level. Submerged terraces are usually better preserved than are elevated marine terraces because, being underwater, they are less affected by processes of erosion and landscape degradation. Because they are obscured from view and more difficult to study than are terraces on land, less is known about them. The most distinctive submerged wave-cut platforms are present at depths of −130 to −200 feet off the southwest coast of Great Britain, at −560 to −660 feet off northern Australia, and a sequence of at least five terraces at depths of −500 to −700 feet off the Southern California coast and its nearshore islands from Santa Barbara to San Diego. Study of marine terraces at these anomalous elevations worldwide has been used to document sea-level variations and the tectonic uplift and subsidence of coastal landmasses through time and to estimate the magnitude of these changes.

There are several types of marine terraces, differing in their mode of formation. Processes of coastal erosion form marine terraces on coasts where the shoreline is backed by a steep, irregular cliff, where the supply of sediment is too limited to build beaches, and where nearshore waters are so shallow that waves break directly on the sea cliff. The base of the cliff is continuously pounded by the full force of the waves, resulting in very large impact pressures. Beach cobbles and abrasive sandladen water are hurled at the base of the cliff. The force with which these stones attack the cliff is difficult to exaggerate; there are reports of beach cobbles thrown 150 feet above sea level and of others raining down on the roof of a lighthousekeeper's cottage.

Wave-cut platforms result from the rapid horizontal cutting away of rock at the base of a sea cliff. Wave erosion quarries a notch into the base of the sea cliff, which undercuts and oversteepens the cliff until landsliding causes the cliff to collapse; this loose rubble is rapidly carried away by the waves and is a source of new rocks to continue wave attack. In some places, notch formation is aided by the bioabrasion activities of marine organisms that live attached to and feed on rocky shores and that secrete chemicals to help dissolve the rock or abrade it away with rasping feeding appendages. Dissolution of the rock is sometimes aided by groundwater seeping out of the base of the sea cliff. The net effect of this process is the landward retreat of the sea cliff and formation of a platform that slopes gently seaward (0-3 degrees). The boundary between the "step" of a wave-cut terrace and the steeper "riser" of the sea cliff that terminates its landward edge is called the inner edge point. The elevation of this point is the position of highest sea level during formation of the feature, and it is used to calculate the height of ancient sea levels in marine terraces.

The platform is continually abraded by the waves transporting sand down the coast, and its width is determined by the amount of landward sea-cliff erosion. The rate of sea-cliff retreat can be surprisingly rapid. Average rates on the California coast exceed 6 inches per year for hard rock cliffs and up to 3 feet per year in cliffs composed of soft, unconsolidated sediment. The coast of East Anglia, Great Britain, experiences up to 13 feet of sea-cliff retreat annually. Continuous records are available for the Huntcliff coast of Yorkshire, Great Britain, at a former Roman signal station where the cliff has retreated 100 feet in eight hundred years.

Overlying many elevated wave-cut platforms is a mantle of unconsolidated sediment, referred to as coverhead. This material, deposited by a combination of terrestrial and marine sedimentary processes, buries the abraded platform surface and may be as much as 100 feet thick. Typical coverhead deposits include a basal layer of well-sorted beach sand, rounded beach cobbles, and marine fossils—the shoreline sediments left on the platform as sea level retreated. These basal strandline sediments are often covered by a heterogeneous deposit composed of poorly sorted rock debris, soil, and stream gravel deposited by sediment washing or landsliding down from the sea-cliff slopes above and by coastal streams building their alluvial fans toward the coast.

Marine terraces are also formed by sediment deposition, which usually occurs on relatively flat coasts with a wide continental shelf where the energy of approaching waves is dissipated by friction with the shallow sea bottom. In areas such as the Atlantic and Gulf coasts of North America and the Netherlands, rivers transport large quantities of mud and sand into coastal waters. Waves and wind currents pile sand into a long, narrow strandline feature known as a barrier island, a common landform found fringing low-lying coasts worldwide. Once formed during a stillstand of several thousand years, a barrier island will migrate landward as sea level rises. Waves and tidal currents cause barrier sand to wash over the top of the island or around its ends, moving it grain-by-grain landward with the rising sea. When sea level begins to fall, the barrier does not migrate seaward but is left behind inland of the new strandline, where it remains as a record of the former high stand of sea level; this feature is usually referred to as a beach ridge. Subsequent to sea level falling and the strandline moving seaward, river and marsh sediment is often deposited on the seaward side of the abandoned barrier. When sea level again rises to its former position, a new barrier island migrates with it; this island is, in turn, abandoned on the coast at the highest position of sea level. If the new barrier migrates far enough inland to reach the older remnant barrier, it will be welded onto it, forming a wide composite barrier island composed of two or more barrier islands.

Throughout the Pleistocene, sediment deposition has widened the Atlantic and Gulf coastal plains, and repeated cycles of sea-level rise and fall have formed concentric arcs of abandoned barrier islands stretching inland more than 30 miles from the present-day coast. These beach ridge barrier islands are a type of marine terrace, each recording an ancient high stand of sea level. One of the most prominent is Trail Ridge in southern Georgia, a sand ridge more than 36 miles long that encloses a low swampy area on its landward side known as the Okefenokee Swamp. Submarine beach ridges are less common than their inland counterparts but have been reported from western Brittany at depths of up to −660 feet.

In tropical areas where the coastal waters are warm all year and clear of suspended mud, corals will flourish and often form massive reefs. Reefs that grow during a long stillstand will become large and well developed, with a gently sloping top that corresponds to sea level. These coastal depositional features are also found stranded above the shoreline when sea level falls or the landmass rises, resulting in reef terraces. This type of marine terrace is most common on island coasts but is also present on the coasts of the Mediterranean and Red Seas and on the Yucatan Peninsula of Mexico.

Methods of Study

The location, elevation, and shape of marine terraces are determined by standard field-mapping and land-surveying techniques and by study of aerial photographs and satellite imagery. The time of formation, or age, of marine terraces is routinely determined by analyzing fossils found in terrace sediments with any of several reliable methods. Some terraces, particularly older ones, cannot be accurately dated because they do not contain well-preserved fossils. Carbon 14 is the most commonly used radioactive isotope for dating organic remains. The carbon 14 technique is highly accurate if done on pristine fossil material, and it has been used to date relatively young terraces containing corals, mollusks, and fossilized plant matter found within cover-head or reef deposits. This technique is expensive, and when samples have been chemically altered since they were formed, the technique can be unreliable. A less precise but useful and less expensive method has been developed using the amino acids present in the protein "glue" (called conchiolin) found within the calcium carbonate shell structure of marine mollusks. The amino acid dating technique takes advantage of the alteration of amino acid molecules within the conchiolin as they age. Because the rate of alteration varies with climate and other local factors, a standard dating curve must be constructed for each region in which the method will be used. This curve is usually calibrated using carbon 14 dates, and all other samples are compared to it to ascertain their age. Both of these methods are limited by the short half-life of the carbon 14 isotope (5,730 years) and can be used only on terraces less than about 60,000 years old. Older terraces are dated using other radioactive isotopic techniques. The uranium decay series is particularly effective and is widely employed on both coral and mollusk fossils. Using the thorium-ranium and protactinium-uranium ratios, marine terraces as old as 640,000 years have been documented.

Knowledge of marine terrace ages has proved useful in land-use planning applications, such as evaluating whether particular earthquake faults or coastal landslides have been recently active, but their greatest contribution has been in the study of sea-level change during the Pleistocene. Once the terraces from a single stretch of coast are dated, a simple graph is constructed using the present-day elevation of the inner edge point (for wave-cut terraces) or its upper surface (for reef and beach ridge terraces). The points on the graph show the location of the strandline relative to the coastal

The Palos Verdes Peninsula coastline of Southern California shows thirteen distinct terrace levels. *(R. Kent Rasmussen)*

land surface during sea-level stillstands at various times during the last several hundred thousand years; the curve that connects these points is an estimate of the relative sea-level variations for that location.

Context

Because marine terraces at successively higher elevations are increasingly older, a simple conclusion is that ancient high stands of sea level stood at elevations much higher than sea level today and that sea-level oscillations have become less and less pronounced with time. When the marine terrace sea-level curves from many areas worldwide are compared with one another, however, there is virtually no correlation between the various curves. This finding shows that the world's coastal landmasses are rising and subsiding independent of one another and of global sea-level oscillations. Studies of late Pleistocene sea-level history using independent evidence indicate that the highest sea-level stillstands within the last 400,000 years were only about 5-10 feet higher than present sea level. This demonstrates that coastal areas with late Pleistocene marine terraces are being tectonically uplifted at rates exceeding the rate of sea-level rise.

There have been many attempts to separate the confusing and often complicated effects of coastal uplift from the Pleistocene oscillations of sea level by analysis of marine terrace sea-level curves. The best results have come from studies on coasts that have experienced steady tectonic uplift during the Pleistocene. In this situation, once a terrace forms at a high stand of sea level, it is raised up above the destructional forces of the sea as the uplifted coast steadily emerges. The island of Barbados in the eastern Caribbean and the Huon Peninsula in New Guinea both have spectacular flights of well-preserved marine terraces. The marine terrace sea-level curves derived from these coasts show cycles of sea-level rise and fall superimposed on a fairly simple uplift history. As expected, the terraces were arranged in a pattern of increasing elevation with increasing age. Using reliable independent evidence of Pleistocene glacial-ice and ocean-volume changes derived from oxygen isotopes in deep-sea cores, average rates of constant tectonic uplift were first assumed and later were calculated for both areas. When the amount of tectonic uplift was subtracted for each terrace, the resulting oscillations of the marine terrace sea-level curves showed a cyclic pattern that corresponds remarkably well with global climatic fluctuations during the last 400,000 years. They also show that sea level has not been significantly higher than its present-day elevation for the last 150,000 years. The timing of these sea-level cycles corresponds very well with variations in the earth's orbital movements about the sun, a suspected cause of Pleistocene glacial cycles (ice ages) and sea-level change. Because much is known about the history of Pleistocene sea-level variations from marine terrace data and other types of studies, dates of marine terraces are also used to determine the tectonic uplift or subsidence history of coastal areas. In tectonically active areas, such as the coast and offshore islands of Southern California, coastal uplift rates several feet per century have been documented by the combined relationships of the elevation and age of marine terraces.

Bibliography

Davis, J. L. *Geographical Variation in Coastal Development*. London: Longman Group, 1977. A general text comparing geomorphic features in coastal regions worldwide.

Hearty, P. J., and P. Aharon. "Amino Acid Chronostratigraphy of Late Quaternary Coral Reefs: Huon Peninsula, New Guinea, and the Great

Barrier Reef, Australia." *Geology* 16 (July, 1988): 579-583. A technical article describing the application of the amino acid dating technique to the classic coral reef terraces of the Huon Peninsula and a comparison of the results with uranium-series ages. *Geology* is a monthly publication of the Geological Society of America.

Moore, W. S. "Late Pleistocene Sea-Level History." In *Uranium Series Disequilibrium: Applications to Environmental Problems*, edited by M. Ivanovich and R. S. Hannah. Oxford, England: Clarendon Press, 1982. Half of this chapter in an advanced textbook is a description of state-of-the-art terrace dating techniques written in a technical style. Included is a summary of all known dates for fossil-bearing marine terraces all over the world that is slightly easier to read.

Oaks, R. Q., Jr., and N. K. Coch. "Pleistocene Sea-Levels: Southeastern Virginia." *Science* 140 (May 13, 1963): 979. A technical article containing an excellent description of uplifted beach ridges on the Atlantic coastal plain of the United States. Also describes ages and dating techniques and their use in determining sea-level history.

Pethic, J. *An Introduction to Coastal Geomorphology*. Baltimore: Edward Arnold, 1984. Written primarily for the Western European region, this advanced text contains several good sections on the formation of marine terraces, their use in sea-level and tectonic uplift studies, and the correlation of terrace data with other studies.

Sharp, R. P. *Coastal Southern California*. Dubuque, Iowa: Kendall/Hunt, 1978. Written by a legendary Southern California field geologist and teacher, this guide is easily understood and used by the nongeologist. Contains the route and descriptions for several day trips of geological investigation that any active and curious individual would find worthwhile.

Sheldon, J. S. *Geology Illustrated*. San Francisco: W. H. Freeman, 1966. Well illustrated with dramatic black-and-white photographs (both ground level and aerial), this book uses an encyclopedia-like format to describe a wide variety of geological processes and landforms. Sheldon is known all over the world for his excellent aerial photography and his clear, concise treatment of geological information in his educational films and materials.

Shepard, F. P. *Geological Oceanography*. New York: Crane, Russak, 1977. A classic and very readable text, with descriptions and explanations of a variety of coastal features by one of the greatest oceanographers in the history of the science.

"Uplifted Marine Terraces." *Science*, December 5, 1986: 1225-1229.

James L. Sadd

Cross-References

Coastal Processes and Beaches, 71; Continental Glaciers, 93; Reefs, 518; Transgression and Regression, 646.

Metamorphic Rocks

Metamorphic rocks bear witness to the instability of the earth's surface. They reveal the long history of interaction among the plates that comprise the surface and of deep-seated motions within the plates. Among the metamorphic rocks are found many ores and stones of value to human civilization.

Field of study: Petrology

Principal terms

CONTACT METAMORPHISM: metamorphism characterized by high temperature but relatively low pressure, usually affecting rock in the vicinity of igneous intrusions

FOLIATION: a texture or structure in which mineral grains are arranged in parallel planes

METAMORPHIC FACIES: an assemblage of minerals characteristic of a given range of pressure and temperature; the members of the assemblage depend on the composition of the protolith

METAMORPHIC GRADE: the degree of metamorphic intensity as indicated by characteristic minerals in a rock or zone

METAMORPHISM: changes in the structure, texture, and mineral content of solid rock as it adjusts to altered conditions of pressure, temperature, and chemical environment

PELITIC ROCK: a rock whose protolith contained abundant clay or similar minerals

PRESSURE-TEMPERATURE REGIME: a sequence of metamorphic facies distinguished by the ratio of pressure to temperature, generally characteristic of a given geologic environment

PROTOLITH: the original igneous or sedimentary rock later affected by metamorphism

REGIONAL METAMORPHISM: metamorphism characterized by strong compression along one direction, usually affecting rocks over an extensive region or belt

TEXTURE: the size, shape, and relationship of grains in a rock

Summary

A significant part of the earth's surface is made up of rocks quite different from sedimentary or igneous rocks. Many of them have distinctive textures and structures, such as the wavy, colored bands of gneiss or the layered mica flakes of schist. They often contain certain minerals not found or not common in igneous or sedimentary rocks, such as garnet and staurolite. Studies of their overall chemical composition and their relations to other rocks in the field show that they were once igneous or sedimentary rocks, but, being subjected to high pressure and temperature, they have been distorted and altered or recrystallized through a process called metamorphism. Metamorphism involves both mechanical distortion and recrystallization of minerals present in the original rock, the protolith. It can cause changes in the size, orientation, and distribution of grains already present, or it can cause the growth of new and distinctive minerals built mostly from materials provided by the destruction of minerals that have become unstable under the changed conditions. The chemical components in the rock are simply reorganized into minerals that are more stable under higher pressure and temperature.

Metamorphic rocks can be classified in a purely descriptive fashion according to their textures and dominant minerals. Because the growing understanding of metamorphic processes can be applied to interpret the origin and history of the rocks, they are also classified according to features related to these processes. The most common classification schemes, in addition to the purely descriptive, categorize the rocks by general metamorphic processes, or metamorphic environments; by the original rocks, or protoliths; by metamorphic intensities, or grades; by the general pressure and temperature conditions, called facies; and by the ratios of pressure to temperature, called pressure-temperature regimes.

The oldest classification is purely descriptive, based on the rock texture (especially foliation) and mineral content. Foliation is an arrangement of mineral grains in parallel planes. The most common foliated rocks are slate, schist, and gneiss. In slate, microscopic flakes of mica or chlorite are aligned so that the rock breaks into thin slabs following the easy cleavage of the flakes. Schist contains abundant, easily visible flakes of mica, chlorite, or talc arranged in parallel; it breaks easily along the flakes and has a highly reflective surface. Gneiss contains little mica, but its minerals (commonly quartz, feldspar, and amphibole) are separated into different-colored, parallel bands, which are often contorted or wavy. The foliated rocks can be described further by naming any significant minerals present, such as "garnet schist." Nonfoliated metamorphic rocks lack parallel structure and are usually named after their dominant minerals. Common types are quartzite (mostly quartz), marble (mostly calcite), amphibolite (with dominant amphibole), serpentinite (mostly serpentine), and hornfels (a mixture of quartz, feldspar, garnet, mica, and other minerals). Quartzite breaks through its quartz grains, whose fracture surfaces give the break a glassy sheen. Marble breaks mostly along the cleavage of its calcite crystals, so that each flat cleavage surface has its own glint. Hornfels often exhibits a smooth fracture with a luster reminiscent of a horn.

Metamorphic Rock Classification
Based on Texture and Composition

Texture	Nonfoliated			Foliated			
	Nonlayered			Layered	Nonlayered		
	fine to coarse grained	fine to coarse grained	fine grained	coarse grained	coarse grained	fine grained	very fine grained
Composition	calcite					chlorite	
			mica			mica	
				quartz			
				feldspar			
					amphibole		
					pyroxene		
Name	MARBLE	QUARTZITE	HORNFELS	GNEISS	SCHIST	PHYLLITE	SLATE

The second classification of metamorphic rocks is based on the general processes that formed them or the corresponding environments in which they are found. The recognized categories are usually named regional, contact, cataclastic, burial, and hydrothermal metamorphism.

Regional metamorphism is characterized by compression along one direction that is stronger than the pressure resulting from burial. The compression causes foliation, typified by the foliated rocks slate, schist, and gneiss. These rocks are found in extensive regions, often in long, relatively narrow belts parallel to folded mountain ranges. According to the theory of plate tectonics, folded mountain ranges like the Appalachians and the Alps began as thick beds of sediments deposited in deep troughs offshore from continents. The sediments were later caught up between colliding continents, strongly compressed, and finally buckled up into long, parallel folds. The more deeply the original sediment is buried, the more intense is the metamorphism. Clay in the sediment recrystallizes to mica, oriented with the flat cleavage facing the direction of compression. Thus, pelitic or clay-bearing sediments become slates and schists with foliation parallel to the folds of the mountains. At higher temperatures, the mica recrystallizes into feldspar, and the feldspar and quartz migrate into light-colored bands between bands of darker minerals so that the rock becomes gneiss. At yet higher temperatures, some of the minerals melt (a process called anatexis), and the rock, called a migmatite, becomes more like the igneous rock granite.

Contact metamorphic rocks are commonly found near igneous intrusions. Heat from the intrusive magma causes the surrounding rock, called country rock, to recrystallize. Though the rock is under pressure because of burial, there is usually no tendency toward foliation because the pressure is equal from all directions. Some water may be driven into the rock through fine cracks or, conversely, water may be driven from the rock by the heat. Especially mobile atoms such as potassium can migrate into the rock and combine with its minerals to form new crystals (a process called metasomatism). Usually, however, the chemical content is not greatly altered, and recrystallization chiefly involves atoms from smaller crystals or the cement migrating into larger, crystals or forming new minerals. In quartz sandstone, for example, the quartz crystals grow to fill all the pore space in a tight, polygonal network called crystalloblastic texture, and the rock becomes quartzite. Similarly, the tiny crystals in limestone or dolomite grow into space-filling calcite crystals, forming marble or, if other minerals are present, the mixed rock called skarn. Pelitic rocks recrystallize to hornfels, containing a variety of minerals, such as quartz, feldspar, garnet, and mica.

Cataclastic (or dynamic) metamorphism occurs along fault zones, where both the rock and individual grains are intensely sheared and smeared out by stress. In deep parts of the fault, the sheared grains recrystallize to a fine-grained, finely foliated rock called mylonite.

Burial metamorphism occurs in very deep sedimentary basins, where the pressure and temperature, along with high water content, are sufficient to form fine grains of zeolite minerals among the sedimentary grains. The process is intermediate between diagenesis, which makes a sediment into a solid rock, and regional metamorphism, in which the texture of the rock is modified.

Hydrothermal metamorphism (which many prefer to call "alteration") is caused by hot water infiltrating the rock through cracks and pores. It is most common near volcanic and intrusive activity. The water itself, or substances dissolved in the water, may be incorporated into the crystals of certain minerals. One important product is serpentine, formed by the addition of water to olivine and pyroxene, which is significant in sub-sea-floor metamorphism.

A third classification is based on the original rocks, or protoliths. This classification is possible because relatively little material is added to or lost from the rock during metamorphism, except for water and carbon dioxide, so the assemblage of minerals present depends on the overall chemical composition of the original rock. The most abundant protoliths are pelitic rocks (from clay-rich sediments, usually with other sedimentary minerals), basaltic igneous rocks, and limestone or other carbonate rocks. Each kind of protolith recrystallizes into a different characteristic assemblage of minerals. The categories can be named, for example, metapelites, metabasalts, and metacarbonates.

Metamorphic intensity, or grade, is the basis for a fourth classification scheme. As the pressure and temperature increase, certain minerals become unstable, and their chemical components reorganize into new minerals more stable in the surrounding conditions. The presence of certain minerals, called index minerals, therefore indicates the intensity of pressure and temperature. The grades most commonly used are named for index minerals in pelitic rocks; in the late nineteenth century, they were described by

George Barrow in zones of metamorphic rocks in central Scotland. These Barrovian grades, in order of increasing intensity, are marked by the first appearances of chlorite, biotite, garnet, staurolite, kyanite, and sillimanite.

A fifth classification scheme categorizes the rocks according to the intensity of pressure and temperature, or facies, without reference to protoliths. The concept of facies was developed by Penti Eskola, working in Finland about 1915, who enlarged on the work of Barrow in Scotland and V. M. Goldschmidt in Norway. Eskola realized that each protolith has a characteristic mineral assemblage within a given facies, or range of pressure and temperature. The facies are named for one of the assemblages within a specified range of conditions; for example, the greenschist facies, named for low-grade metamorphosed basalt, refers to equivalent low-grade assemblages from other protoliths as well. Other examples are the amphibolite facies, the range of pressures and temperatures that would give the staurolite and kyanite grades in pelitic rocks; and the granulite facies, which corresponds to extreme conditions bordering on anatexis.

Finally, the facies themselves, or the metamorphic assemblages in them, can be classified according to ratios of pressure to temperature, called pressure-temperature regimes. The greenschist, amphibolite, and granulite facies include mineral assemblages of increasing metamorphic intensity whose pressure and temperature rise together approximately as they would with increasing depth under most areas of the earth's surface. This sequence is sometimes referred to as the Barrovian pressure-temperature regime because Barrow's metamorphic grades in pelitic rocks fall in these facies. In another regime, called the Abukuma series after an area in Japan, the temperature is much cooler for any given pressure. The blueschist facies, characterized by blue and green sodium-rich amphiboles and pyroxenes, is typical of this series. The converse situation, a regime in which temperature rises much faster than pressure, corresponds to contact metamorphism, and is called the hornfels facies.

Methods of Study

Initial studies of metamorphic rocks are almost always done in the field. The tectonic or structural nature of the region suggests the processes to which the rock has been subjected. For example, an area of folded mountains can be expected to exhibit regional metamorphism; a volcanic area, some contact metamorphism; and a fault zone, some cataclastic metamorphism. Some features are obvious at the scale of an outcrop, such as banding and foliation, or the halo of recrystallized country rock abutting an igneous intrusion. Some textural features, such as foliation, large crystals such as the garnet in garnet schist, or the luster of a fracture surface as in quartzite or marble, are easily seen in a hand specimen. Similarly, a preliminary estimate of mineral content can be made from a hand specimen.

Many features, however, are best seen in thin section under a petrographic microscope. Usually all but the finest grains can be identified. From the relative abundance of the various minerals and their known chemical compositions the overall chemical composition of the rock can be calculated. The protolith can then be identified by comparing the calculated composition to the known compositional ranges of igneous and sedimentary rocks.

Textures seen under the microscope reveal much about the history of the rock. Foliation, for example, usually indicates regional metamorphism. A space-filling, polygonal texture can show contact or hydrothermal recrystallization, and a crumbled, smeared-out cataclastic texture indicates faulting.

Certain minerals such as staurolite or kyanite indicate the grade of metamorphism. Metamorphism is a slow process, however, especially at low temperatures, and conditions sometimes change too rapidly for the mineral assemblage to come to equilibrium. It is not uncommon to find crystals only

---

REACTIONS THAT FORM METAMORPHIC ROCKS

1.
 - a. cherty dolostone + vapor → marble + vapor
 - b. 5 dolomite + 8 quartz + water → tremolite + 3 calcite + 7 carbon dioxide
 - c. $5CaMg(CO_3)_2 + 8SiO_2 + H_2O \rightarrow Ca_2Mg_5Si_8O_{22}(OH)_2 + 3CaCO_3 + 7CO_2$

2.
 - a. peridotite + vapor → verde antique marble
 - b. 4 olivine + 4 water + 2 carbon dioxide → serpentine + 2 magnesite
 - c. $4Mg_2SiO_4 + 4H_2O + 2CO_2 \rightarrow Mg_3Si_2O_5(OH)_4 + 2MgCO_3$

3.
 - a. peridotite + vapor (with dissolved silica) → serpentinite
 - b. 3 olivine + 4 water + silica → 2 serpentine
 - c. $3Mg_2SiO_4 + 4H_2O + SiO_2 \rightarrow 2Mg_3Si_2O_5(OH)_4$

4.
 - a. cherty dolostone + vapor → soapstone + vapor
 - b. 3 magnesite + 4 quartz + water → talc + 3 carbon dioxide
 - c. $3MgCO_3 + 4SiO_2 + H_2O \rightarrow Mg_3Si_4O_{10}(OH)_2 + 3CO_2$

5.
 - a. high-aluminum shales → kyanite schist
 - b. kaolinite-clay → 2 kyanite + 2 quartz + 4 water
 - c. $Al_4Si_4O_{10}(OH)_8 \rightarrow 2Al_2SiO_5 + 2SiO_2 + 4H_2O$

6.
 - a. cherty limestone → marble + vapor
 - b. calcite + quartz → wollastonite + carbon dioxide
 - c. $CaCO_3 + SiO_2 \rightarrow CaSiO_3 + CO_2$

7.
 - a. sodium-rich igneous felsite → blueschist
 - b. albite (feldspar) → jadeite + quartz
 - c. $NaAlSi_3O_8 \rightarrow NaAlSi_2O_6 + SiO_2$

8.
 - a. sedimentary clay-rich shale → corundum-bearing garnet schist
 - b. 6 staurolite → 4 garnet + 12 kyanite + 11 corundum + 3 water
 - c. $6Fe_2Al_9Si_4O_{23}(OH) \rightarrow 4Fe_3Al_2Si_3O_{12} + 12Al_2SiO_5 + 11Al_2O_3 + 3H_2O$

partially converted into new minerals or to find lower-grade minerals coexisting with those of higher grade. At low grade, some structures of the original rock, such as bedding, may be preserved. Even the outlines of earlier crystals may be seen, filled in with one or more new minerals. High-grade metamorphism usually destroys earlier structures.

All metamorphic rocks available for study are at surface conditions, so the pressures and temperatures that caused them to recrystallize have been relieved. If conditions were relieved slowly enough, and especially if water was available, the rock may have undergone retrograde metamorphism, reverting to a lower grade and thus adjusting to the less intense pressure and temperature. Retrograde metamorphism is usually not very complete, and some evidence of the most intense conditions almost always remains. For example, the distinct outline of a staurolite crystal might be filled with crystals of quartz, biotite mica, and iron oxides.

More recent methods of investigation sometimes applied to metamorphic rocks are X-ray diffraction and electron microprobe analysis. The pattern of X rays scattered from crystals depends on the exact arrangement and spacing of atoms in the crystal structure, which is useful for identifying minerals. X-ray diffraction can be used to identify crystals that are too small or too poorly formed to be identified with a microscope. The microprobe can analyze the chemical composition of crystals even of microscopic size. Determination of exact composition or of variation in composition within growth zones of a single crystal can be especially useful for identifying variations in conditions during crystal formation.

Context

Most people encounter metamorphic rocks while traveling through mountains and other scenic regions. Recognizing these rocks is easier if one has a general idea of where the various types occur and how they appear in outcrops.

Regional metamorphic rocks are best exposed in two kinds of localities: the continental shields and the eroded cores of mountain ranges. Ancient basement rocks of the continental platform, composed of regionally metamorphosed and igneous rocks, are exposed in shields without a cover of sedimentary rock. The Canadian Shield, extending from northern Minnesota through Ontario and Quebec to New England, is the major shield of North America. Similar shields are exposed in western Australia and on every other continent. The old, eroded mountains of Scotland and Wales contain abundant outcrops in which some of the pioneering studies of metamorphic rocks were conducted. The Appalachians have even larger exposures, extending from Georgia into New England. Somewhat smaller outcrops occur in many parts of the Rocky Mountains and the Coast Ranges. The foliated rocks schist, slate, and gneiss make up the bulk of these exposures. Rock cleavage parallel to the foliation is an important clue to recognizing outcrops of these rocks; slopes parallel to foliation tend to be fairly smooth and straight, while slopes eroded across the foliation are ragged and steplike. Bare slopes of schist can reflect light strongly from the many parallel flakes of mica. Slate is usually dull and dark-colored but characteristically splits into ragged slabs. The colorful, contorted bands of gneiss are easily recognized.

Contact metamorphic rocks are much less widespread and are generally confined to areas of active or extinct volcanism. The Cascades and the Sierra Nevada show many examples, but some of the best exposures are found near ancient intrusives in the Appalachians and in New England. Contact metamorphic rocks are more of a challenge to recognize because they generally lack foliation. The best clue is physical contact with a body of igneous rock. They often form a shell or halo around an igneous body, most intensely recrystallized at the contact and extending outward a few centimeters to a few kilometers (depending mostly on the size of the igneous body), until they eventually merge into the surrounding unaltered country rock. The halo is generally similar to the country rock but, because it is recrystallized, it is usually harder, more compact, and more resistant to erosion. Broken surfaces can be distinctive. Depending on the nature of the country rock, one might look for the glassy sheen of fractured quartzite, the glinting cleavage planes of marble, or the smooth, hornlike fracture of hornfels.

Hydrothermally altered rocks from sub-sea-floor metamorphism, when finally exposed on land, are found among regional metamorphic rocks and appear much like them except for the greenish-gray colors of chlorite, serpentine, and talc. Terrestrial hydrothermal alteration is most easily recognized in areas of recent volcanic activity. Good examples are exposed in the southwestern United States, such as the so-called porphyry copper deposits. Many such areas contain valuable deposits, and so may have been mined or prospected. The outcrops are often much fractured and veined near the intrusion. Where alteration is most intense, the rock may appear bleached; father from the intrusion, it may have a greenish hue because of low-grade alteration. Quartz and sulfide minerals such as pyrite are common in the veins, but weathering often leaves a rusty-looking, resistant cap called an iron hat or gossan over the deposit.

Rocks formed by cataclastic and burial metamorphism require specialized equipment for their recognition. Zones many miles wide containing mylonite, the product of cataclastic metamorphism, are exposed along the Moine fault in northwestern Scotland (where mylonite was first studied) and along the Brevard fault, extending along the Appalachians from Georgia into North Carolina. Examples of burial metamorphic rocks are found under the Salton Sea area of California and the Rotorua area of New Zealand.

Bibliography

Bates, Robert L. *Geology of the Industrial Rocks and Minerals.* Mineola, N.Y.: Dover, 1969. Somewhat technical but readable and practical descriptions of the geological occurrence and production of metamorphic rocks and minerals (among others), listing their principal uses. Representative rather than comprehensive. Topical bibliography and good index.

Compton, Robert Ross. *Manual of Field Geology.* New York: John Wiley & Sons, 1962. A standard undergraduate text with a chapter devoted to the interpretation of metamorphic rocks and structures in the field. Some knowledge of mineralogy is assumed. The illustrations are helpful.

Ehlers, Ernest G., and Harvey Blatt. *Petrology: Igneous, Sedimentary, and Metamorphic.* New York: W. H. Freeman, 1982. Undergraduate text in elementary petrology for readers with some familiarity with minerals and chemistry. Thorough, readable discussion of most aspects of metamorphic rocks. Abundant illustrations and diagrams, good bibliography, and thorough indices.

Pough, Frederick H. *A Field Guide to Rocks and Minerals.* 4th ed. Boston: Houghton Mifflin, 1976. The best of the most widely available field guides, authoritative but easy to read. Color plates of representative mineral specimens, and sufficient data to be useful for distinguishing minerals. Very brief description of rocks, including metamorphic rocks. Elementary crystallography and chemistry are presented in the introduction.

Strahler, Arthur N. *Physical Geology.* New York: Harper & Row, 1981. The chapter on metamorphic rocks is a good intermediate-level approach to classification and metamorphic processes. Related chapters on geological environments may interest the reader. Excellent bibliography for the beginning student, with thorough glossary and index.

Tarbuck, Edward J., and Frederick K. Lutgens. *The Earth.* 2d ed. Westerville, Ohio: Charles E. Merrill, 1987. One of the better earth science texts for beginning college or advanced high school readers. Good elementary treatment of metamorphic rocks and, in other chapters, of related environments. Color pictures throughout are excellent. Bibliography, glossary, and a short index.

James A. Burbank, Jr.

Cross-References

Plate Tectonics, 505.

Meteorite and Comet Impacts

The hypervelocity impact of a meteorite or comet on a planetary surface destroys the impacting body and produces a crater that is many times larger than the impacting body. Impact has been responsible for shaping many planetary surfaces and may have caused climatic and biotic changes on the earth.

Field of study: Structural geology

Principal terms

ASTEROID: a small, rocky body in orbit around the sun; a minor planet

COMET: a small body in orbit around the sun that has a fuzzy appearance and often develops a long tail as a result of release of gas and dust as the solid core is heated by sunlight

COMPLEX CRATER: an impact crater of large diameter and low depth-to-diameter ratio caused by the presence of a central uplift or ring structure

HYPERVELOCITY IMPACT: impact involving an object that is traveling faster than the speed of sound in the impacted material and producing compressive shock waves

METEORITE: a rock of extraterrestrial origin that has fallen on the earth's surface; most meteorites are believed to be fragments of asteroids

SHOCK WAVE: a compressional wave formed when a body undergoes a hypervelocity impact; it produces abrupt changes in pressure, temperature, density, and velocity in the target material as it passes through

SIMPLE CRATER: a small impact crater with a simple bowl shape

TARGET MATERIAL: material, generally rock, that is hit by a projectile or impacting body

Summary

Most objects swept up by the earth as it orbits the sun are vaporized, or at least appreciably slowed down by the earth's atmosphere; however, large objects such as asteroids or comets, with masses greater than about 100 metric tons, can hit the earth's surface still traveling at close to their velocity in space (generally between 10 and 40 kilometers per second). Such objects can explode on impact and produce a depression in the earth's surface called an impact crater. Although impact craters are relatively rare on the earth's surface, studies of other planetary bodies, such as Mercury and Mars, as well as the Moon, indicate that impact cratering is a major geological process in the solar system. The relatively low number of impact craters on the earth is a result of the earth being a geologically active planet; as a consequence, impact craters are generally quickly destroyed by other processes, such as gradation, tectonism, or volcanism.

Impact craters have been widely recognized as impact features only in the latter part of the twentieth century. They can be confused with other features, such as volcanic craters; many older impact craters have been so deeply eroded that only a circular area of deformation remains. The first crater to be identified as having an impact origin is Meteor (or Barringer) Crater in Arizona. This crater was first suggested to be an impact crater in 1891 by G. K. Gilbert, a well-known American geologist, and Marcus Baker. Gilbert also proposed that the lunar craters were formed by meteoroid impact. He later changed his mind about Meteor Crater and suggested that it was formed by volcanism. D. M. Barringer, however, supported the view that the crater had been formed by the impact of a large nickel-iron meteorite, fragments of which were found scattered around the crater. Barringer attempted to develop the crater as a commercial enterprise for the recovery of nickel-iron. He was not successful in using the crater as a source of iron, but the crater is now widely accepted as an impact crater largely as a result of his publications in the early 1900's. By 1928, only two impact craters had been recognized, and both were associated with meteorite fragments.

In the 1960's, shock metamorphic effects were recognized, and their identification became an important criterion for identifying impact craters. Shock metamorphism refers to the changes that take place in rock material as a result of high temperature and pressure generated during a high-velocity impact event. The recognition of shock metamorphism as a criterion for identifying impact structures has led to the discovery of more than one hundred impact structures since 1960.

The size and morphology of an impact crater depend initially on the size and velocity of the impacting body and on the nature of the surface or target material. The forward motion of small meteoroids is stopped by the earth's atmosphere, and the object falls to the earth, where it can be collected as a meteorite. Such objects do little damage on impact. Larger objects can retain a portion of their forward velocity and can penetrate into the ground, particularly if the meteorite lands in soft material such as soil or sediment. The resulting feature is called a penetration funnel . A number of craters of this type are found in Campo del Cielo, Argentina, associated with fragments of a nickel-iron meteorite.

Large impact craters or explosion craters are produced during hypervelocity, or hypersonic, impact events. During such an event, the impacting body explodes and a crater is produced that is many times the diameter of

Barringer Crater in Arizona. *(PhotoDisc)*

the impacting body. Hypervelocity impacts occur when the impacting body hits the surface, traveling at a velocity that exceeds the velocity of sound in the target material—that is, greater than several kilometers per second for most rock material. Upon impact, the kinetic energy of the impacting body is transferred to the earth's surface in the form of intense shock waves. The kinetic energy of a moving body is equal to one-half its mass times its velocity squared. Thus, a nickel-iron meteorite with a diameter of 20 meters, weighing about 32,000 metric tons, traveling at a velocity of 20 kilometers per second, would impart about 6.4×10^{15} joules of energy into the target rock. This amount of energy is equivalent to approximately one and a half million tons of dynamite and would be about seventy-five times more powerful than the atom bomb that was dropped on Hiroshima.

Three stages are recognized in the formation of an impact crater: compression, excavation, and postcratering modification. The compression stage is initiated when the projectile (the meteorite or comet) makes contact with the ground. Two compressional shock waves are produced: One travels back through the impacting body, destroying it, and one moves through the ground. At the initial point of contact, pressures exceeding a megabar can be produced, which results in deformational stresses that are orders of magnitude greater than the strength of rock material. High temperatures are generated by the shock wave and increase with increasing shock pressure. At shock pressures greater than about 250 kilobars, most rock types will be crushed; above about 500 kilobars, rock material will begin to melt. At pressures exceeding 1,000 kilobars, rock will be vaporized. The high pressure and temperature produced during the passage of the shock wave distinguish impact from other geological processes.

The compressional shock wave moves out radially into the ground, and the pressure and temperature drop off exponentially from the initial point

409

of contact. Thus, near the initial point of contact, the pressure and associated temperature are high enough to vaporize rock material completely. Surrounding that, a larger volume of rock is melted; surrounding that, a much larger volume of rock is crushed by the passage of the shock wave. For impact events that form craters larger than about 1 or 2 kilometers in diameter, the shock pressures and temperatures are great enough to melt and/or vaporize the impacting body completely.

The excavation stage begins with the development of rarefaction waves that follow the compressional shock wave. During passage of the rarefaction waves, the rock material is decompressed and set into motion. Rock material moves upward and outward, leading to ejection of rock and the development of a transient cavity. Rock material near the point of impact is ejected first. This material is subjected to the highest pressure and temperature and is therefore in the form of vapor and melt. It also has the highest velocity and thus is thrown farthest from the crater. As the cavity grows, the ejected material (ejecta) is derived from progressively greater distances from the initial point of contact and is thus subjected to progressively lower pressure and temperature and has progressively lower ejection velocity. Thus, the rock excavated from the greatest depth experiences the lowest pressure and temperature and lands nearest the crater.

The postcratering modification stage involves changes that occur after the crater is formed. These changes include slumping of the crater walls, isostatic adjustments, and erosion and infill of the crater. This stage can continue over a long period of time and eventually leads to obliteration of the crater. Old impact scars left after the crater has been removed by erosion have been called astroblemes (star wounds).

The ejecta forms an ejecta blanket and secondary craters surrounding the primary impact crater. A continuous blanket of ejecta extends out from the crater a distance of one to two crater diameters. The ejecta blanket is thickest next to the crater and thins outward. Rocks from shallow depths in the crater are deposited first, followed by rocks from progressively deeper depths, resulting in a reversed stratigraphy in the ejecta blanket. Beyond the ejecta blanket, the ejecta is discontinuous and large blocks of rock form low-velocity craters, called secondary craters, which often occur in clusters or loops.

The surface around the impact crater is elevated above the surrounding surface to form a rim. The raised rim is a result of a variety of processes. The rock adjacent to the crater is crushed by the shock wave and thus takes up more volume, which causes the land surface next to the crater to be elevated. In addition, the ejecta blanket is thickest next to the crater. Furthermore, in some impact craters, the surface rock originally inside the crater was lifted up and thrown back over itself to form an overturned flap. This also adds to the elevation of the land surface immediately adjacent to the crater.

Some of the melt stays inside the crater, and some of the ejecta falls back inside the crater. The melt flows to the floor of the crater and is covered with an ejecta layer that insulates it and prevents it from cooling rapidly. As a result, the melt layer crystallizes and forms a layer of melt rock that has texture similar to that of an igneous rock. The melt rock can be distinguished from normal igneous rocks by the presence of inclusions showing evidence of shock metamorphism, by their unusual compositions corresponding to a mixture of the target rocks, and, in some cases, by contamination from the impacting meteorite. The ejecta that falls back into the crater forms a layer of fragmental rock called a fallback breccia.

In map view, most impact craters are circular; in cross section, the profile of a crater varies with size. Smaller craters have simple bowl shapes with a depth-to-diameter ratio of about 1:5. Larger craters are more complex and have lower depth-to-diameter ratios because of the presence of a central uplift or interior rings. On earth, the transition from simple to complex craters takes place at a crater diameter of about 2 to 4 kilometers. The central uplift may be the result of postshock rebound of the crater floor, gravitational collapse of crater rim material, or both.

More than one hundred impact craters have been identified on the earth. Known terrestrial impact craters range in diameter from less than 1 kilometer to about 140 kilometers and in age from less than 100,000 to nearly 2 billion years. Twelve impact craters still have meteorite fragments associated with them, and in every case the meteorite is an iron or stony iron. All the craters with associated meteorite fragments are small (less than 2 kilometers in diameter) and young (less than 1 million years old). Meteoritic contamination of the melt rocks in some of the larger impact structures indicates that at least some of them were probably formed by stony meteorites.

The rate of crater formation has decreased with time. Immediately after the planets were formed, the cratering rate was exceedingly high, but the rate decreased exponentially with time until about 3 billion years ago. Since then, the rate has been fairly uniform. It is estimated that for the last 100 million years, a crater larger than 10 kilometers was formed on the average of about every 100,000 years, while a crater larger than 100 kilometers was formed about every 50 million years. Most of the craters were probably formed by meteorites rather than by comets, but many of the larger craters may have been formed by comets.

Methods of Study

Knowledge of impact craters and the cratering process comes from the study of known impact craters, experiments with human-made explosions, laboratory experiments with high-velocity projectiles, and theoretical studies. Many of the known impact craters have been studied and mapped extensively in order to develop criteria for recognizing impact structures. Meteor Crater in Arizona is the best-known and most studied impact crater. Many shock metamorphic features were first recognized there and then later used as criteria for identifying other impact structures that no longer have meteorite fragments associated with them.

High-speed motion pictures of explosions, using chemical explosives and nuclear devices, have been studied in order to understand the cratering process. Crater formation can also be observed on a smaller scale by observing the impact of missiles and on an even smaller scale by impacting projectiles in the laboratory. Such studies can be used to relate impact-cratering effects to known conditions of impact velocities, projectile properties, and target conditions. These studies indicate that under average conditions, the diameter of the impact crater is about ten to twenty times the diameter of the impacting body. Such information cannot be obtained from the study of impact craters, as the velocity and mass of the projectiles are not known.

The formation of large impact craters (greater than 1 kilometer in diameter) requires more energy than the largest nuclear bomb explosion. Thus, these large events cannot be studied directly, but computer programs have been developed in an attempt to model what might happen during a highly energetic impact. Some scientists have suggested that during a large impact event, involving kilometer-sized impacting bodies, a fireball is formed that can carry material up through the atmosphere and eject it into space at velocities close to or exceeding the earth's escape velocity.

Studies of lunar impact craters have been used to calibrate crater formation rates over time. The lunar highlands were formed about 4.5 billion years ago, while the different mare (lunar sea) surfaces were formed between 3 and 4 billion years ago. Since their formation, these surfaces have not been modified by degradation processes or tectonism. By determining the number of impact craters per unit area on these various surfaces, which have been radiometrically dated, scientists have been able to determine the rate of formation of impact craters through time.

Context

If scientists' current theories concerning the origin of the planets is correct, then the planets were formed by the accretion of asteroid-sized bodies called planetesimals. The early history of the planets and their satellites was dominated by meteoroid bombardment. The surfaces of the Moon, Mercury, parts of Mars, and many of the satellites of the outer planets attest an early history of intense bombardment. The number of craters per unit area has been used to determine the relative ages of different regions on a planetary surface. The large circular basins on the Moon and on planetary bodies are believed to be the results of asteroid impacts. Some scientists have even argued that the earth's moon was formed as a result of a Mars-sized body colliding with the earth. Most of the atmospheres and hydrospheres of the terrestrial planets (Earth, Venus, Mars) may have been contributed during the late stage of accretion by cometary impacts. Some scientists have speculated that the first organic molecules that led to the development of life on the earth were brought in by impacting meteorites or comets.

Impact has played a major role in the earth's history. During the first half-billion years of the earth's history, impact cratering was the dominant process in shaping the face of the globe. Impact may have also played an important role in the geologic history of the earth over the last few billion years. It has been suggested that large impacts might have triggered mantle plumes that led to extensive volcanism and may have caused the breakup of continents and initiated continental drift. Furthermore, impacts are probably responsible for the formation of a group of unusual glass objects called tektites.

It has even been suggested that large impacts might be the cause of climatic changes, mass extinctions, and geomagnetic reversals. Large impacts might cause extinctions in a variety of ways. Nitrogen oxide can be produced by shock heating of the atmosphere by an impacting body, which would lead to the development of acid rains. The impact of comets could introduce large quantities of gases or chemicals, such as ammonium, that could poison the environment. Large impacts can throw huge quantities of dust into the atmosphere that would block out sunlight, preventing photosynthesis and possibly even triggering an ice age. The impact of a 10-kilometer-diameter asteroid at the end of the Mesozoic era, about 65 million years ago, is thought to have been responsible for a mass extinction. It has been proposed that periodic increases in comet impacts have been responsible for mass extinctions that occur approximately every 26 million years. Evidence suggests that at least three large impact events might be associated with reversals of the earth's magnetic field.

Astronomers have shown that a large number of asteroids in earth-crossing orbits could in the future collide with the earth. These objects range in diameter from less than 1 kilometer to about 10 kilometers. The impact of a 10-kilometer body would produce a crater more than 100 kilometers in diameter and would destroy everything within a few hundred kilometers of the crater. It has even been suggested that a large impact event might be mistaken for a nuclear bomb attack and that such an event could trigger a war.

Some large impact structures have economic value. The 140-kilometer-diameter Sudbury impact structure in Ontario, Canada, is the primary source of most nickel and is also a source of copper and other metals. Several billion dollars worth of metals have been recovered from this structure since it was discovered in 1883. At least one buried impact crater has provided a trap for petroleum. The Boltysh structure in the Ukraine is a buried 25-kilometer-diameter impact structure that contains about 3 billion tons of oil shale. Several oil-producing structures in North America are also suspected as being of impact origin.

Bibliography

Classen, J. "Catalogue of 230 Certain, Probable, Possible, and Doubtful Impact Structures." *Meteoritics* 12 (1977): 61. As indicated by the title, this article is essentially a list of certain and possible impact craters. The author gives data on crater location, size, age, and evidence for an impact origin, where present. He also gives an extensive reference list.

French, Bevan M., and Nicholas M. Short, eds. *Shock Metamorphism of Natural Materials.* Baltimore: Mono Book, 1968. A collection of papers dealing with cratering mechanics and shock metamorphism. The papers vary in degree of difficulty.

Horz, Friedrich, ed. "Meteorite Impact and Volcanism." *Journal of Geophysical Research* 76 (August 10, 1971). A collection of papers dealing with impact structures and shock metamorphism. Most of the papers are fairly technical.

Roddy, D. J., R. O. Pepin, and R. B. Merrill, eds. *Impact and Explosion Cratering.* Elmsford, N.Y.: Pergamon Press, 1977. A collection of papers on cratering mechanics and impact craters on various planetary bodies. The papers vary in degree of difficulty, but some should be suitable for advanced high school students.

Shoemaker, E. M. "Impact Mechanics at Meteor Crater, Arizona." In *The Solar System.* Vol. 4, *The Moon, Meteorites and Comets,* edited by B. M. Middlehurst and G. P. Kuiper. Chicago: University of Chicago Press, 1963. A classic paper on Meteor Crater and its origin. Suitable for high school readers.

Silver, Leon T., and Peter H. Schultz, eds. *Geological Implications of Impacts of Large Asteroids and Comets on the Earth.* Special Paper 190. Denver, Colo.: U.S. Geological Survey, 1982. A collection of papers on impact cratering and the effects of large impacts on the climate and life with emphasis on the events that occurred about sixty-five million years ago, at the end of the Mesozoic era (the age of the dinosaurs). These papers vary in degree of difficulty, but some are suitable for advanced high school students and undergraduates.

Billy P. Glass

Cross-References

Catastrophism, 53; The Cretaceous-Tertiary Boundary, 118; The Fossil Record, 257.

Minerals: Physical Properties

Minerals are naturally occurring, inorganic solids with a definite chemical composition and a definite crystal structure. Many minerals are readily identified by their physical properties, but identification of other minerals may require instruments designed to examine details of their chemical composition or crystal structure. The characteristic physical properties of some minerals, such as hardness, malleability, ductility, and electrical properties, make them commercially useful and economically valuable.

Field of study: Mineralogy and crystallography

Principal terms

CLEAVAGE: the tendency for minerals to break in smooth, flat planes along zones of weaker bonds in their crystal structure

CRYSTAL: a solid bounded by smooth planar surfaces that are the outward expression of the internal arrangement of atoms; crystal faces on a mineral result from precipitation in a favorable environment

CRYSTALLINE: a characteristic of a solid, indicating that it has a regular periodic arrangement of atoms in a three-dimensional framework

DENSITY: in an informal sense, the relative weight of mineral samples of equal size; it is defined as mass per unit volume

LUMINESCENCE: the emission of light by a mineral

LUSTER: the reflectivity of the mineral surface; there are two major categories of luster: metallic and nonmetallic

MINERAL: a naturally occurring, inorganic, homogeneous solid with a definite chemical composition and a definite crystal structure

MOHS HARDNESS SCALE: a series of ten minerals arranged in order of increasing hardness, with talc as the softest mineral known (1) and diamond as the hardest (10)

ROCK: an aggregate of one or more minerals

TENACITY: the resistance of a mineral to bending, breakage, crushing, or tearing

Summary

Minerals are the building blocks of rocks. A mineral is a naturally occurring, inorganic, homogeneous solid with a definite chemical composition (or definite range of compositions), which can be expressed by a chemical formula, and a definite atomic crystal structure, which is reflected in its crystal shape and other physical properties. By definition, man-made materials cannot be minerals and, therefore, synthetic gemstones produced in the laboratory are not considered minerals. Also, minerals are inorganic, meaning that they are not produced by organisms and are not composed of organic (carbon-based) molecules. As a result, pearls are not minerals. Coal and hydrocarbons are not minerals because they are not homogeneous and because they contain organic molecules. Minerals must be solids, which means that liquids are not minerals (although liquid mercury is sometimes considered to be a mineral; mercury is a solid at temperatures below −40 degrees Celsius).

The chemistry of some minerals varies within a particular range because of substitution of one element for another in the crystal structure (ionic substitution). For example, iron (Fe^{2+}) and magnesium (Mg^{2+}) may substitute for each other in the olivine minerals. In the plagioclase feldspars, calcium (Ca^{2+}) may substitute for sodium (Na^{1+}), with a concomitant substitution of aluminum (Al^{3+}) for silicon (Si^{4+}) so that the mineral remains electrically neutral.

Crystal structure is the internal ordering of atoms in a specific three-dimensional geometric framework, called a crystal lattice. Crystal faces are the outward expression of this internal ordering. All specimens of a particular mineral have the same internal crystal structure and are referred to as crystalline, regardless of whether crystal faces are present. Volcanic glass is not a mineral, because it lacks an orderly internal atomic crystal structure; it is a noncrystalline, amorphous material.

Minerals have diagnostic physical properties resulting from their chemistry and crystal structure. Physical properties of minerals include color, hardness, streak, luster, crystal shape, cleavage, fracture, and density, or specific gravity. Several minerals have additional diagnostic physical properties, including tenacity, magnetism, taste, reaction to hydrochloric acid, luminescence, and radioactivity.

Color is an obvious physical property, but it is one of the least diagnostic for mineral identification. In some minerals, color results from the presence of major elements in the chemical formula; in these minerals, color is a diagnostic property. For example, malachite is always green, azurite is always blue, and rhodochrosite and rhodonite are always pink. In other minerals, color is the result of trace amounts of chemical impurities or of defects in the crystal lattice structure. Depending on the impurities, a particular species of mineral can have many different colors. For example, pure quartz is colorless, but quartz may be white (milky quartz), pink (rose quartz), purple (amethyst), yellow (citrine), brown (smoky quartz), green, blue, or black. Similarly, fluorite may be colorless, white, pink, purple, blue, brown, green, or yellow. The following is a partial list of colors and trace chemical impurities which may cause them: Pink coloration may be caused by iron or titanium; green coloration may be caused by iron, chromium, or vanadium; blue coloration may be caused by titanium, titanium and iron, or a combination of iron with different valences; yellow coloration may be caused by nickel, and greenish yellow coloration may be caused by iron. The color of sphalerite varies from white to yellow to brown to black, with increasing amounts of iron substituting for zinc. Defects in the crystal lattice structure can also cause color. For example, the purple color of fluorite and amethyst and the brown color of smoky quartz result from defects in the crystal structure. Some structural effects (and colors) can be caused by exposing minerals to radiation. Some minerals have different colors depending on the direction in which the mineral is oriented as light passes through it. This property is known as pleochroism and is best seen in thin sections or small mineral grains using a petrographic microscope. Hornblende, hypersthene, tourmaline, and staurolite are pleochroic.

Hardness is the resistance of a mineral to scratching or abrasion. Hardness is a result of crystal structure; the stronger the bonding forces between the atoms, the harder the mineral. The Mohs hardness scale, devised by a German mineralogist, Friedrich Mohs, in 1822, is a series of ten minerals arranged in order of increasing hardness. The minerals on the Mohs hardness scale are talc (the softest mineral known), gypsum, calcite, fluorite, apatite, potassium feldspar (orthoclase), quartz, topaz, corundum, and diamond (the hardest mineral known). A mineral with higher hardness number can scratch any mineral of equal or lower hardness number. The relative hardness of a mineral is easily tested using a number of common materials, including the fingernail (a little over 2), a copper coin (about 3), a steel nail or pocket knife (a little over 5), a piece of glass (about 5.5), and a steel file (6.5). The hardness of some minerals varies with crystallographic direction. For example, kyanite can be scratched with a knife parallel to the direction of elongation, but it cannot be scratched perpendicular to the elongation.

Streak is the color of the mineral in powdered form. Streak is more definitive than mineral color, because although a mineral may have several color varieties, the streak will be the same for all. Streak is best viewed after rubbing the mineral across an unglazed porcelain tile. The tile has a hardness of approximately 7, so minerals with a hardness of greater than 7 will not leave a streak, although their powdered color may be studied by crushing a small piece.

Luster refers to the reflectivity of the mineral surface. There are two major categories of luster: metallic and nonmetallic. Metallic minerals include metals (such as native copper and gold), as well as many metal sulfides, such as pyrite and galena. Another metal sulfide, sphalerite, commonly has a submetallic luster, although in some samples, its luster is nonmetallic. Nonmetallic lusters can be described as vitreous or glassy (characteristic of quartz and olivine), resinous (resembling resin or amber, characteristic of sulfur and some samples of sphalerite), adamantine or brilliant (diamond), greasy (appearing as if covered by a thin film of oil, including nepheline and some samples of massive quartz), silky (in minerals with parallel fibers, such as malachite, chrysotile asbestos, or fibrous gypsum), pearly (similar to an iridescent pearl-like shell, such as talc), and earthy or dull (as in clays).

Crystal shape is the outward expression of the internal three-dimensional arrangement of atoms in the crystal lattice. Crystals are formed in a cooling or evaporating fluid as atoms begin to slow down, move closer, and bond together in a particular geometric gridwork. If minerals are unconfined and free to grow, they will form well-shaped, regular crystals. On the other hand, if growing minerals are confined by other, surrounding minerals, they may have irregular shapes. Because crystal faces are related to the internal structure of a mineral, the crystal faces of several specimens of the same mineral will be similar. This fact was observed as early as 1669 by Nicolaus Steno, who observed that "the angles between equivalent faces of crystals of the same substance, measured at the same temperature, are constant," which is commonly referred to as the Law of Constancy of Interfacial Angles. Although different crystals of the same mineral may be different sizes, the angles between corresponding crystal faces are always the same. Some of the common shapes or growth habits of crystals include acicular (or needlelike, such as natrolite), bladed (elongated and flat like a knife blade, such as kyanite), blocky (equidimensional and cubelike, such as galena and fluorite), and columnar or prismatic (elongated or pencil-like, such as quartz and tourmaline). Other crystal shapes are described as pyramidal, stubby, tabular, barrel-shaped, or capillary.

Cleavage is the tendency for minerals to break in smooth, flat planes along zones of weaker bonds in their crystal structure. Cleavage is one of the most important physical properties in identifying minerals because it is so closely related to the internal crystal structure. Cleavage is best developed in minerals that have particularly weak chemical bonds in a given direction. In other minerals, differences in bond strength are less pronounced, so cleavage is less well developed. Some minerals have no planes of weakness in their crystal structure; they lack cleavage and do not break along planes. Cleavage can occur in one direction (as in the micas, muscovite, and biotite) or in more than one direction. The number and orientation of the cleavage planes are always the same for a particular mineral. For example, orthoclase feldspar has two directions of cleavage at right angles to each other. The amphiboles (hornblende, tremolite, actinolite, and the like) also have two directions of cleavage, but at angles of approximately 56 and 124 degrees. Cubic cleavage (three directions at right angles) is a characteristic feature of halite (table salt) and galena. Rhombohedral cleavage (three directions not at right angles) is characteristic of calcite. Fluorite has four directions of cleavage, which produces octahedral (eight-sided) cleavage fragments with triangular cleavage faces. Sphalerite has six directions of cleavage (dodecahedral cleavage). Cleavage is an outstanding criterion for identification of minerals in which it is well developed.

Fracture is irregular breakage that is not controlled by planes of weakness in minerals. Conchoidal fracture is a smooth, curved breakage surface, commonly marked by fine concentric lines, resembling the surface of a shell. Conchoidal fracture is common in broken glass and quartz. Fibrous or splintery fracture occurs in asbestos and sometimes in gypsum. Hackly fracture is jagged with sharp edges and occurs in native copper. Uneven or irregular fracture produces rough, irregular breakage surfaces.

Density is defined as mass per unit volume (typically measured in terms of grams per cubic centimeter). In a very informal sense, density refers to the relative weight of samples of equal size. Quartz has a density of 2.6 grams per cubic centimeter, whereas a "heavy" mineral such as galena has a density of 7.4 grams per cubic centimeter (about three times as heavy). Specific gravity (or relative density) is the ratio of the weight of a substance to the weight of an equal volume of water at 4 degrees Celsius. The terms "density" and "specific gravity" are sometimes used interchangeably, but density requires the use of units of measure (such as grams per cubic centimeter), whereas specific gravity is unitless. Specific gravity is an important aid in mineral identification, particularly when studying valuable minerals or gemstones, which might be damaged by other tests of physical properties. The specific gravity of a mineral depends on the chemical composition (type and weight of atoms) as well as the manner in which the atoms are packed together.

Tenacity is the resistance of a mineral to bending, breakage, crushing, or tearing. A mineral may be brittle (breaks or powders easily), malleable (may be hammered out into thin sheets), ductile (may be drawn out into a thin wire), sectile (may be cut into thin shavings with a knife), flexible (bendable, and stays bent), and elastic (bendable, but returns to its original form). Minerals with ionic bonding, such as halite, tend to be brittle. Malleability, ductility, and sectility are diagnostic of minerals with metallic bonding, such as gold. Chlorite and talc are flexible, and muscovite is elastic.

Magnetism causes minerals to be attracted to magnets. Magnetite and pyrrhotite are the only common magnetic minerals, and they are called ferromagnetic. Lodestone, a type of magnetite, is a natural magnet. When in a strong magnetic field, some minerals become weakly magnetic and are attracted to the magnet; these minerals are called paramagnetic. Examples of paramagnetic minerals include garnet, biotite, and tourmaline. Other minerals are repelled by a magnetic field and are called diamagnetic minerals. Examples of diamagnetic minerals include gypsum, halite, and quartz.

Some minerals are easily identified by their taste. Most notable is halite, or common table salt. Another mineral with a distinctive taste is sylvite, which is distinguished from halite by its more bitter taste. Some minerals, such as calcite, effervesce or fizz when they come into contact with hydrochloric acid; the bubbles of gas released are carbon dioxide. Other minerals that effervesce in hydrochloric acid include aragonite, dolomite if powdered, and malachite. Most of the carbonate minerals will react with hydrochloric acid, although some will not react unless the acid is heated.

Luminescence is the term for emission of light by a mineral. Minerals that glow or luminesce in ultraviolet light, X rays, or cathode rays are fluorescent minerals. The glow is the result of the mineral changing invisible radiation

PHYSICAL PROPERTIES OF MINERALS

Property	Explanation
Chemical composition	Chemical formula that defines the mineral
Cleavage	Tendency to break in smooth, flat planes along zones of weak bonding; depends on structure
Color	Depends on presence of major elements in the chemical composition; may be altered by trace elements or defects in structure; often not definitive
Crystal shape	Outward expression of the atomic crystal structure
Crystal structure	Three-dimensional ordering of the atoms that form the mineral
Density	Mass per unit volume (grams per cubic centimeter)
Electrical properties	Properties having to do with electric charge; quartz, for example, is piezoelectric (emits charge when squeezed)
Fracture	Tendency for irregular breakage (not along zones of weak bonding)
Hardness	Resistance of mineral to scratching or abrasion; measured on a scale of 1-10 (Mohs hardness scale)
Luminescence	Emission of electromagnetic waves from mineral; some minerals are fluorescent, some thermoluminescent
Luster	Reflectivity of the surface; may be either metallic or nonmetallic
Magnetism	Degree to which mineral is attracted to a magnet
Radioactivity	Instability of mineral; radioactive minerals are always isotopes
Specific gravity	Relative density: ratio of weight of substance to weight of equal volume of water at 4 degrees Celsius
Streak	Color of powdered form; more definitive than color
Taste	Salty, bitter, etc.; applies only to some minerals
Tenacity	Resistance to bending, breakage, crushing, tearing: termed as brittle, malleable, ductile, sectile, flexible, or elastic

to visible light, which happens when the radiation is absorbed by the crystal lattice and then reemitted by the mineral at lower energy and longer wavelength. Fluorescence occurs in some specimens of a mineral but not all. Examples of minerals which may fluoresce include fluorite, calcite, diamond, scheelite, willemite, and scapolite. Some minerals will continue to glow or emit light after the radiation source is turned off; these minerals are phosphorescent. Some minerals glow when heated, a property called thermoluminescence, present in some specimens of fluorite, calcite, apatite, scapolite, lepidolite, and feldspar. Other minerals luminesce when crushed, scratched, or rubbed, a property called triboluminescence, present in some specimens of sphalerite, corundum, fluorite, and lepidolite and, less commonly, in feldspar and calcite. Luminescence is generally caused by the presence of certain rare earth or lanthanide elements, or uranium or thorium. The fluorescence of specimens of calcite and willemite from Franklin, New Jersey, is attributed to the presence of manganese.

Radioactive minerals contain unstable elements that alter spontaneously to other kinds of elements, releasing subatomic particles and energy. Some elements come in several different forms, differing by the number of neutrons present in the nucleus. These different forms are called isotopes, and one isotope of an element may be unstable (radioactive), whereas another isotope may be stable (not radioactive). Radioactive isotopes include potassium 40, rubidium 87, thorium 232, uranium 235, and uranium 238. Examples of radioactive minerals include urananite and thorianite.

Some minerals also have interesting electrical characteristics. Quartz is a piezo-electric mineral, meaning that when squeezed, it produces electrical charges. Conversely, if an electrical charge is applied to a quartz crystal, it will change shape and vibrate as internal stresses develop. The oscillation of quartz is the basis for its use in digital quartz watches.

Methods of Study

In many cases, the physical properties of minerals can be studied using relatively common, inexpensive tools. For example, hardness may be determined by attempting to scratch the mineral in question with a number of common objects, including a fingernail, copper coin, pocket knife or nail, piece of glass, or steel file. The relative hardness of a mineral may also be determined by attempting to scratch one mineral with another, thereby bracketing the unknown mineral's hardness between that of other minerals on the Mohs hardness scale. Streak may be determined by rubbing the mineral across an unglazed porcelain tile to observe the color (and sometimes odor) of the streak, if any. Color and luster are determined simply by observing the mineral.

The angles between adjacent crystal faces may be measured using a goniometer. There are several types of goniometers, but the simplest is a protractor with a pivoting bar, which is held against a large crystal so that the angles between faces can be measured. There are also reflecting goniometers, which operate by measuring the angles between light beams reflected from crystal faces.

Density can be determined by measuring the mass of a mineral and determining its volume (perhaps by measuring the amount of water it displaces in a graduated cylinder), then dividing these two measurements. In order to determine the specific gravity (SG) of a mineral, it is necessary to have a specimen that is homogeneous, without cracks, and without bits of other minerals attached. Mineral specimens with a volume of about 1 cubic centimeter are most useful. Specific gravity is usually determined by first weighing the mineral in air and then weighing it while it is immersed in water. When immersed in water, it weighs less because it is buoyed up by a force equivalent to the weight of the water displaced. A Jolly balance, which works by stretching a spiral spring, can measure specific gravity. For tiny mineral specimens weighing less than 25 milligrams, a torsion balance (or Berman balance) is useful for accurate determinations. Heavy liquids, such as bromoform (SG = 2.89) and methylene iodide (SG = 3.33), are also used to determine the specific gravity of small mineral grains. In heavy liquids, grains of quartz (SG = 2.65) will float, whereas heavy minerals, such as zircon (SG = 4.68) or garnet (SG = 3.5 to 4.3), will sink. The mineral grain is placed into the heavy liquid and then acetone is added to the liquid until the mineral grain neither floats nor sinks (that is, until the specific gravity of the mineral and the liquid are the same). Then, the specific gravity of the liquid is determined using a Westphal balance.

An ultraviolet light source (with both long and short wavelengths) is used to determine whether minerals are fluorescent or phosphorescent. A portable ultraviolet light can be used to prospect for fluorescent minerals. Thermoluminescence can be triggered by heating a mineral to 50-100 degrees Celsius. Radioactivity is measured using a Geiger counter or scintillometer.

Minerals may be identified using a number of techniques and instruments, including X-ray diffraction, differential thermal analysis, thermogravimetric analysis, the electron microscope, and the petrographic microscope. X-ray diffraction works by bombarding a crystal or powdered mineral with X rays. When the X rays strike the mineral, they are scattered or diffracted; the X rays penetrate below the mineral surface and are reflected from layers of atoms in the crystal lattice. Minerals can then be identified by comparing the positions and intensities of the reflections produced with those listed in standard tables of mineral data. Differential thermal analysis (DTA) identifies temperatures at which one mineral is transformed into another; it is used primarily to identify clay minerals. Thermogravimetric analysis, which is commonly used in conjunction with DTA, measures the change in weight of minerals as they are heated. The electron microscope allows the mineralogist to study extremely small minerals. The transmission electron microscope is used to gain insights into the structure of minerals. The scanning electron microscope is primarily used to examine surface details of minerals. The energy dispersive analyzer can be used in conjunction with both types of electron microscope and can provide semiquantitative chemical analyses of selected mineral particles.

A petrographic microscope with a polarized light source is required for studying optical properties of minerals, which can be extremely useful in identification. Optical properties can be studied using crushed or powdered samples of minerals (called grain mounts) or with thin sections of rocks. The optical properties are determined by observing changes in polarized light passing through the mineral. Optical properties include index of refraction, birefringence, relief, pleochroism, extinction angle, and interference figures. A universal stage may be used to change the orientation of the mineral for more accurate measurements of optical properties. Index of refraction refers to the ratio between the velocity of light in a vacuum and to the velocity of light in the mineral. In general, high refractive index is related to high specific gravity. Index of refraction is commonly measured by placing crushed grains of an unknown mineral in oils of known refractive index for comparison. Birefringence, which is present in minerals that have more than one index of refraction, splits a beam of light passing through the mineral into two beams of unequal velocities. Birefringence is studied by its color in crossed polarized light on minerals of known thickness. Relief is the difference between the refractive index of the mineral and the refractive index of the surrounding material (usually a mounting medium such as Canada balsam or epoxy); it is studied by observing how clearly the mineral stands out from the mounting medium. Pleochroism is a difference in the color of a mineral as it is rotated on the microscope stage in plane polarized light.

Chemical analyses of minerals may be performed using wet chemistry (which requires dissolution of mineral specimens), blowpipe analysis (an old technique for studying the behavior of a mineral in a flame), neutron activation analysis (usually performed in a nuclear reactor, with neutrons bombarding the mineral, producing characteristic radioisotopes that are studied with gamma-ray spectroscopy), atomic absorption spectrometry (which involves dissolution of the mineral and studying how light emitted by a cathode-ray tube is absorbed by a flame produced by burning the liquid sample), X-ray fluorescence spectrometry (XRF, which bombards a mineral with X rays and examines the wavelengths of radiation emitted from the mineral), the electron microprobe (which works like XRF—but with an electron beam to bombard the atoms in the mineral—and can be used to study the chemical composition of very small areas of minerals), and cathodouminescence (which activates minerals using an electron beam, thus exciting certain ions and producing luminescence).

Context

The physical properties of minerals affect their usefulness for commercial applications. Minerals with great hardness, such as diamonds, corundum (sapphire and ruby), garnets, and quartz, are useful as abrasives and in cutting and drilling equipment. Other minerals are useful because of their softness, such as calcite (hardness 3), which is used in cleansers because it will not scratch the surface being cleaned. Also, calcite in the form of marble is commonly used for sculpture because it is relatively soft and easy to carve. Alabaster, a form of gypsum (hardness 2), is also commonly carved and used for decorative purposes. Soapstone, made of the mineral talc (hardness 1), was quarried and carved by American Indians in eastern North America. Talc is also used in talcum powder because of its softness.

Some metals, such as copper, are ductile, which makes them useful for the manufacture of wire. Copper is one of the best electrical conductors. Copper is also a good conductor of heat and is often used in cookware. Gold is the most malleable and ductile mineral. Because of its malleability, gold can be hammered into sheets so thin that 300,000 of them would be required to make a stack 1 inch high. Because of its ductility, 1 gram of gold (about the weight of a raisin) can be drawn into a wire more than a mile and a half long. Pure gold is too soft to be used in coins or jewelry and is almost always alloyed with other metals (such as silver, or copper) to harden it. Because of its high specific gravity (SG = 19.3 when pure), gold can be collected by panning in streams. Gold is the best conductor of heat and electricity known, but it is generally too expensive to use as a conductor.

Other minerals are valuable because they do not conduct heat or electricity. They are used as electrical insulators or for products subjected to high temperatures. For example, kyanite, andalusite, and sillimanite are used in the manufacture of spark plugs and other high-temperature porcelains. Muscovite is also useful because of its electrical and heat-insulating properties; sheets of muscovite are often used as an insulating material in electrical devices. The "isinglass" used in furnace and stove doors is muscovite. Sheets of muscovite were used as windowpanes in medieval Europe because of the mineral's cleavage, transparency, near absence of color, and glassy luster. Because of its glassy luster, ground muscovite is commonly used to give wallpaper a shiny finish.

The fibrous cleavage, flexibility, incombustibility (flame resistance), and low conductivity of heat that are characteristic of asbestos (chrysotile, crocidolite, and other asbestos minerals) make it ideally suited for fireproofing, heat and electrical insulation, roofing materials, and brake linings. Asbestos can also be woven into fireproof fabrics because of its flexible fibers. The use of asbestos is declining, however, because of health concerns; the tiny, needlelike fibers may become lodged in the lungs, leading to various asbestos-related diseases.

Cleavage is the property responsible for the use of graphite as a dry lubricant and in pencils. Graphite has perfect cleavage in one direction, and is slippery because microscopic sheets of graphite slide easily over one another. A "lead" pencil is actually a mixture of graphite and clay; it writes by leaving tiny cleavage flakes of graphite on the paper. Conchoidal fracture is the property which permits arrowheads and stone tools to be made from quartz, cryptocrystalline quartz (flint and chert), and obsidian (volcanic glass). The sharp edges are produced where many conchoidal fractures merge.

The color of the streak (or crushed powder) of many minerals makes them valuable as pigments. Hematite has a red streak and is used in paints and cosmetics. Lazurite has a blue powder and was once used as the paint pigment ultramarine blue (now produced artificially). Silver is commonly used in photographic films and papers because in the form of silver halide, it is light-sensitive and turns black. After developing and fixing, metallic silver remains on the film to form the negative. Silver is also one of the best conductors of heat and electricity and is often used in electronics.

The uranium-bearing minerals (urananite, carnotite, torbernite, and autunite) are used as sources of uranium, which is important because its nucleus is susceptible to fission (splitting or radioactive disintegration), producing tremendous amounts of energy. This energy is used in nuclear power plants for generating electricity and in atom bombs. Pitchblende, a variety of urananite, is a source of radium, which is used as a source of radioactivity in industry and medicine. The high specific gravity of barite makes it a useful additive to drilling muds to prevent oil well gushers or blowouts. It is also opaque to X rays and is used in medicine for "barium milkshakes" before patients are X-rayed so that the digestive tract will show up clearly.

Many minerals are economically valuable because of their unique physical properties. Others are useful as a result of their chemical properties. These include halite, or table salt; sylvite, which is used as a salt substitute for persons with high blood pressure and as a source of potassium for fertilizers; apatite, which is used as a source of phosphate for fertilizers; borax minerals, which are used in the manufacture of boric acid and detergents; gypsum, which is used in plaster of paris and walboard; and numerous minerals that are ores for iron, lead, copper and zinc, and other metals.

Bibliography

Blackburn, W. H., and W. H. Dennen. *Principles of Mineralogy*. Dubuque, Iowa: Wm. C. Brown, 1988. This book is divided into three parts: The first part is theoretical and includes crystallography and crystal chemistry, along with a section on the mineralogy of major types of rocks. The second part is practical and includes chapters on physical properties of minerals, crystal geometry, optical properties, and methods of analysis. The third part contains systematic mineral descriptions. Designed for an introductory college course in mineralogy, but should be useful for amateurs as well.

Bloss, F. D. *An Introduction to the Methods of Optical Crystallography*. New York: Holt, Rinehart and Winston, 1961. This book is for the advanced student of mineralogy who is interested in the ways in which the crystal structure of a mineral changes the characteristics of a beam of light passing through it, as studied with the petrographic microscope. Theoretical, it may be of interest to persons with a background in physics or geology.

Chesterman, C. W., and K. E. Lowe. *The Audubon Society Field Guide to North American Rocks and Minerals*. New York: Alfred A. Knopf, 1988. This book contains 702 color photographs of minerals, grouped by color, as well as nearly a hundred color photographs of rocks. All the mineral photographs are placed at the beginning of the book, and descriptive information follows, with the minerals grouped by chemistry. Distinctive features and physical properties are listed for each of the minerals, and information on collecting localities is also given. A section at the back of the book discusses various types of rocks. A glossary is also included. Suitable for the layperson.

Desautels, P. E. *The Mineral Kingdom*. New York: Ridge Press, 1968. An oversize, lavishly illustrated coffee-table book with useful text supplementing the color photographs. It covers how minerals are formed, found, and used and includes legends about minerals and gems as well as scientific data. Provides a broad introduction to the field of mineralogy. Intended for the amateur, it also makes fascinating reading for the professional geologist.

Frye, Keith. *Modern Mineralogy*. Englewood Cliffs, N.J.: Prentice-Hall, 1974. This book addresses minerals from a chemical standpoint and includes chapters on crystal chemistry, structure, symmetry, physical properties, radiant energy and crystalline matter, the phase rule, and mineral genesis. Designed as an advanced college textbook for a student who already has some familiarity with mineralogy. Includes short descriptions of minerals in a table in the appendix.

Kerr, P. F. *Optical Mineralogy*. New York: McGraw-Hill, 1977. This book is designed to instruct advanced mineralogy students in the study and identification of minerals using a petrographic microscope. The first part concerns the basic principles of optical mineralogy, and the second part details the optical properties of a long list of minerals.

Klein, C., and C. S. Hurlbut, Jr. *Manual of Mineralogy*. 20th ed. New York: John Wiley & Sons, 1985. One of a series of revisions of the original mineralogy textbook written by James D. Dana in 1848. The first part of the book is dedicated to crystallography and crystal chemistry, with shorter chapters on the physical and optical properties of minerals. The second part of the book provides a classification and detailed, systematic description of various types of minerals, with sections on gem minerals and mineral associations. Considered to be the premier mineralogy textbook for college-level geology students; many parts of it will be useful for amateurs.

Mottana, A., R. Crespi, and G. Liborio. *Simon & Schuster's Guide to Rocks and Minerals*. New York: Simon & Schuster, 1978. This book is fully illustrated with color photographs of 276 minerals. It provides background information on physical properties, environment of formation, occurrences, and uses of each mineral. A sixty-page introduction to minerals provides sophisticated technical coverage that will be of interest to both mineralogy students and amateurs. The last part of the book illustrates and describes one hundred types of rocks. A glossary is also included.

Pough, F. H. *A Field Guide to Rocks and Minerals.* Boston: Houghton Mifflin, 1976. This well-written and well-illustrated book is suitable for readers of nearly any age and background. One of the most readable and accessible sources, it provides a fairly complete coverage of the minerals. Designed for amateurs.

Zussman, J., ed. *Physical Methods in Determinative Mineralogy.* New York: Academic Press, 1967. A reference book that describes technical methods used in the study of rocks and minerals, including transmitted and reflected light microscopy, electron microscopy, X-ray fluorescence spectroscopy, X-ray diffraction, electron microprobe microanalysis, and atomic absorption spectroscopy. Written for the professional geologist or advanced student.

Pamela J. W. Gore

Cross-References

Minerals: Structure, 428.

Minerals: Structure

The discovery of the internal structures of minerals by the use of X-ray diffraction was pivotal in the history of mineralogy and crystallography. X-ray analysis revealed that the physical properties and chemical behavior of minerals are directly related to the highly organized arrangements of their atoms, and this knowledge has had important scientific and industrial applications.

Field of study: Mineralogy and crystallography

Principal terms

CLEAVAGE: the capacity of crystals to split readily in certain directions; in well-defined structures, cleavage planes pass between sheets of atoms

CRYSTAL: externally, a solid material of regular form bounded by flat surfaces called faces; internally, a substance whose orderly structure results from a periodic three-dimensional arrangement of atoms

ION: an electrically charged atom or group of atoms

IONIC BOND: the strong electrical forces holding together positively and negatively charged ions (for example, sodium and chloride ions in common salt)

MINERAL: a naturally formed inorganic substance with characteristic physical properties, a definite chemical composition, and, in most cases, a regular crystal structure

X RAY: radiation interpretable in terms of either very short electromagnetic waves or highly energetic photons (light particles)

Summary

From earliest times, people have been fascinated by distinctively formed and attractively colored crystals conspicuously embedded in drab rocks. These striking minerals, with their flat faces meeting in straight edges, stimulated the human imagination and challenged the intellect. Primitive humans saw these patterned stones, despite their flaws, as special, and they used them as decorations, talismans, and medical aids. Ancient natural philosophers reasoned that minerals were imperfect realizations of some ideal geometric shape; they also accepted the idea that minerals were alive and grew in caves, as if they were a kind of plant. This animistic idea of minerals lasted into the Renaissance, but by the sixteenth century, the new mechanical philosophy that undergirded the scientific revolution caused certain scientists to think of crystals in terms of the arrangements of material particles. Atomism began to rival animism in the explanation of mineral structure. For example, Thomas Harriot, an English mathematician, proposed that the densest minerals were those in which every atom touched its twelve nearest "neighbors" (today, scientists would say that this structure represents the closest packing of uniform spheres). In the early seventeenth century, Johannes Kepler, best known for his work in astronomy, explained the myriad patterns of snowflakes by the association of spherical particles.

Although students of crystals had long surmised that crystals possessed regular internal structures, it was not until late in the seventeenth century that some scientists began to formulate specific theories. In 1678, Christiaan Huygens, the Dutch physicist and astronomer, became interested in calcite, a transparent mineral that exhibited an extraordinary property: An object viewed through this mineral appeared as a double image. By assuming calcite to be made up of small ellipsoidal particles packed in a regular array, Huygens used his wave theory of light to show how these particles caused the formation of double rays of light.

Important steps toward the understanding of mineral structure occurred through the work of Nicolaus Steno, a Danish physician at the court of the Medicis in Florence. He helped to discredit the idea of the vegetative growth of minerals by showing how a tiny seed crystal could grow by the superimposition of particles on its faces. In 1669, he noted that quartz crystals, whatever their size and origin, had constant angles between corresponding faces. Despite this important discovery, Steno did not generalize his observations, and it was not until a century later that the law of constancy of angles for all crystals of a given substance was established.

During the seventeenth and eighteenth centuries, natural philosophers used two basic ideas to explain the external structures of minerals: particles, in the form of spheres, ellipsoids, or various polyhedra; and an innate attractive force, or the emanating "glue" needed to hold particles together. These attempts to rationalize mineral structures still left a basic problem unanswered: How does one explain the heterogeneous physical and chemical properties of minerals with homogeneous particles? This problem was not answered satisfactorily until the twentieth century, but the modern answer grew out of the work of scientists in the eighteenth and nineteenth centuries. The most important of these scientists was René-Just Haüy, often called the "father of crystallography."

Haüy, a priest who worked at the Museum of Natural History in Paris, helped to make crystallography a science. Before Haüy, the science of crystals had been a part of biology, geology, or chemistry; after Haüy, the science of crystals was an independent discipline. His speculations on the nature of the crystalline state were stimulated when he accidentally dropped a calcite specimen, which shattered into fragments. He noticed that the fragments split along straight planes that met at constant angles. No matter what the shape of the original piece of calcite, he found that broken fragments were rhombohedra (slanted cubes). He reasoned that a rhombo-hedron, similar to the ones he obtained by cleaving the crystal, must be preformed in the inner structure of the crystal. For Haüy, then, the cleavage planes existed in the crystal like the mortar joints in a brick wall. When he discovered similar types of cleavage in a variety of substances, he proposed that all crystal forms could be constructed from submicroscopic building blocks. He showed that there were several basic building blocks, which he called primitive forms or "integral molecules," and they represented the last term in the mechanical division of a crystal. With these uniform polyhedra, he could rationalize the many mineral forms observed in nature. Haüy's building block was not, however, the same as what later crystallographers came to call the unit cell, the smallest group of atoms in a mineral that can be repeated in three directions to form a crystal. The unit cell is not a physically separable entity, such as a molecule; it simply describes the repeat pattern of the structure. On the other hand, for Haüy, the crystal was a periodic arrangement of equal molecular polyhedra, each of which might have an independent existence.

In the nineteenth century, Haüy's ideas had many perceptive critics. For example, Eilhardt Mitscherlich, a German chemist, discovered in 1819 that different mineral substances could have the same crystal form, whereas Haüy insisted that each substance had a specific crystal structure. As Haüy's system came under attack, scientists suggested new models of crystal struc-ture, and as these models met with difficulties, the theories that took their place became increasingly abstract and idealized. Some crystallographers shunned the concrete study of crystals (leaving it to mineralogists), and they defined their science as the study of ordered space. This mathematical analysis bore fruit, for crystallographers were able to show that, despite great variety of possible mineral structures, all forms could be classified into six crystal systems on the basis of certain geometrical features, usually axes. The cube is the basis of one of these systems, the isometric, in which three identical axes intersect at right angles. Symmetry was another factor in describing these crystal systems. For example, a cube has fourfold symmetry around an axis passing at right angles through the center of any of its faces. As some crystallographers were establishing the symmetry relationships in crystal systems, others were working on a way to describe the position of crystal faces. In 1839, William H. Miller, a professor of mineralogy at the University of Cambridge, found a way of describing how faces were ori-ented about a crystal, similar to the way a navigator uses latitude and longitude to tell where his ship is on the earth. Using numbers derived from axial proportions, Miller was able to characterize the position of any crystal face.

Friedrich Mohs, a German mineralogist best remembered for his scale of the hardness of minerals, was famous in his lifetime for his system of mineral classification, in which he divided minerals into genera and species, similar to the way biologists organized living things. His system was based on geometrical relationships that he derived from natural mineral forms. He wanted to transform crystallography into a purely geometrical science, and he showed that crystal analysis involved establishing certain symmetrical groups of points by the rotation of axes. When these point groups were enclosed by plane surfaces, crystal forms were generated. The crystallographer's task, then, was to analyze the symmetry operations characterizing the various classes of a crystal system.

Beginning in 1848, Auguste Bravais, a French physicist, took the same sort of mathematical approach in a series of papers dealing with the kinds of geometric figures formed by the regular grouping of points in space, called lattices. Bravais applied the results of his geometric analysis to crystals, with the points interpreted either as the centers of gravity of the chemical molecules or as the poles of interatomic electrical forces. With this approach, he demonstrated that there is a maximum of fourteen kinds of lattices, which differ in symmetry and geometry, such that the environment around any one point is the same as that around every other point. These fourteen Bravais lattices are distributed among the six crystal systems. For example, the three isometric Bravais lattices are the simple (with points at the vertices of a cube), body-centered (with points at the corners along with a point at the center of a cube), and face-centered (with corner points and points at the centers of the faces of the cube). With the work of Bravais, the external symmetry of a mineral became firmly grounded on the idea of the space lattice, but just how actual atoms or molecules were arranged within unit cells remained a matter of speculation.

In the latter part of the nineteenth century, various European scientists independently advanced crystallography beyond the point groups of the Bravais lattices by recognizing that the condition of translational equivalence was a restriction justified only by an external consideration of points. The condition of translational equivalence means that if one found oneself within a lattice and could move from point to point, one would find the same view of one's surroundings from each position. Leonhard Sohncke in Germany recognized that other symmetry elements could bring about equivalence. For example, he introduced the screw axis, in which a rotation around an axis is combined with a translation along it. In the late 1880's, the Russian mineralogist Evgraf Federov introduced another symmetry element, the glide plane, in which a reflection in a mirror plan is combined with a translation without rotation along an axis. Using various symmetry elements, Federov derived the 230 space groups, which represent all possible distributions that atoms can assume in minerals. Shortly after the work of Federov, William Barlow, an English chemist, began to consider the problem of crystal symmetry from a more concrete point of view. He visualized crystals not in terms of points but in terms of closely packed spherical atoms with characteristic diameters. In considering atoms to be specifically sized spheres, he found that there are certain geometric arrangements for packing them efficiently. One can appreciate his insight by thinking about arranging coins in two dimensions. For example, six quarters will fit around a central quarter, but only five quarters will fit around a dime (illustrating the importance of size in determining coordination number). Barlow showed that similar constraints hold for the three-dimensional packing of spherical atoms of different diameters.

The most revolutionary breakthrough with regard to mineral structure occurred in 1912, when Max von Laue, a German theoretical physicist working with two experimentalists, discovered that X rays passing through a crystal produced a diffraction pattern on a photographic plate. Before 1912, scientists had little genuine knowledge of the internal structure of minerals, but after 1912, scientists used X rays scattered from crystals to determine the precise arrangement of atoms in even extremely complicated minerals. The scientists principally responsible for extending the technique of X-ray diffraction to minerals were the British father-and-son team of William Henry Bragg and William Lawrence Bragg . Most notably, Lawrence Bragg interpreted the spots on diffraction photographs as reflections of the X rays from the planes of atoms in the crystal, and he derived a famous equation based on this understanding of X-ray diffraction. During the years 1913 and 1914, the Braggs worked assiduously in determining the structures of many minerals, for example, diamond (carbon), fluorite (calcium fluoride), pyrite (iron disulfide), and calcite (calcium carbonate), as well as sphalerite and wurzite (two forms of zinc sulfide). World War I delayed the spread of the X-ray diffraction technique, but after the war, the number of research workers increased, first in the United States, then in the Scandinavian countries.

As scientists determined more and more mineral structures, they became convinced that minerals are basically composed of spherical atoms or ions, each of characteristic size, packed closely together. For example, the silicate minerals were of central concern to William Lawrence Bragg in England and Linus Pauling in the United States. The basic unit in these minerals is the tetrahedral arrangement of four oxygen atoms around a central silicon atom. Each tetrahedral unit has four negative charges, and so one would expect that electric repulsion would force these tetrahedral building blocks to fly apart. In actual silicate minerals, however, these units are linked, in chains or rings or sheets, in ways that bring about charge neutralization and stability. These tetrahedra may also be held together by such positively charged metal ions as aluminum, magnesium, and iron. These constraints lead to a fascinating series of structures. Pauling devised an enlightening and useful way of thinking both about these silicate structures and about complex inorganic substances in general. In the late 1920's, he proposed a set of principles (now known as Pauling's Rules) that govern the structures of ionic crystals, that is, crystals in which ionic bonding predominates. The silicate minerals provide striking examples of his principles. One of his rules deals with how a positive ion's electrical influence is spread among neighboring negative ions; another rule states that highly charged positive ions tend to be as far apart as possible in a structure. Pauling's Rules allowed him to explain why certain silicate minerals exist in nature and why others do not. During the 1930's, the X-ray analysis of minerals was largely the province of physicists in Europe and chemists in the United States. Very few mineralogists, with their lack of advanced training in physics and mathematics, were equipped to make contributions in this new field. It was not until the 1960's that the structure determination of minerals became an important activity in some large geology departments. By this time, the slow and cumbersome early methods of working out mineral structures had been surpassed by computerized X-ray crystal-

Linus Pauling, whose investigations of silicates revealed much about mineral structure. *(The Nobel Foundation)*

lography, through which it was possible to determine, quickly and elegantly, the exact atomic positions of highly complex minerals. In the 1970's and 1980's, scientific interest shifted to the study of minerals at elevated temperatures and pressures. These studies often showed that temperature and pressure changes cause complex internal structural modifications of the mineral, including shifts in distances between certain ions and in their orientation to others. New minerals continue to be discovered and their structures determined. Structural chemistry has played an important role in deepening understanding both of these new minerals and of old minerals under stressful conditions. This knowledge of mineral structure has benefited not only mineralogists, crystallographers, and structural chemists, but also inorganic chemists, solid-state physicists, and many earth scientists.

Methods of Study

Early scientists identified minerals by such readily observable physical properties as hardness, density, cleavage, and crystal form. In the seventeenth and eighteenth centuries, chemists were able to determine the chemical compositions of various minerals by gravimetric methods, which involved dissolving samples in appropriate solvents and then precipitating and weighing individual constituents. In this way, they found that such metals as cobalt, nickel, and manganese were part of the fabric of certain minerals. The first methods of examining the external structure of minerals were quite primitive. In the seventeenth century, Nicolaus Steno cut sections from crystals and traced their outlines on paper. A century later, Arnould Carangeot invented the contact goniometer. This device, which enabled crystallographers to make systematic measurements of interfacial crystal angles, was basically a flat, pivoted metal arm with a pointer that could move over a semicircular protractor. William Hyde Wollaston, an English metallurgist, invented a more precise instrument, the reflecting goniometer, in 1809. This device used a narrow beam of light reflected from a mirror and directed against a crystal to make very accurate measurements of the angles between crystal faces. The reflecting goniometer ushered in a period of quantitative mineralogy that led to the multiplication of vast amounts of information about the external structure of minerals.

The discovery of the polarization of light in the nineteenth century led to another method of mineral investigation. Ordinary light consists of electromagnetic waves oscillating in all directions at right angles to the direction of travel, but a suitable material can split such light into two rays, each vibrating in a single direction (this light is then said to be plane polarized). Various inventors, mainly in England, perfected the polarizing microscope, a versatile instrument using plane-polarized light to identify minerals and to study their fine structure. Even the darkest minerals could be made transparent if sliced thin enough. These transparent slices produced complex but characteristic colors because of absorption and interference when polarized light passed through them.

In the twentieth century, X-ray diffraction provided scientists with a tool vastly more powerful than anything previously available for the investigation of internal mineral structures. Before the development of X-ray methods, the internal structure of a mineral could be deduced only by reasoning from its physical and chemical properties. After X-ray analysis, the determination of the detailed internal structures of minerals moved from speculation to precise measurement. The phenomenon of diffraction had been known since the seventeenth century. It can be readily observed when a distant street light is viewed through the regularly spaced threads of a nylon umbrella, causing spots of light to be seen. In a similar way, Max von Laue reasoned that the closely spaced sheets of atoms in a crystal should diffract X rays, with closely spaced sheets diffracting X rays at larger angles than more widely spaced ones. William Lawrence Bragg then showed how this technique could be used to provide detailed information about the atomic structure of minerals. In terms of determinations done and papers produced, X ray methods have continued to outnumber other techniques that later proliferated for studying mineral structure.

The powder method of X-ray diffraction proved to be a useful and easy way to identify minerals and to obtain structural information. Albert W. Hull at General Electric and Peter Debye and Paul Scherrer at Göttingen University independently discovered this method, which consists of grinding a mineral specimen into a powder that is then formed into a rod by gluing it to a thin glass fiber. As X rays impinge on it, this rod is rotated in the center of a cylindrical photographic film. The diffraction pattern on the film can then be interpreted in terms of the arrangement of atoms in the mineral's unit cell.

Although X rays have been the most important type of radiation used in determining mineral structures, other types of radiation, in particular infrared (with wavelengths greater than those of visible light), have also been effective. Infrared radiation causes vibrational changes in the ions or molecules of a particular mineral structure, and that permits scientists to map its very detailed atomic arrangement. The technique of neutron diffraction makes use of relatively slow neutrons from reactors to determine the locations of the light elements in mineral structures (the efficiency of light elements in scattering neutrons is generally quite high).

In recent decades, scientists have continued to develop sophisticated techniques for exploring the structure of minerals. Each of these methods has its strengths and limitations. For example, the electron microprobe employs a high-energy beam of electrons to study the microstructure of minerals. This technique can be used to study very small amounts of minerals as well as minerals in situ, but the strong interaction between the electron beam and the crystalline material produces anomalous intensities, and thus electron-microprobe studies are seldom used for a complete structure determination. Many new techniques have helped scientists to perform structural studies of minerals in special states—for example, at high pressures or at temperatures near the melting point—but the most substantial advancements in determining mineral structures continue to involve X-ray analysis.

Context

A central theme of modern mineralogy has been the dependence of a mineral's external form and basic properties on its internal structure. Because the arrangement of atoms in a mineral provides a deeper understanding of its mechanical, thermal, optical, electrical, and chemical properties, scientists have determined the atomic arrangements of many hundreds of minerals by using the X-ray diffraction technique. This great amount of structural information has proved to be extremely valuable to mineralogists, geologists, physicists, and chemists. Through this information, mineralogists have gained an understanding of the forces that hold minerals together and have even used crystal-structure data to verify and correct the formulas of some minerals. Geologists have been able to use the knowledge of mineral structures at high temperatures and pressures to gain a better understanding of the eruption of volcanoes and other geologic processes. Physicists have used this structural information to deepen their knowledge of the solid state. Through crystal-structure data, chemists have been able to expand their understanding of the chemical bond, the structures of molecules, and the chemical behavior of a variety of substances.

Because minerals often have economic importance, many people besides scientists have been interested in their structures. Rocks, bricks, concrete, plaster, ceramics, and many other materials contain minerals. In fact, almost all solids except glass and organic materials are crystalline. That is why a knowledge of the structure and behavior of crystals is important in nearly all industrial, technical, and scientific enterprises. This knowledge has, in turn, enabled scientists to synthesize crystalline compounds to fill special needs: for example, high-temperature ceramics, electrical insulators, semiconductors, and many other materials.

Bibliography

Bragg, William Lawrence. *Atomic Structure of Minerals*. Ithaca, N.Y.: Cornell University Press, 1937. Bragg wrote this book while he was Baker Professor at Cornell University in the spring semester of 1934. Primarily a discussion of mineralogy from the perspective of the vast amount of new data generated by the successful application of X-ray diffraction analysis to crystalline minerals. Because of its provenance in a series of general lectures, the text is highly readable, though a knowledge of elementary physics and chemistry is presupposed. Of use to mineralogists, physicists, chemists, and all other scientists interested in the physical and chemical properties of minerals.

Bragg, William Lawrence, G. F. Claringbull, and W. H. Taylor. *The Crystalline State*. Vol. 4, *The Crystal Structures of Minerals*. Ithaca, N.Y.: Cornell University Press, 1965. The X-ray analysis of crystals generated so much data that proved to be of interest to workers in so many branches of science that Bragg needed several collaborators and volumes to survey the subject; this volume is a comprehensive compilation of crystal-structure information on minerals. Because each collaborator wrote on that aspect of the subject of which he had expert knowledge, the analyses of structures are authoritative. Can be appreciated and used by anyone with a basic knowledge of minerals, as crystallographic notation is kept to a minimum and the actual structures take center stage.

Evans, Robert Crispin. *An Introduction to Crystal Chemistry*. 2d ed. New York: Cambridge University Press, 1964. In this book, Evans, a Cambridge chemist, analyzes crystal structures in terms of their correlation with physical and chemical properties. His approach is not comprehensive; rather, he discusses only those structures that are capable of illustrating basic principles that govern the behavior of these crystals. Though the author's approach demands some knowledge of elementary chemistry and physics on the part of the reader, there is no need for detailed crystallographic knowledge.

Lipson, Henry S. *Crystals and X-Rays*. New York: Springer-Verlag, 1970. Lipson wrote this book, which is part of the Wykeham Science series, to give advanced high school students and college undergraduates an inspiring introduction to the present state of X-ray crystallography. Though many scientists treat crystallography as a mathematical subject, Lipson stresses the observational and experimental, for example, by showing how the X-ray diffraction technique was used to determine the structures of some simple minerals.

Pauling, Linus. *The Nature of the Chemical Bond and the Structure of Molecules and Crystals: An Introduction to Modern Structural Chemistry.* 3d ed. Ithaca, N.Y.: Cornell University Press, 1960. Pauling's first scientific paper was the X-ray analysis of a mineral, molybdenite, and he went on to determine many other mineral structures. He used both crystal structures and quantum mechanics to develop a classic theory of the chemical bond. This book grew out of his own work and his tenure as Baker Professor at Cornell University during the fall semester of 1937. The beginner will encounter difficulties, but readers with a good knowledge of chemistry will find this book informative and inspiring.

Sinkankas, John. *Mineralogy for Amateurs.* New York: Van Nostrand Reinhold, 1966. As the title suggests, Sinkankas intended his book primarily for the amateur mineralogist. Because of its simplified presentation of many complex ideas, it has become popular with nonprofessionals. Includes a good chapter on the geometry of crystals, in which the basic ideas of mineral structure are cogently explained. Very well illustrated with photographs and drawings (many of the latter done by the author).

Smith, David G., ed. *The Cambridge Encyclopedia of Earth Sciences.* New York: Crown, 1981. This volume is part of a Cambridge series of reference works dedicated to the sciences. The various sections of this encyclopedia (on geology, mineralogy, oceanography, seismology, and the physics and chemistry of the earth) were written by authorities from England and the United States. Though primarily a reference work, this book is both readable and informative in most sections; for example, part 2 contains a good analysis of the internal structure of minerals. Some knowledge of elementary physics and chemistry is needed for a full understanding of most sections. Profusely illustrated with helpful diagrams and photographs.

Robert J. Paradowski

Cross-References

Minerals: Physical, 415.

Mountain Belts

Mountain belts are products of plate tectonics, produced by the convergence of crustal plates. Topographic mountains are only the surficial expression of processes that profoundly deform and modify the crust. Long after the mountains themselves have been worn away, their former existence is recognizable from the structures that mountain building forms within the rocks of the crust.

Field of study: Tectonics

Principal terms

CONTINENTAL SHELF: the submerged offshore portion of a continent, ending where water depths increase rapidly from a few hundred to thousands of meters

EPEIROGENY: uplift or subsidence of the crust within a region, without the internal disturbances characteristic of orogeny

GABBRO: a silica-poor intrusive igneous rock consisting mostly of calcium-rich feldspar and iron and magnesium silicates; its volcanic equivalent is basalt

GRANITE: a silica-rich intrusive igneous rock consisting mostly of quartz, potassium- and sodium-bearing feldspar, and biotite or hornblende

IGNEOUS: from the Latin *ignis* (fire), a term referring to rocks formed from the molten state or to processes that form such rocks

METAMORPHISM: the change in the mineral composition or texture of a rock because of heat, pressure, or the chemical action of fluids in the earth

OROGENY: the profound disturbance of the earth's crust, characterized by crustal compression, metamorphism, volcanism, intrusions, and mountain formation

PLATE TECTONICS: the theory that the crust of the earth consists of large moving plates; orogeny occurs where plates converge and one plate overrides the other

SEDIMENTARY ROCKS: rocks that form by surface transport and deposition of mineral grains or chemicals

SUBDUCTION: the sinking of a crustal plate into the interior of the earth; subduction occurs at subduction zones, where plates converge

Summary

Mountains have many origins. They can be volcanic, like Mount Vesuvius, or they can be formed by vertical movements along faults, as the Sierra Nevada or the Ruwenzori of central Africa were. Some mountains are the result of relatively gentle epeirogenic (deformational) uplift of the crust—for example, the Black Hills or the Adirondack Mountains. The causes of epeirogeny, however, are poorly understood. The great mountain chains of the earth, such as the Andes, the Rocky Mountains, or the Himalayas, formed not only from uplift but also from internal deformation of the crust, volcanic activity, metamorphism, and the intrusion of vast quantities of molten rock into the crust, especially granite and related rocks. These processes are collectively called orogeny, and mountain chains that form from such processes are called orogenic belts. Orogeny is one of the most important consequences of plate tectonics. It occurs when plates collide and one plate overrides the other—a process known as subduction—in response to compressional forces and heating generated by the plate collision.

The earth's crust consists of two types, called continental and oceanic by geologists. The continents and the adjacent continental shelves are underlain by granitic crust, averaging about 40 kilometers thick. The ocean floors are made of gabbro and basalt, averaging about 5 kilometers thick. The true edge of a continent is not the shoreline, which is constantly changing, but the boundary between continental and oceanic crust. The edge of the continental shelf coincides closely with this boundary.

Many mountain belts form at subduction zones with a continental overriding plate and an oceanic descending plate. The downward bending of the plate creates a deep, narrow trench on the ocean floor, sometimes more than 10 kilometers below sea level. The descending oceanic plate sinks into the earth's interior, eventually to be reabsorbed. The overriding plate experiences orogeny. All orogenic belts differ in detail, but most have certain major features in common. A typical orogenic belt consists of parallel zones, which may be defined by distinctive rock type, type of metamorphism, level of igneous activity, or type of deformation of the rocks. The zones, generally parallel to the boundary where the two plates collide, are the result of different crustal conditions and processes at different distances from the plate boundary. It is useful to regard orogenic belts as having an "outer" side, adjacent to the plate boundary, and an "inner" side within the overriding plate.

The first zones recognized in orogenic belts were those defined by environment of deposition: an outer zone of thick, deep-water sedimentary rocks and volcanic rocks and an inner zone of thinner, shallow-water sedimentary rocks without abundant volcanic rocks. The nineteenth century American geologist James Hall first described these zones, which he envisioned as parallel troughs formed by downward folding of the crust. Because these troughs were viewed as immense versions of ordinary downward folds,

439

or synclines, James Dana later called the troughs geosynclines. The outer trough was called eugeosyncline, and the inner trough was called miogeosyncline.

The original concept of the geosyncline disturbed many geologists, because it did not quite match the structure of active mountain belts. In 1964, Robert Dietz reexamined the geosyncline concept and showed that the rocks need not have accumulated in troughs. With this insight, it became clear that the rocks of miogeosynclines corresponded closely to those of the continental shelves, while eugeosynclines were a good match to the rocks of many volcanic island chains. Later, it became clear that the rocks of the eugeosyncline often formed separately from the miogeosyncline and were later juxtaposed by plate motions.

Because of these revisions of the original geosyncline concept, many geologists have abandoned the original terms and prefer the terms "geocline," "eugeocline," and "miogeocline" instead. The eugeocline is the outer belt of deep-water sedimentary rocks and volcanic rocks. The miogeocline is the inner belt of shallow-water sedimentary rocks. Beyond the miogeocline is the platform, where thin shallow-water or terrestrial rocks were deposited on the stable interior of the overriding plate. In addition, most orogenic belts have an inner belt of coarse sedimentary rocks deposited late in the history of the orogenic belt. This belt, called the molasse basin, consists of debris eroded from the mountains and deposited at their base.

The Alps are an example of a mountain belt produced by recent epeirogenic uplift. *(Robert McClenaghan)*

Molasse basins consist mostly of rocks deposited in shallow-water or land environments.

Much of the structure of a mountain belt is related to processes in the descending plate. As the descending plate reaches a depth of about 100 kilometers, it begins to melt, and molten rock, or magma, invades the overriding plate. In general, volcanic rocks in orogenic belts become progressively richer in silica with increasing distance from the plate boundary, because the rising magma has more time to react with silica-rich continental crust. Also, volcanic and intrusive rocks in mountain belts tend to become more silica-rich over time. Most orogenic belts have a main axis of igneous activity, the igneous arc, where volcanic and intrusive activity are concentrated. The igneous arc is generally on the inner side of the eugeocline. In deeply eroded orogenic belts, intrusive rocks of the igneous arc, usually granitic in composition, are exposed as great masses known as batholiths.

Different thermal conditions in different parts of the overriding plate give rise to two distinct zones of metamorphism. Adjacent to the descending plate, rocks are carried downward to great depths, often 20 kilometers or more, but because the rocks are in contact with the still-cool descending plate, they remain unusually cool. Temperatures in this zone generally average 200-300 degrees Celsius, instead of the 500-600 degrees Celsius that might be expected at 20 kilometers depth. This low-temperature, high-pressure metamorphism is known as blueschist metamorphism, because many of the minerals that form often impart a bluish color to the rocks.

Generally coinciding with the igneous arc is an inner belt of metamorphism where temperatures are high but pressures are moderate. Peak temperatures commonly exceed 600 degrees Celsius, with pressures typically reflecting depths of 5-10 kilometers. Such conditions are called amphibolite metamorphism. Adjacent to the region of highest temperature is a region of lower-temperature metamorphism, where temperatures of 400-500 degrees Celsius prevail. This type of metamorphism is called greenschist metamorphism, from the greenish color of many of the minerals formed. The outer zone of blueschist metamorphism generally occupies the outer part of the geocline. Amphibolite metamorphism generally coincides with the igneous arc and the inner part of the eugeocline. Greenschist metamorphism commonly extends into the miogeocline.

The deformation of rocks in the overriding plate depends on the nature of the rocks, stress, temperature, and confining pressure. Orogenic belts display several zones of distinctive structures. The most important of these belts are the accretionary prism, zone of basement mobilization, and the foreland fold-and-thrust belt. The eugeocline in general is a region of intense deformation; deformation in the miogeocline is less intense. The outermost edge of the orogenic belt is occupied by the accretionary prism, and it often forms much of the eugeocline. Where the colliding plates meet,

441

sediment is scraped off the descending plate. Other sediment is eroded from the continent and pours into the trench. The sediment from the continent is deposited rapidly, with little weathering or sorting, to form impure sandstone called graywacke. Submarine landslides and slumps are common in the unstable setting of the trench; the resulting complex of chaotically deposited graywacke is called flysch. A wedge of intensely deformed sediment accumulates on the edge of the continent, much the way a wedge of snow accumulates ahead of a snowplow. This wedge of sediment is the accretionary prism. Frequently, fragments of oceanic crust break off the descending plate and are incorporated into the accretionary prism. These slices of oceanic crust, called ophiolites, are of enormous geologic value. Not only do they mark the location of former subduction zones, but they also provide otherwise unobtainable cross sections of oceanic crust exposed on dry land. The actual contact between the two plates is marked by mélange, a chaotic mixture of broken rock with fragments ranging from microscopic to kilometers in size.

The high temperatures in the igneous arc and amphibolite zone of metamorphism can make the rocks of the crust plastic. That is, the rocks flow like stiff fluids, even though they do not melt. Because the hot rocks are less dense than the cooler crust around them, they rise upward. A mass of rock that flows upward in this manner is called a diapir. The process of heating deep crust (or "basement") so that it rises is called basement mobilization, and evidence of it occurs in many mountain belts. The mobilized crust appears as intensely deformed and highly metamorphosed rock called gneiss. Often the rising mass of gneiss appears to have shouldered the overlying rocks aside, so that the rocks are arched upward with a central core of gneiss. Such a structure is called a gneiss dome.

Compressional forces arising from plate convergence thicken the crust of the overriding plate in a number of ways. Within the accretionary prism, sheets of sedimentary rock are thrust downward beneath the overriding plate, resulting in a stack of faulted slices of rock. These slices may thicken internally by fracturing along small faults and stacking the resulting small slices one above the other, a process called duplexing. Within the igneous arc, crustal thickening occurs when magma invades the crust, increasing its volume. Magma can also be added to the base of the crust, a process called underplating. Heating of the crust within the igneous arc makes much of the lower crust plastic, permitting the plastic crust to be squeezed upward by compression. This process probably assists the upward movement of gneiss domes.

The thickening of the crust results in the uplift of the surface and the formation of topographic mountains. In the foreland, which basically coincides with the miogeocline, deformation is largely a response to events in the active core of the mountain belt. Some of the deformation in the foreland seems to be driven by rising masses of mobilized basement. Rocks of the miogeocline and some of the underlying crust fracture into sheets that are shoved over the rocks beneath. Fractures or faults where one mass of rock overrides another are called thrust faults. These rocks may also be buckled into folds by compressive forces. Rocks nearer the surface often slide off the rising mountain belt. The rocks may break into thin sheets called nappes that stack one atop the other or may crumple into folds. Often the folded rocks have detached from the rocks beneath, much like a carpet slides and folds when a piece of furniture is pushed over it. This process of detachment is called décollement. Because this deformation involves only the surface layers of rock and not the underlying basement rocks, it is called thin-skinned deformation.

Other kinds of plate collisions result in different combinations of structures. Oceanic-oceanic subduction zones have somewhat simpler orogenic belts. When the descending plate begins to melt, magma rises and breaks through the surface to create a volcanic island arc such as is found in the Aleutian Islands or the Lesser Antilles. Since both plates are oceanic crust, made largely of basalt, the magma also is basalt. Erosion strips sediment off the volcanic islands and dumps much of it in the trench to form an accretionary prism. Over a very long time, the island arc may be built up into a continuous belt of intensely deformed volcanic rocks and sedimentary rocks derived from them. The Greater Antilles and the Isthmus of Panama probably formed this way. Such orogenic belts consist essentially of a eugeocline and igneous arc, with the associated metamorphic zones and deformation structures.

Continent-continent collisions start out as continent-ocean subduction zones, but eventually the convergence of plates brings two continents together, one of which is pushed beneath the other. Continent-continent collisions include the Himalaya, where India is being pushed beneath Tibet, and the Persian Gulf, where the Arabian Peninsula is being pushed beneath Iran. Because continental crust is thick and relatively light, the descending plate cannot be subducted. Instead, one continent rides onto the other, creating a double thickness of crust. Eventually, resistance to further movement may cause plate motions to change on a regional or even global scale.

Usually the overriding continent has had a long history of orogeny before the collision, whereas the other continent may have had none. Orogeny results in such a wide range of structures that it is usually immediately obvious which of the continents is or was the overriding plate. The collision boundary between the two continents, called a suture, may display relics of the former accretionary prism, including mélange, fragments of ophiolites, or evidence of blueschist metamorphism.

Often a small block of crust, called a terrane, collides with a larger plate. The terrane may be a volcanic island chain or a small fragment of continental crust. The northern coast of New Guinea is an area where terranes (in this case volcanic island chains) are colliding with a continent. The addition of terranes to a larger plate is called accretion. Terranes are recognizable as

443

distinct blocks of crust separated from adjacent rocks by major faults. In many cases, eugeoclines did not form near their corresponding miogeoclines but as separate terranes. Repeated accretion of terranes can add large areas to a continent. Roughly 1,000 kilometers of the western United States was accreted to North America in the last 500 million years.

After orogeny ceases, it is common for mountain belts to experience a period of crustal extension and faulting. Once the compressional forces that uplifted the mountains subside, many mountain ranges simply cannot support their own mass and begin to spread under their own weight. It requires about 20 million years for erosion to level a mountain range. Nevertheless, long after the topographic mountains are gone, the structures created by orogeny remain. The most conspicuous markers of ancient orogenies are usually the eugeocline and miogeocline, igneous arc, and molasse deposits.

If mountains are worn away within a few tens of millions of years, it follows that the present Urals and Appalachians, the products of continental collisions more than 200 million years ago, cannot be remains of the original mountains. In the case of the Appalachians, this point is clear because rivers such as the Potomac flow across the structures in the mountain belt. Clearly, the Appalachians were once level enough (or buried) that rivers could flow across them. The Appalachians, the Urals, and even the Alps are the result of recent epeirogenic uplift after erosion had largely, or entirely, leveled the original mountains. Why mountain belts sometimes experience renewed periods of uplift long after orogeny ceases is unknown.

Methods of Study

Since the earliest days of geology, the formation of mountain belts has been regarded as one of the central problems of geology. Interpreting the structure and origin of mountain belts makes use of virtually every methodology used in the earth sciences. One of the most conspicuous features of geology as a science is its emphasis on spatial and temporal relationships. The relationships in space and time between rock units are the key pieces of information required to interpret the history of the earth. The task facing the geologist on the small scale is to integrate hundreds of local observations into a coherent picture that describes the distribution and history of the rocks in a small region. The task facing the geologist on the large scale is to integrate many local studies into a coherent history of a large region (possibly the entire earth).

Geologic mapping is the major tool for summarizing the spatial relationships of rock units. In modern geologic mapping, the geologist in the field plots observations on a base map or aerial photograph, using various symbols for rock type and orientation of bedding or other structures. Time and sequence relationships can be determined at contacts between different rock units, using methods that are essentially common sense. The fundamental principle is that later events are overprinted on earlier ones. The principle of superposition, for example, holds that the lowest layers in a series were deposited before the higher layers. Observations at rock contacts allow the geologist to piece together the sequence of events in an area.

Geologic mapping in orogenic belts presents particular problems. Rocks are often folded or tilted, displaced by faults, and shuffled out of their normal sequence by thrust faults. The fundamental principles described above still apply. Folded rocks must have been folded after the youngest rocks in the fold were deposited, faults must be younger than the rocks they displace, and so on. Additional geologic methods can provide further information. Many structures in sedimentary rocks, such as ripple marks formed by gentle wave action, are all essentially governed by gravity and have a distinct "right way up." They help the geologist determine the original orientation of tilted rocks. Fossils can be used to correlate rock layers and determine that rocks have been faulted out of their normal order. Radiometric dating can provide additional information on the sequence and timing of events.

Comparisons between regions are invaluable in working out the history of an orogenic belt. Processes that are hard to recognize in some regions may be very obvious in others. Thrust faults and exotic terranes are major features of the Appalachians, for example, but both types of structures were first recognized in rugged mountain ranges where the rocks are much better exposed than in the low and forested Appalachians.

The internal structure of mountain ranges can be studied by a variety of methods. In ranges such as the Alps or the Himalaya, several kilometers of internal structure may be directly exposed in deep valleys, yet even this much relief amounts to less than a tenth of the thickness of the crust. Drilling, usually for petroleum, also provides direct access to the interior of orogenic belts. Seismic studies, where natural or artificial seismic waves are detected after traveling through the earth's crust, are an additional means of probing deep beneath mountain ranges. Gravity and magnetic studies, in which buried rock masses can be detected by their gravitational attractions or magnetic fields, provide indirect access to the deep interior of mountain ranges.

Context

To preindustrial ages, mountains were places of danger filled with natural hazards, wild animals, and possibly hostile peoples. To a technological age with a vastly greater command of nature, mountains are places of beauty that generate income. Switzerland is the supreme example of a land where mountains yield wealth for human beings. Apart from their value as centers for recreation, mountains are economically valuable as sources of mineral wealth. The geologic processes that create mountains also concentrate mineral deposits. The Andes of Chile contain some of the world's largest copper deposits, the Andes of Colombia produce emeralds, and the foothills of the Sierra Nevada are famous for their gold. Understanding the processes that create mountains also improves understanding of how mineral resources form and where to seek them.

Yet mountains can be actively dangerous. The processes that create them also result in earthquakes and volcanic eruptions, while the processes that erode them create floods and landslides. Many of the most mountainous nations, where mountain building is a continuing process, are also underdeveloped nations with vulnerable, rapidly expanding populations.

Bibliography

Cook, Frederick A., L. D. Brown, and J. E. Oliver. "The Southern Appalachians and the Growth of Continents." *Scientific American* 243 (October, 1980): 156-168. Seismic probing has revealed the fault where ancient Africa rode onto North America to create the southern Appalachians. *Scientific American*, which is aimed at the scientifically informed but nonspecialist reader, roughly at college level, is probably the best source of information on recent advances in science for nonspecialists. Its coverage of advances in the earth sciences since 1970 has been especially thorough.

Dietz, Robert S. "Geosynclines, Mountains, and Continent-Building." *Scientific American* 226 (March, 1972): 30-38. Dietz's reanalysis of the traditional geosyncline concept bridged the conceptual gap between modern concepts of plate tectonics and the evolution of ancient mountain belts.

Howell, David G. "Terranes." *Scientific American* 253 (November, 1985): 116-125. Terranes are small blocks of crust added to a mountain belt by plate collisions. This article describes types of terranes and what happens when they collide.

James, David E. "The Evolution of the Andes." *Scientific American* 229 (August, 1973): 60-69. A summary of a relatively simple mountain chain whose formation is related to the continuing convergence of the Pacific and South American plates. Such mountain belts are a good starting point for understanding more complex mountains.

Jones, David L., A. Cox, P. Coney, and Myrl Beck. "The Growth of Western North America." *Scientific American* 247 (November, 1982): 70-84. The westernmost 1,000 kilometers of North America are a mosaic of at least 200 small blocks added by plate collision in the last 500 million years. This article was written by some of the scientists who were most influential in discovering this process.

Molnar, Peter. "The Structure of Mountain Ranges." *Scientific American* 255 (July, 1986): 70-79. A summary of discoveries about the internal structure of mountain ranges. The apparently solid crust of a mountain range is actually brittle and unable to support its own weight without external compressive forces.

Molnar, Peter, and P. Tapponier. "The Collision Between India and Eurasia." *Scientific American* 236 (April, 1977): 30-41. The mechanical paradox of plate tectonics driving India northward into Eurasia has been resolved by the deformation of the Eurasian crust far beyond the immediate collision zone. This article illustrates one way that continental collisions can result in large-scale changes in crustal behavior.

Spencer, Edgar W. *Introduction to the Structure of the Earth*. 3d ed. New York: McGraw-Hill, 1988. A college-level textbook on structural geology. Chapters 1-3 survey plate tectonics and its role in deforming the crust. Chapter 19 describes mountain ranges that form from continent-ocean plate collision, particularly the structure of western North America. Chapter 20 describes mountains that form by continent-continent collision, with emphasis on the Appalachians and Alps.

Suppe, John. *Principles of Structural Geology*. Englewood Cliffs, N.J.: Prentice-Hall, 1985. A college-level textbook on the structures that form when rocks deform. Chapter 1 is a survey of plate tectonics and its role in deforming the crust. Chapter 12 describes the structure of the Appalachians, and Chapter 13 describes the structure of western North America.

Steven I. Dutch

Cross-References

Continental Crust, 78; Continental Growth, 102; Folds, 248; Island Arcs, 332; Plate Tectonics, 505; Stress and Strain, 607; Subduction and Orogeny, 615.

Normal Faults

Normal faults are common features that occur when the earth's crust is subjected to tensional forces. The sense of movement is primarily vertical and results in an extension of the crust. These faults are generally associated with broad-flexed or uplifted areas and are an integral part of the modern concept of plate tectonics.

Field of study: Structural geology

Principal terms

DIP: the angle of inclination of a fault, measured from a horizontal surface; the dip direction is perpendicular to strike

FAULT: a break in the earth's crust that is characterized by movement parallel to the surface of the fracture

FAULT DRAG: the bending of rocks adjacent to a fault

FOOTWALL: the crustal block underlying the fault

GRABEN: a long, narrow depressed crustal block bounded by normal faults that may form a rift valley

HANGING WALL: the crustal block that overlies the fault

HORST: a long, narrow elevated crustal block bounded by normal faults that may result in a fault-block mountain

SLICKENSIDES: fine lines or grooves along a faulted body that usually indicate the direction of latest movement

STRESS: the forces acting on a solid rock body within a specified surface area

THROW: the vertical displacement of a rock sequence or key horizon measured across a fault

Summary

A normal fault is a fracture that separates two crustal blocks, one of which has been displaced downward along the fractured surface. Some workers use the term "gravity faults" to indicate an apparent normal fault if genesis, rather than geometry, is implied. Crustal blocks overlying or underlying a normal or reverse fault are commonly designated as the hanging wall and footwall blocks, respectively. These are old descriptive terms that were used in the early English coal-mining districts. Faults that were inclined toward the downdropped side were common in the area and the term "normal fault" was applied. At places where the movement was in the opposite direction, the breaks in the rock were designated as reverse faults. The displacement of normal faults, which can be intermittent, ranges from less than a meter to thousands of meters. The inclination or dip of the fault can be from nearly horizontal to vertical, but generally ranges from 45 to 60 degrees. In some areas, the angle of dip decreases with depth and results in a curved surface that is concave upward. This curved surface is termed a listric normal fault and is a common type of fracture in the Gulf Coast region of the United States.

Normal faults are the product of a dynamic process that results in conditions of changing stress (force per unit area) along a plane of weakness in the earth's crust. The fault develops from a point along this plane. According to Lamoraal de Sitter, the stress is minimum at the starting point along the surface and maximum at the edges. Because of the edge conditions, the plane steepens and splits into several divergent smaller faults, or splays. These small segments may join to form a larger normal fault with a scalloped trace. Subparallel normal faults with smaller displacements generally accompany the large faults. At places, the adjacent beds are systematically fractured without significant displacement. These fractures are termed joints and generally have a high density (close spacing) near the fault.

The deforming forces can be related to three mutually perpendicular but unequal axes designated as maximum principal stress (σ_1), intermediate principal stress (σ_2), and minimum principal stress (σ_3). In the case of normal faults, the primary deforming force (σ_1) is vertical or nearly vertical. The least stress (σ_3) is horizontal. The normal faults are actually steeply inclined shear fractures that formed in response to forces promoting the sliding of adjacent blocks past each other. These fractures generally form at an angle of 30 degrees from the maximum principal stress. The orientation of the maximum principal stress is horizontal for thrust faults and for wrench (transform) faults.

Normal faults are classified according to the type of displacement of fault blocks relative to a known point. Based on the slip or actual movement of formerly adjacent points on opposing fault blocks, three types are commonly designated: strike slip, or movement along the trend of the fault; dip slip, or movement directly down the fault surface; and oblique slips, or diagonal movement down the fault surface. The movement along all these examples is translational; consequently, no rotation of the blocks in respect to each other has occurred outside a disturbed zone adjacent to the fault. If the actual displacement is not known, the term "separation" is used by most geologists to indicate the apparent movement on a map or cross section. Heave and throw are the horizontal and vertical components, respectively, of the dip separation as measured along a vertical profile that is at right angles to the trend of the fault.

There are several varieties of normal faults. Detachment (denudation) faults have a low angle of dip (usually less than 30 degrees) and are common features in the western United States. Growth or contemporaneous faults are listric normal faults that are active during sediment accumulation. Layered rocks on the downthrown side of the fault are thicker than equivalent beds on the upthrown side. Smaller subsidiary or antithetic faults commonly form on the downthrown side of the main fault but dip in a direction opposite to the master fault. These are common features along the Gulf Coast of the United States.

Some special faults may result from the same stress orientation as normal faults; that is, the maximum principal stress is vertical. These closely related faults, however, are characterized by rotational movement between blocks. For example, hinge faults increase in displacement along the length of the fault; linear features that were parallel before faulting are not all parallel after faulting. A pivotal (scissor) fault is another example of a rotational fault. In this type, the fault blocks pivot about an axis that is at right angles to the fault surface; the movement on the downthrown side is in opposite directions (up and down) along the length of the fault.

Major steeply dipping normal faults occur in the Colorado Plateau (Arizona and New Mexico) where these features are closely associated with regional flexures called monoclines. The western part of the plateau along the Colorado River is divided into large structural blocks by three north-trending faults. One of these faults, the Hurricane fault of Arizona and Utah, dips to the west and has a maximum displacement of 3,048 meters.

At some places, normal faults bound narrow blocks that have been displaced up or down. An uplifted or elevated block is called a horst; the depressed block is termed a graben. Topographically, these structural features may be represented by a series of mountain ranges and intervening valleys, respectively. The Basin and Range system of the western United States is a good example of this horst-and-graben type complex. In this region, both low- and high-angle normal faults have shaped an area that extends from southern Oregon southward to northern Mexico; the area has been broadly uplifted and the crust stretched in an east-west direction by the normal faulting. Some estimates of the total extension across the region are more than 100 percent. The displacement along these large faults ranges up to 5,486 meters. In Europe, the Rhine graben is a classic example of a well-developed rift system. This narrow structural trough trends northward for nearly 300 kilometers through West Germany and controls the path of the upper Rhine River.

Methods of Study

Normal faults are recognized in the field or on vertical aerial photographs and satellite images by identifying features characteristic of this type of fault. On the earth's surface, these faults occur as geological lines that are revealed by a sharp, curvilinear line in the bedrock that is usually accentuated by vegetation, a sharp contact in adjacent rock types in section or map view, a marked change in structural style, an abrupt change in topography, or an anomalous drainage pattern. Normal faults are usually recognized in vertical drill holes by an omission of rock layers; comparison of rock samples (drill cuttings) or mechanical logs from adjacent borings will generally reveal the part of the rock column that is missing. Caution, however, must be exercised to make sure that the strata have not been eroded, have not thinned, or have not changed character laterally.

There are many distinctive geometric, mineralogic, or physiographic features that are associated with large normal faults. Some of these features, however, are also characteristic of other types of faults and should thus merely be considered as "clues" in recognizing normal faults. Many normal faults are expressed at the surface by low cliffs or scarps that reflect the minimum displacement along the fault; however, these straight slopes are usually modified by erosion to form faultline scarps. The scarps may be notched by streams crossing the upthrown block at a high angle to the fault trend; continued erosion by these side streams may result in triangular-shaped bedrock facets on the footwall with fan-shaped stream deposits on the downthrown hanging wall block. Movement along the fault usually disturbs rocks adjacent to the break and results in beds along the fault being bent up or down; in other words, the rocks bend before rupture takes place. This phenomenon is called fault drag. Normal drag occurs when rocks on the upthrown side are bent down into the fault and rocks on the down-thrown side are bent upward. Reverse drag occurs when beds on both sides of the fracture are bent down. Because movement along the fault produces an irregular surface between the fault blocks at some places, subsurface fluids are provided an avenue to the surface. Both hot- and cold-water springs are common occurrences along large normal faults. Solutions moving along the fault may also deposit minerals such as calcite or quartz between the blocks. These fillings are usually stained yellow or reddish brown by iron oxide. Movement along the fault is usually recorded by fine lines or by narrow grooves on the fault surface called slickensides. These features, however, may only indicate the latest movement along the fault. Impressions of the slicken lines are sometimes preserved on the outer surfaces of the mineral fillings. A series of larger scale (several centimeters or more of relief), parallel grooves and ridges may produce an undulating fault surface. The movement of large blocks along the fault usually produces low, step-like irregularities on the surface that are steeply inclined in the direction of movement. These features can be used to identify the direction of movement when a fault surface is poorly exposed. As the fault blocks slide past each other, angular rock fragments are dislodged and may accumulate to form a tectonic breccia; a microscopic breccia, or mylonite, may also result from movement. In some cases, the dislodged rock may be ground to a pliable, claylike substance called gouge. At places, a large fragment of bedrock, called a horse, is caught along a normal fault.

Normal faults generally occur in definite region patterns that are easily represented on geologic maps. Zones of overlapping, or *en echelon*, normal faults are common in the Gulf Coast region of the United States. In Texas, individual faults within the Balcones and Mexia-Luling fault systems are not continuous along strike but overlap with adjacent faults that have a similar trend. These fault zones generally follow the path of the buried Ouachita fold belt and mark the boundary between the geologically stable area of Texas and the less stable Gulf Coast region. Parallel or subparallel faults in an area also form a distinctive map pattern; if most of the faults are downthrown in the same direction, these structural features are designated as step faults. Radial fault patterns are common over or around central uplifts or domed areas of the crust. These faults are generally associated with local stretching of the crust that results from the emplacement of salt masses (plugs) or igneous intrusions.

Some normal faults are also closely related to the development of plunging (inclined) folds and form characteristic map patterns. According to de Sitter, steeply dipping normal cross faults may form nearly at right angles to the trend of concentric (formed of parallel layers) folds. These faults are parallel to the principal deforming force and occur during the folding process. The maximum displacement occurs along the highest part (crest) of the fold; these faults usually die out along the flanks. Longitudinal crest faults may occur parallel to the trend of the folds. These normal faults are perpendicular to the principal deforming force and probably form as the compressional forces diminish.

Context

Faults have played a significant role in shaping the earth's surface throughout geologic time. The occurrence of normal faults is closely tied to modern plate tectonics, a unifying concept for the geological sciences. The faults are generally associated with modern and ancient divergent lithospheric plate boundaries, both on continents and in the ocean basins. The regions adjacent to modern plate margins, which are characterized by high heat flow and shallow-focus earthquake activity, are places where new oceanic-type crust is generated. In modern ocean basins, inferred normal faults bound narrow downdropped blocks (grabens) along the axis of the mid-ocean ridge system, the longest continuous geologic feature on earth. Topographically, these structural troughs form deep valleys along the ridge crest. Individual troughs range up to 30 kilometers wide and are filled or partly filled with sediment. The Mid-Atlantic Ridge, a mountain range along the midline of the Atlantic basin, extends northward from Antarctica to Iceland. In Iceland, measurements across the boundary between the North American and Eurasian plates indicate that the crustal blocks are currently moving apart at the rate of a few centimeters per year.

In the Middle East, along the Red Sea, steeply dipping normal faults are associated with a large dome or uplift over a plumelike hot spot in the earth's mantle. Near the Afar region of Ethiopia, the uplift has been subdivided by three radial fault systems that intersect at angles of about 120 degrees. These systems are characterized by large, high-angle normal faults that initially formed a series of downdropped blocks, or grabens. The three-pronged structural feature represents a "triple junction" that separates the African, Arabian, and Indian-Australian lithospheric plates. The East African rift system, which consists of both east and west zones, forms the second prong; it trends northward for nearly 5,000 kilometers and is marked by a series of elongate lakes. The maximum displacement on the bordering faults is nearly 2,500 meters at some places. The east-trending third arm of this large feature is a rift that is partly occupied by the Gulf of Aden.

Earth scientists have also been able to identify historical divergent lithospheric plate boundaries from regional geologic and geophysical (application of physics to geological problems) studies. In the modern Appalachian mountain chain in the eastern part of the United States, a series of elongate structural troughs (grabens and half-grabens) occur along the axis of the range. These structural features, which extend from Nova Scotia in Canada southwestward to North Carolina, contain thick deposits of Triassic (period of geologic time ranging from about 200 to 245 million years ago) sedimentary rocks with associated igneous rocks. Internally, steeply dipping normal faults divide the troughs into narrow tilted blocks that range up to 10 kilometers wide. Some of the border faults were active during Triassic deposition and have a cumulative displacement of nearly 4,000 meters. The formation of these troughs probably marked the separation of North America and Europe about 200 million years ago.

Normal faults are also economically important. These faults serve as traps for hydrocarbons at many places. Migrating oil and gas moving updip from a place of origin, usually a sedimentary basin, are trapped against the fault, which acts as an impermeable barrier or seal. If the fault is not completely sealed, however, it may serve as an avenue for fluids to move to a higher level. Most commercial hydrocarbon deposits occur on the upthrown side of the fault where "rollover" of the rock layers has provided a suitable site for the accumulation of hydrocarbons. The faults are also the locus of metallic mineral deposits. Mineralization may occur in the openings along the fault or in the adjacent fractured rock. Drag ore, related to fault drag, occurs at some places. Also, rich ore bodies are moved downward along younger normal faults. A classic example is at the United Verde extension mine near Jerome, Arizona. There, a rich copper deposit was displaced more than 500 meters vertically.

Bibliography

Allison, I. S., R. F. Black, J. M. Dennison, R. K. Fahnestock, and S. M. White. *Geology: The Science of a Changing Earth.* 5th ed. New York: McGraw-Hill, 1974. An interesting and well-written book on physical geology that is designed for high school or introductory college students. An excellent section on geotectonics that details features associated with normal faults.

Billings, M. P. *Structural Geology*. 3d ed. Englewood Cliffs, N.J.: Prentice-Hall, 1972. A popular college textbook that presents basic concepts and structural features in a clear, concise, and understandable manner. Emphasizes a field approach to recognizing and solving geological problems. Suitable for upper-level undergraduate geology students.

Cloos, E. "Experimental Analysis of Gulf Coast Fracture Patterns." *American Association of Petroleum Geologists Bulletin* 52 (1968). This journal article describes the results of experimental work with clay and dry sand models. Model grabens bounded by normal faults as well as single normal faults accompanied by fault drag are produced by applying tensional forces in a pressure box. These model fractures are compared to the Texas Gulf Coast fracture pattern. Suitable for high school-level readers and college students who are interested in geological models.

Coble, Charles R., E. C. Murray, and D. R. Rice. *Earth Science*. Englewood Cliffs, N.J.: Prentice-Hall, 1986. A general textbook designed for junior high school students and interested laypersons. In the structural section of the text, challenging scientific questions are presented for the reader. Also, the activities of specialists in the structural career field are summarized.

Davis, G. H. *Structural Geology of Rocks and Regions*. New York: John Wiley & Sons, 1984. This book is very readable and takes a practical approach to regional tectonics. Basic concepts and principles of structural geology are emphasized. Suitable for upper-level undergraduate geology students.

De Sitter, L. U. *Structural Geology*. New York: McGraw-Hill, 1959. This book effectively relates geological theory with practice; it compares similar geological phenomena, both on a small and large scale, from different parts of the world. An advanced book designed primarily for students with a good background in geology.

Judson, Sheldon, Marvin E. Kauffman, and L. Don Lee. *Physical Geology*. 7th ed. Englewood Cliffs, N.J.: Prentice-Hall, 1987. An interesting and well-written book on physical geology designed for introductory college students. Includes a section on geotectonics that details features associated with normal faults. An excellent glossary of technical terms.

Park, R. G. *Foundations of Structural Geology*. New York: Methuen, 1983. An excellent reference for high school or college students specifically interested in structural geology. Good coverage of the relationship between geologic structures and plate tectonics.

Shelton, J. W. "Listric Normal Faults: An Illustrated Summary." *American Association of Petroleum Geologists Bulletin* 68 (1984). This article provides the reader with details about the characteristic features of a specific type of normal fault. The paper also discusses the general geometry, causes, and occurrences of most types of normal faults. A number of illustrations, mostly cross sections, are included in the text as an aid to understanding the concepts presented.

Donald F. Reaser

Cross-References

Earthquakes, 148; Thrust Faults, 631; Transform Faults, 638.

Ocean Basins

Ocean basins contain basaltic crust produced by sea-floor spreading at mid-ocean ridges, which may be covered with a thin layer of oceanic sediments. Sea-floor sediments and rocks in the oceans may contain a record of the history of the development of ocean basins. Ocean basin deposits have provided evidence supporting the theories of sea-floor spreading and plate tectonics.

Field of study: Tectonics

Principal terms

BASALT: a dark-colored, fine-grained rock erupted by volcanoes, which tends to be the basement rock underneath sediments in the ocean basins

BIOGENIC SEDIMENTS: the sediment particles formed from skeletons or shells of microscopic plants and animals living in seawater

DEPOSITION: the process by which loose sediment grains fall out of seawater to accumulate as layers of sediment on the the sea floor

LITHOSPHERE: the outermost layers (the crust and outer mantle) of the earth, which are arranged in distinct rigid plates that may be moved across the earth's surface by sea-floor spreading and plate tectonics

MAGNETIC ANOMALIES: linear areas of ocean crust that have unusually high or low magnetic field strength; magnetic anomalies are parallel to the crest of the mid-ocean ridges

MID-OCEAN RIDGE: a continuous mountain range of underwater volcanoes, located along the center of most ocean basins; volcanic eruptions along these ridges drive sea-floor spreading

RIFTING: the splitting of continents into separate blocks, which move away from one another across the earth's surface

SEA-FLOOR SPREADING: a theory that the continents of the earth move apart from one another by rifting of continental blocks, driven by the eruption of new ocean crust in the rift

SEISMIC REFLECTION: study of the layered sediments in ocean basins by bouncing sound waves sent into the sea floor off the different rock layers

SEISMIC REFRACTION: examination of the deep structure of the ocean crust, using powerful sound waves that are bent into the crustal layers rather than being immediately reflected back to the ocean surface

Summary

Ocean basins make up one-third of the earth's surface, and the rocks and sediments in these basins may preserve an important record of the past history of the oceans. Earth materials in the ocean basins consist of a layer of volcanic basalts produced by sea-floor volcanic eruptions at the mid-ocean ridges, which may be covered by layers of marine sediments and sedimentary rocks. The shape of individual ocean basins may be changed as a result of the interactions of lithospheric plates, such as plate collisions, plate accretion, and plate destruction. Ocean basin rocks and sediments may contain valuable deposits of metals and other economic minerals, which may represent important natural resources that could be extracted by humans at some time in the future.

Eruption of volcanic basalts in ocean basins makes up an important part of sea-floor spreading. The creation of new oceanic crust by volcanic eruptions along the mid-ocean ridges provides the driving force to move blocks of continental lithosphere across the earth's surface. For example, the separation of South America from Africa during the past 200 million years has been driven by the creation of the South Atlantic Ocean by sea-floor spreading along the Mid-Atlantic Ridge between these two continents.

Ocean basin shapes may be altered by plate interactions. As lithospheric plates are rifted and move apart from one another, new ocean basins are created between the continental landmasses. In contrast, lithospheric plates may run into one another, and plate collisions cause the ocean basin between continents to be destroyed by subduction, in which crustal slabs are forced downward into the mantle and are remelted. An example is seen in southern Europe, where the collision of the northward-moving African plate with the Eurasian plate has caused the Mediterranean Sea to become shallower and narrower at the same time that crumpling of the edges of the continents has caused mountain building of the Alps in Europe and the Atlas Mountains in northern Africa.

The volcanic basement rocks in ocean basins may preserve a record of the earth's magnetic field during the past, through the record of oriented magnetic minerals contained within basalts. When basalt erupted at mid-ocean ridges cools as a result of exposure to cold seawater, magnetic minerals within the igneous rock are aligned with the earth's magnetic field and are "locked" into position by the crystallization of adjacent mineral grains. Thus, the alignment of magnetic minerals within sea-floor basalts in the ocean basins acts as an enormous magnetic tape recorder, which preserves a record of the alternating reversals of the magnetic field of the earth. Oceanographers investigating the magnetism of the sea floor during the 1950's discovered the existence of long, straight areas of ocean crust with unusual magnetic properties. These linear magnetic anomalies were parallel to the mid-ocean ridges but offset from the ridge crests. The anomalies are symmetrical about the mid-ocean ridges: Anomaly records on both sides form identical "mirror images" of each other. This finding supports the theory of sea-floor spreading, which predicted that creation of new oceanic crust at mid-ocean ridges would cause rifting and separation of previously cooled basalts to either side of the ridge.

Sea-floor spreading theory further predicted that the older the anomaly, the farther it has been pushed away from the ridge crest. This prediction was proved by deep-ocean drilling, which drilled into and determined the age of sediments immediately atop specific magnetic anomalies in the ocean crust. Estimates of the rate of creation of new oceanic crust along the ridges may be calculated from the distance between the crest of the mid-ocean ridge and specific magnetic anomalies whose age has been determined. These calculations have proven that creation of new oceanic crust does not occur at a constant rate through time and that there have been episodes of rapid sea-floor spreading and of slower spreading during the past.

Examination of the chemistry and mineral composition of the volcanic rocks of an ocean basin may provide a record of the history of volcanic eruptions at the mid-ocean ridges and help geologists to determine the chemistry and type of igneous rocks being erupted at any point in the past. Understanding the mineral content of sea-floor crust erupted at specific times in the past allows geochemists to make predictions about the nature of the deeper portions of the earth's crust. Chemical changes in sea-floor basalts may reflect similar changes occurring in the lower crust or upper mantle of the planet.

In addition to preserving historical information in the harder igneous rock basement, ocean basins provide a record of sediment deposition during the past. These regions of the ocean floor are among the flattest areas on the surface of the planet and have minimal relief: Most ocean basins are smooth and nearly flat, with less than 1 meter of vertical altitude change in 1 kilometer of horizontal distance. Their smoothness is the result of burial of the blocky, irregularly faulted volcanic basement rocks beneath layers of slowly accumulating mixtures of biogenic sediments, turbidites, and other sediment particles derived from continental sources. Newly erupted basement rocks are gradually covered by oceanic sediments, so there is an overall correlation between crustal age and total sediment thickness within an ocean basin. Verification of this relationship by deep-ocean drilling provided further support for sea-floor spreading.

Sedimentary rock layers provide information on the history of deposition in the ocean basins by their structure and by the fossils preserved in sediment layers. Marine geologists examine the types, sizes, and sorting of the individual grains that compose sea-floor sediments. Geologists attempt to determine the sources of sediment particles deposited in ocean basins and to analyze both the changes in sediment particles as they fall through the water column and changes occurring on the sea floor after the sediments are deposited. Fine-grained particles derived from continental sources may be carried far out to sea by the winds to be deposited in the deep ocean basins. Also, biogenic particles either may dissolve as they sink through the oceans or may be dissolved on the sea floor by deep-ocean water masses.

Paleontologists study the fossils buried within layers of sedimentary rock. Marine sediments deposited in water depths shallower than the carbonate compensation depth have abundant microscopic fossil remains of ancient one-celled plants and animals (plankton) that lived in the shallow water of the oceans during past geologic time. As these organisms died, their remains sank to the sea floor, to become an important part of the sedimentary rock layers. Examining the record of fossils preserved in sea-floor sediments is like reading the pages of a book containing the history of the ocean basin: Changes in the type and number of ancient fossil organisms and fossil assemblages may be preserved in the microfossils contained in sea-floor sediment layers.

Perhaps one of the most fascinating aspects of the study of sea-floor sediments in the deep-ocean basins is that these materials contain significant amounts of micrometeorites and extraterrestrial material. Micrometeorites

The topography of the ocean basins is visible in this computer-generated image of earth. *(National Aeronautics and Space Administration)*

may fall on the continents, but their scarcity and small size make them difficult to identify. In deep-ocean basins far from the continents, however, sediment accumulates at a much slower rate as a result of a combination of the distance from continental sediment sources and the dissolution of biogenic sediment particles by corrosive bottom waters. As a result, deep-ocean sediments tend to be fine-grained red clays, which have few to no fossils and which may be deposited at rates as slow as 1 millimeter per million years. In these red clays, extraterrestrial materials may make up a significant portion of the sediment particles because of the extremely slow sediment deposition rates.

Methods of Study

Because the ocean basins contain a variety of geologic materials, including igneous, sedimentary, and metamorphic rocks, a number of different techniques are used in the study of ocean basins, depending upon the specific feature of interest. Methods that are suitable for the study of one aspect of the ocean basins may be completely useless for obtaining information about other features of the basin. Some of the methods used include acoustic profiling, seismic reflection and refraction studies, dredging, sediment coring, and deep-ocean drilling. In addition, information on ocean basins may be derived from ancient sea-floor deposits that have been uplifted and are presently found above sea level on the continents.

The overall shape of the ocean basin and the water depths of individual parts of the basin may be studied by acoustic profiling, or echo sounding. This technique uses an acoustic transponder (a sound source) mounted on the hull of an oceanographic vessel to emit sound waves, which travel down through the water until they are reflected back by the sea floor up to a shipboard recorder, which measures the total time between emission of the sound pulse and its return. The water depth is equal to one-half the total time (sound must go down, then up again), multiplied by the speed of sound in seawater. Profiles of the ocean basin's shape are obtained by continuously running the echo sounder while the ship is sailing across the basin.

While acoustic profiling gives the water depth of the ocean basin, the energies of the sound waves are insufficient to provide information about the buried structure of the ocean floor. The shape and thickness of the basement rocks and the sediment cover on the floor of an ocean basin may be examined by seismic reflection profiling, which is somewhat similar to echo sounding. In seismic reflection studies, a large energy source (such as the explosion of a dynamite charge) is released in seawater to create high-energy sound waves, which move down through the ocean with sufficient energy to penetrate the sediment layers of the sea floor before they are reflected back up to the vessel by the different sub-bottom layers. Reflection profiling is also made by continuously producing these high-energy sound waves while sailing across a basin, to obtain a record of the thickness and geometry of the sediment layers and the harder basement rocks of the sea floor.

The deep structure of basement rocks is investigated by seismic refraction studies, which may be made by one oceanographic vessel and a stationary floating recorder (sonobuoy) or by two vessels. In seismic refraction studies, extremely large energy sources are released in the ocean to create powerful sound waves, which have the ability to penetrate through sea-floor sediment layers into the deeper layers of igneous basement rocks. Sound waves penetrating layers in the sea floor are bent (refracted) into the layer and travel through it for a certain distance before they are refracted back up to the ocean surface. An acoustic recorder at the surface measures the depth of sound penetration below the sea floor and the time elapsed since the explosion. Refraction profiles may be done by exploding charges off the stern of a moving vessel, using a stationary sonobuoy as the recording device, or refraction profiles may be made in a "two-ship" experiment, where one vessel acts as the "shooter" and the second vessel is the stationary recorder. By alternately "leapfrogging" past each other, two ships may make a much longer continuous reflection profile than is possible with only one vessel and a sonobuoy.

The history of sediment deposition preserved by marine sediments in ocean basins may be studied by obtaining long cores of sea-floor sediments, either by sediment coring or by ocean-floor drilling programs. Once a long sediment core is obtained from the ocean floor, the sediment particles and fossils within the sediments are studied layer by layer in order to examine the sedimentation history of the basin. Younger sediments are placed atop older layers, so by beginning with the uppermost layers of the sediment core and continuing into deeper layers, the geologist can examine the record of progressively older deposits in the ocean basin.

Direct examination of the basement rocks of the ocean basins may be made by dredging rocks from the mid-ocean ridges, by drilling through the sediment cover to take cores of the volcanic basalts, or by studying portions of the ocean floor that have been uplifted above sea level by tectonic activity. Dredging uses a wire mesh bag attached to a rigid iron frame, which is towed on a long cable behind a vessel to obtain rock samples from the sea floor. As the ship moves across the surface, it drags the dredge along the bottom, and sea-floor rocks are broken off by the frame and caught in the wire mesh bag attached to the rear of the dredge.

Deep-ocean drilling programs have provided long basalt cores that have been drilled from the sea floor in different ocean basins. The deepest sea-floor borehole drilled in the oceans by the Deep Sea Drilling Project, site 504B, located near the Galápagos Islands in the eastern Equatorial Pacific, has been extended to a depth of 1,350 meters below the surface of the sea floor. Drilling at this location recovered 275 meters of sea-floor sediments, and more than a kilometer of sea-floor basalts have been penetrated, with the possibility of further deepening of this hole into the ocean crust by later drilling operations at this location.

Information about the ocean basins has also been provided by uplifted sections of sea floor, located in areas of plate collisions. These ancient sea-floor deposits, or ophiolite sequences, are found on the island of Newfoundland in eastern Canada, on the island of Cyprus in the eastern Mediterranean Sea, and on the island of Oman in the Persian Gulf, among other locations. Rocks in the ophiolite sequences are an important natural resource, because they contain copper and many other valuable metals interspersed between basalts and igneous rocks. These ancient ocean-floor deposits, which have been uplifted above sea level by the collision of two lithospheric plates, may contain enormous reserves of rare metallic minerals. The metallic deposits in ophiolites were originally deposited as vein minerals between the volcanic basalts in the deeper portions of ocean crust. By understanding the factors controlling the formation of ophiolites, scientists may be able to predict other locations where these rocks may be found, and humans will be better able to utilize these valuable minerals in the future.

Context

Volcanic basement rocks and sedimentary rock layers in the ocean basins have provided evidence supporting the theories of sea-floor spreading and plate tectonics, and these geologic materials have preserved a record of earth history. Evidence for sea-floor spreading derives from the linear magnetic anomaly patterns of ocean crust and from the symmetry of features about the mid-ocean ridges. Sea-floor sediment thicknesses increase with greater distance from the ridge crest, and both the sediment thickness patterns and the magnetic anomaly patterns are symmetrical about the mid-ocean ridges, as predicted by the spreading hypothesis. Eruption of sea-floor basalts at the mid-ocean ridges provides the driving mechanism to move blocks of lithosphere around the surface of the earth.

A history of the chemistry and mineral content of volcanic rocks erupted along the mid-ocean ridges is preserved in sea-floor basement rocks, and magnetic minerals in these igneous rocks preserve a record of past reversals of the magnetic field of the earth. Sedimentary rocks and the fossils contained within them provide a history of the ancient organisms which were alive in past oceans, and marine sediment particles preserve a record of ancient sources of particles deposited in the oceans far from land and also may contain the history of the effects of differing ocean chemistry on sea-floor deposits.

Finally, rocks and sediments in the ocean basins may represent an untapped resource of economic mineral deposits, based on the wealth of minerals that have been found in ophiolite sequences, sections of ancient sea floor that have been uplifted above sea level by the interactions of moving lithospheric plates. These sea-floor mineral deposits may provide future economic resources, which may be exploited after all the continental mineral deposits are exhausted.

Bibliography

Anderson, Roger N. *Marine Geology: A Planet Earth Perspective.* New York: John Wiley & Sons, 1986. A textbook discussing various aspects of oceanography, whose content is aimed at readers with minimal scientific background.

Busby, Cathy, and William Morris, et al. "Evolutionary Model for Convergent Margins Facing Large Ocean Basins: Mesozoic Baja California." *Geology* 26, no. 3 (March, 1998): 227-231.

Cox, Allan, comp. *Plate Tectonics and Geomagnetic Reversals.* San Francisco, Calif.: W. H. Freeman, 1973. A compilation of the important scientific papers discussing the discovery of linear sea-floor magnetic anomalies and the development of the hypotheses of sea-floor spreading and plate tectonics.

Glen, William. *The Road to Jaramillo: Critical Years of the Revolution in Earth Science.* Stanford, Calif.: Stanford University Press, 1982. This volume covers the history of the examination of the magnetic record of the continents, the discovery of linear sea-floor magnetic anomalies, and their role in the development of the hypotheses of sea-floor spreading and plate tectonics.

LeGrand, H. E. *Drifting Continents and Shifting Theories.* New York: Cambridge University Press, 1988. A review of the history of the "modern revolution in geology," which culminated in the development of the theory of global plate tectonics.

Seibold, Eugen, and Wolfgang H. Berger. *The Sea Floor: An Introduction to Marine Geology.* New York: Springer-Verlag, 1982. A textbook covering geological oceanography, designed for freshman-level college courses for students with minimal backgrounds in science, which covers all the information attainable by studying ocean basins.

Van Andel, Tjeerd H. *New Views on an Old Planet: Continental Drift and the History of Earth.* New York: Cambridge University Press, 1985. A survey of the theories of continental drift and plate tectonics, written for an educated lay audience.

_____. *Science at Sea: Tales of an Old Ocean.* San Francisco, Calif.: W. H. Freeman, 1981. A general-audience book discussing some of the methods used by seagoing oceanographers to study the sediments and rocks of the oceanic crust. It also describes some of the hazards and problems inherent in oceanography at sea.

Dean A. Dunn

Cross-References

Earth's Age, 155; Igneous Rocks, 315; The Oceanic Crust, 471; Plate Tectonics, 505; Subduction and Orogeny, 615.

The Ocean Ridge System

The ocean ridge system is a complex chain of undersea volcanic mountains that are found in all the oceans. These mountains contain rift valleys along their axes, which are believed to be spreading centers from which continental motion takes place. All existing evidence, such as volcanic activity, the flow of heat from within the earth, and various types of faulting and rifting, supports modern theories about sea-floor spreading, tectonic plates, and continental drift.

Field of study: Tectonics

Principal terms

ASTHENOSPHERE: a zone of rock within the mantle that has plastic flow properties attributable to intense heat

BASALT: a heavy, dark-colored volcanic rock

CONVERGING PLATES: a tectonic plate boundary where two plates are pushing toward each other

DIVERGENT PLATES: a tectonic plate boundary where two plates are moving apart

METALLOGENESIS: the process by which metallic ores are formed

SEDIMENTS: the solid fragments of rock that have been eroded from other rocks and then transported by wind or water and deposited

TECTONIC PLATES: according to theory, the seven major plates into which the earth's crust is divided

Summary

The ocean ridge system is a complex chain of mountains about 80,000 kilometers in length that winds through the ocean basins. These mountain ranges vary from a few hundred to a few thousand kilometers in width and have an average relief of 0.6 kilometer. Studies of the sea floor indicate that ocean ridges are found in every major ocean basin. They are composed of basalt and covered by various types of sediments. Many of the ridges have narrow depressions that extend thousands of kilometers along their axes. Heat probes lowered into these rifts indicate much higher temperatures than on the flanks of the ridges. Another very significant finding has been that the rocks that make up the sea floor are much younger than those that make up the continents. This finding countered the pre-1960's belief that the rocks of the ocean basin are more ancient than those of the continents.

In the 1960's, the theory of sea-floor spreading suggested that new sea floor is constantly being added by volcanic activity at the ocean ridges. The theory of plate tectonics proposes that the earth's crust is divided into several major plates. These plates extend down into the earth's mantle to a hot, semimolten zone known as the asthenosphere. Since the rock that composes the tectonic plates is less dense than is the rock that forms the mantle, the plates may be considered to be floating on the asthenosphere. Because the interior of the earth is much hotter than the surface, a flow of heat toward the surface is a constant process. This method of heat transfer, known as convection, is a density type of current. Material such as molten rock or hot air or water is less dense than is the same material in a cooler state. As a result, it flows upward with cooler material filling in below. When the hot material reaches a higher level and releases its heat, it too returns to the depths to be reheated. It is this process, along with the relatively low density of the tectonic plates, which causes the plates to drift apart.

The location and symmetry of the mid-ocean ridges, especially in the Atlantic and Indian oceans, suggests the configuration of the continents before they began to drift apart. The modern theories of plate tectonics and sea-floor spreading are in basic agreement with the theory of continental drift as proposed by Alfred Wegener in 1915; however, modern ideas regarding the mechanics of drift differ. The basic concept of sea-floor spreading at the oceanic ridges proposes that tension cracks form in the crust at these spreading centers. Molten rock from the mantle then flows upward through these fissures forming both the volcanic ridges and creating new sea floor. As the fissures widen, new crustal material moves away on both sides and additional new sea floor is created. It is in this way that the mid-ocean ridges such as the Mid-Atlantic Ridge, the East Pacific Rise, the Antarctic Rise, and the Carlsberg Ridge of the Indian Ocean were formed. The spreading away from the oceanic ridges takes place at a rate of about 2 centimeters per year in the Atlantic and about 5 centimeters per year in the Pacific.

As new sea floor is being created at the ridges, old sea floor is being destroyed at continental margins. Here, old sea floor is subducted beneath continental plates. At these converging plate boundaries, old sea floor is forced downward into the mantle, where it undergoes remelting. This molten rock may then find its way back to the surface through cracks or fissures in the overlying rock. Volcanic action is the result of this material reaching the earth's surface. Because old sea floor is destroyed in this manner, it is now understood why rocks that make up the ocean floor are relatively young. Deep-sea-floor drilling projects in the Atlantic and Indian oceans have failed to find rock or fossil samples that are older than the Jurassic period of earth history, which ended about 140 million years ago.

The present ocean ridges show large offsets in some areas. This phenomenon results from transform faulting. As spreading of the plates took place at divergent boundaries, fracture zones developed at right angles to the axes of the ridges. Displacements of the ridges along these faults produced the observed structure.

The narrow rift valleys are believed to have been caused by downfaulting along divergent plate boundaries. Some rift zones, such as the one associated with the Carlsberg Ridge of the northern Indian Ocean, link up with continental rift zones. This rift has been shown to be connected with the African rift zone. Studies conducted during the 1970's indicate that there exist extensive amounts of volcanism and seismic activity along rift zones. Also found along oceanic ridges are hot-water springs. The existence of these springs, known as hydro-thermal vents or smokers, had been predicted, but direct evidence for them was not gathered until the early 1960's, when metal-bearing sediments were discovered on the East Pacific Rise. The first actual observation of a smoker took place in 1980 by the crew of the deep-diving research submarine *Alvin*. The *Alvin* was part of an underwater research program being conducted near the Galápagos spreading center in the eastern Pacific. Researchers were surprised to find a rather extensive plant and animal community living near the smoker at a depth too great for photosynthesis to be a factor. Clams as long as 30 centimeters were found, as well as white crabs and tube worms some 3 meters long.

Smokers have been found to emit great quantities of sulfur-enriched waters. Dissolved within the acidic water are various types of metals. These metals are dissolved from the rocks as the superheated water moves toward the surface. Metals such as copper, zinc, iron, silver, and gold are extracted and concentrated into a supersaturated fluid. When this fluid is discharged from a smoker at the sea floor, it will precipitate to form an ore body if cooling takes place rapidly. These massive sulfide deposits are sometimes made permanent when volcanic eruptions cover them with basalt flows.

Although most of the knowledge regarding the nature of the sea bottom has been gained since the 1960's, the Mid-Atlantic Ridge system has been known to exist since echo-sounding studies were done after World War I. By 1960, it had been determined that the Mid-Atlantic Ridge was continuous with other oceanic ridge systems around the world. The same cannot be said of the rift system that was discovered in the Mid-Atlantic Ridge. Profiles taken across ridge systems do not always indicate such rift depressions. This anomaly can probably be explained by considering the possibility that these rifts have been filled with volcanic material and therefore go undetected.

A rather extensive study was made along the Mid-Atlantic Ridge about 200 miles south of the Azores in 1974. It was found that the rift valley is bordered by a series of steep slopes that appear to be high-angle faults. Many open vertical fissures were observed, some of them as much as 8 meters across. Although no active volcanism was observed during this study, mounds of pillow lava along the fissures indicated that volcanism had taken place. The positioning of the faults and fissures indicated that the Mid-Atlantic Ridge is spreading outward east and west from its central axis.

The East Pacific Rise, which is part of the system of mid-ocean ridges, stands some 2 to 3 kilometers above the ocean floor and is thousands of kilometers wide. The slopes in this area are not as great as those of the Mid-Atlantic system, but the mechanism of ridge formation is the same. It has been suggested that part of the East Pacific Rise extends under the North American continent and that the San Andreas fault may be part of this system. Evidence supporting this possibility is that the rate of displacement along the fault is comparable to the rate of ridge spreading on the East Pacific Rise near the Gulf of California.

Methods of Study

The ocean floor has been investigated by the use of sonar since the early 1900's: Sound waves from a device are sent into the sea in all directions. As these waves strike an object, they are reflected back to the source. By analyzing the reflections, scientists are able to determine the nature of the sea floor. This technique was refined during World War II for the purpose of locating enemy submarines. By the early 1950's, new maps and charts of the sea floor had been made using sonar.

In the 1960's, a plan to drill a deep hole through the crust of the earth to the mantle was proposed. Since this deep hole was to intersect the Mohorovičić discontinuity (named after the Croation geologist who discovered it, Andrija Mohorovičić), the project was called Mohole. Since the crust of the earth is much thinner under the ocean basins than it is under the continents, the ocean was the most logical location in which to drill. Al-

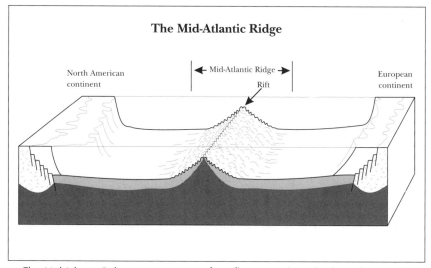

The Mid-Atlantic Ridge is a major site of sea-floor spreading, slowly pushing North America and Europe farther apart.

though the 6-kilometer-deep hole was never drilled, several preliminary holes were. Samples taken from this drilling revealed the presence of gray claylike sediments overlying dark heavy basalts. Project Mohole has since been replaced by a plan to drill many holes at various locations into the rock beneath the ocean depths. This project, which is under the direction of various American oceanographic institutes, is known as the Deep Sea Drilling Project. Much of the drilling was done from the deck of the now-retired *Glomar Challenger* of the company Global Marine, Incorporated. More than six hundred holes have been drilled in the Atlantic, Pacific, and Indian oceans and the Mediterranean Sea, revealing a wealth of data on the nature and evolution of the ocean basins.

In the past, ocean researchers have had to gather data from the decks of ships by lowering various tools to the bottom to gather samples—there was a need for direct observation by the oceanographer. In 1948, the first minisubmarine, or bathyscaphe, made an unmanned dive to 759 meters. In 1954, a manned vessel was taken to a depth of 4,050 meters. By 1960, interest in these submersibles grew quickly in the United States. Since that time, submersibles such as the *Aluminaut* and *Alvin* have been used in various deep-sea research projects. Submersibles have been used for on-site studies of the rift zone of the Mid-Atlantic Ridge and of the hot-spring smokers of the East Pacific Rise.

Another technique known as marine seismology has been employed to study the ocean ridges. Because oil companies have used seismic studies for more than half a century in the exploration for oil deposits on continents, it was only a matter of time before this technology would be adapted for study of the ocean depths. The process involves the making of a sound explosion in the sea, which is accomplished by the use of an air gun. The sound waves are produced by the rapid release of compressed air. The waves then travel through the water to the bottom, where they are reflected to seismometers located on the ocean floor. The data are collected and a seismogram is generated by a computer. This technique has been used to locate active magma chambers in rift zone areas.

Context

Studies of submarine rift areas have revealed the hot-water springs that are referred to as smokers. Smokers discharge water that contains sulfur in solution as well as various types of metals such as copper, iron, zinc, silver, and gold. These metals precipitate to the sea bottom and, if they are cooled rapidly enough, form deposits. If the ore bodies are covered with volcanic material shortly after their deposition, they become protected from the erosive processes of seawater. In time, these deposits drift from oceanic spreading centers to become parts of continents. Many mineral deposits formed in just this manner have been mined since ancient times.

In the late 1950's, metal-rich sediments were first found along mid-ocean ridges. By the late 1960's, metalliferous muds were found at the bottom of the Red Sea, the Gulf of California, and the East Pacific Rise near the Galápagos Islands and south of Baja, California. Since 1977, marine geologists have been able actually to observe the formation of metal deposits while observing smokers from the research submarine *Alvin*. Interest in ridge deposition of metals centers on the processes of metallogenesis. Mining of such deposits, however, even with advanced technology, is not profitable. It will no doubt be considered in the future, however, as the copper deposits in the area of the Galápagos spreading center alone are estimated to be worth in excess of $2 billion.

Bibliography

Anderson, Roger N. *Marine Geology: An Adventure into the Unknown*. New York: John Wiley & Sons, 1986. A well-written, somewhat technical volume on the geology of the oceans. Contains excellent chapters on metallogenesis and mineral deposits. This volume is suitable for the college student of geology or the informed layperson.

Compton, R. R. *Interpreting the Earth*. New York: Harcourt Brace Jovanovich, 1977. A volume suitable for the layperson, it covers general topics in geology. Well illustrated.

Dott, Robert H., and Roger L. Batten. *Evolution of the Earth*. 4th ed. New York: McGraw-Hill, 1988. A well-illustrated volume dealing with topics in historical geology. Suitable for the high school or college freshman student of geology.

Edyll, C. P. *Exploring the Ocean World*. New York: Thomas Y. Crowell, 1972. A collection of papers on topics such as chemistry, physics, and biology of the oceans, underwater archaeology, and marine ecology and pollution. Suitable for the general reader.

Shepard, F. P. *Submarine Geology*. New York: Harper & Row, 1973. A technical volume dealing with such topics as methods and instrumentation of oceanic study, sediments and sedimentation, plate tectonics, erosive processes, and sea-bottom topography, stratigraphy, and deposits. Suitable for the college-level student of geology or oceanography.

Skinner, B. J., and S. C. Porter. *The Dynamic Earth*. New York: John Wiley & Sons, 1989. A well-written, well-illustrated volume dealing with general topics in physical geology such as rocks, minerals, erosion, earth resources, sedimentation, and tectonics. Ideal for the college-level introductory physical geology course.

Weiner, Jonathan. *Planet Earth*. New York: Bantam Books, 1986. A well-illustrated volume dealing with general topics in the earth sciences. The companion to the Public Broadcasting Service (PBS) television series of the same name. Suitable for the general reader.

Wicander, R., and J. S. Monroe. *Historical Geology*. New York: West Publishing, 1989. A very well-illustrated volume dealing with such subjects as geologic time, origin and interpretation of sedimentary rocks, and a detailed account of the historical geology of the earth through various time periods. Excellent for a first-year college course in historical geology.

David W. Maguire

Cross-References

Igneous Rock Bodies, 307; Magmas, 383; Ocean Basins, 456; The Oceanic Crust, 471; Plate Tectonics, 505; Subduction and Orogeny, 615.

The Oceanic Crust

The oceanic crust is that portion of the outer layer of material forming the earth that underlies the world oceans. This crust is a dynamic layer, primarily composed of basalt, where new, submarine mountain ranges are being continuously formed and old ocean floors are being destroyed.

Field of study: Tectonics

Principal terms

FRACTURE ZONES: large, linear zones of the sea floor characterized by steep cliffs, irregular topography, and faults; such zones commonly cross and displace oceanic ridges by faulting

HYDROTHERMAL VENTS: sea-floor outlets for high-temperature, mineralized springs that are associated with sea-floor spreading centers and which are often the site of deep-sea, chemosynthetic biological communities

OCEAN TRENCHES: long, deep, and narrow depressions in the sea floor (greater than 6,000 meters deep), with relatively steep sides; these features mark the boundary between ocean crust and continental crust and are associated with the subduction of oceanic crust

OCEANIC RIDGES: long, narrow elevations of the sea floor, some 2-3 kilometers higher than the surrounding ocean basins, that are associated with the creation of new sea-floor material

PLATE TECTONICS: the theory of mobility within the earth's crust that accounts for mountain building at ocean ridges, spreading of the sea floor, and subduction at ocean trenches by dividing the crust into a series of plates that interact by colliding, rifting, or sliding past one another

SEA-FLOOR SPREADING: the process whereby crustal plates move away from mid-ocean ridges, creating new crustal material as molten rock moves upward through rifts at the ridge crests

SEAMOUNTS: isolated elevations on the sea floor, usually rising to higher than 1,000 meters, which are commonly the shape of an inverted cone reflecting their volcanic origin; a flat-topped seamount is known as a guyot

SEISMIC ACTIVITY: a disturbance of the crust caused by earthquakes or earth movements, often associated with zones of sea-floor subduction and ocean-ridge formation

SUBDUCTION: the process whereby old sea floor that was produced millions of years earlier at the ocean ridges is forced under continental crust in the vicinity of trenches

Summary

The earth is not homogeneous from its center to its surface, but rather it is composed of three concentric layers: the core, the mantle, and the crust. The core of the earth is composed of a dense mixture of nickel and iron, with a solid inner portion and a liquid outer portion. The core is extremely hot, ranging from about 5,000 degrees Celsius at the center to 4,000 degrees Celsius in the outer core. The core extends from a depth of about 2,900 to 6,378 kilometers (center of the earth) and accounts for 31.5 percent of the earth's mass. The next layer, the mantle, consists of less dense rock and holds 68.1 percent of the mass.

The material of the mantle has the properties of iron-magnesium silicate rock rich in olivine and pyroxene minerals. The mantle is about 2,870 kilometers thick and cooler than the core (1,500-3,000 degrees Celsius). The mantle also has two zones. The lower portion is presumed to be essentially rigid, but the upper mantle, or asthenosphere, is more plastic and flows when stressed. The asthenosphere extends to a depth of 700 kilometers and is likely the site of molten magma formation. The outermost layer, the crust, is the less-dense outer shell of the earth. Also known as the lithosphere, this layer consists of granitic continental crust and basaltic oceanic crust. The crust is underlain and likely fused with a layer of heavier mantle rock. The boundary between the crust and mantle is known as the Mohorovičić discontinuity (or Moho); the boundary occurs under continents at depths of 10-70 kilometers but only 5-10 kilometers under the oceans.

Above the Moho, both the oceanic and continental crusts have properties that resemble basalt. Basalt, a common volcanic rock found extensively on the earth's surface, is composed of silicates of calcium, magnesium, and iron. These rocks have an average density of 3 grams per cubic centimeter. Under the continents, but not the oceans, the basalt is overlain by a rock layer with properties similar to those of granite. Granite is a common igneous rock composed of silicates of aluminum and potassium. Such rocks are lighter in color and weight than is basalt, with an average density of 2.8 grams per cubic centimeter. Thus, the continental crust "floats" as massive blocks on a layer of basalt. Because the densities of the continental and oceanic crusts are not greatly different, approximately 93 percent of the continental blocks are submerged in the underlying basaltic crust. Continental blocks are analogous to floating icebergs of various heights in that the Moho is pushed deeper under continental mountain ranges than it is under flat coastal plains. The Moho assumes a shape that reflects the surface of the continent but exaggerated to nine times greater. The bottoms of the continental blocks must rise as material is eroded to the sea thus keeping the exposed-to-submerged ratio constant. This flotation phenomenon is known as isostasy, and the rising process is called isostatic adjustment.

One of the most remarkable characteristics of the oceanic crust is its structural uniformity. Essentially, marine sediments overlie igneous rock, which forms three distinct layers. The sediments vary considerably in thickness and in composition. Shell material and debris from marine plants and animals form dominant sediments around the equator and near the polar seas; detritus from the land and glacial deposits are common near the continents; and chemical precipitates (oozes) are found elsewhere. The oceanic ridge crests are generally free of sediments. From the flanks of the ridges to the continents, the sediments generally increase in thickness to more than 3 kilometers at the continental margins. The three igneous layers are each relatively uniform in composition and thickness. The upper layer has been penetrated by deep ocean drilling and is known to be composed of basaltic lavas, 1-2.5 kilometers thick. The basal layer, directly overlying the mantle, is thin (0.5 kilometer) and presumably formed of layered peridotite. Peridotite is a dense ultrabasic igneous rock consisting mainly of olivine minerals. Similar rocks are thought to be the principal constituent of the mantle. The main (middle) layer is 5 kilometers thick and has properties consistent with a gabbroic composition. Gabbro is a coarse-grained igneous rock consisting mainly of plagioclase feldspar, pyroxene, and olivine minerals. It is the deep-seated equivalent of the overlying, fine-grained basalt. In certain areas, metamorphism and hydrothermal processes have formed other rock and mineral types including amphibolite, greenschists, zeolites, and serpentine. The surface features of the oceanic crust include such interesting and interrelated topographic features as oceanic ridges, fracture zones, seamounts, abyssal plains, deep-sea trenches, and island arcs. The existence of each of these features can be explained by the concepts of plate tectonics and sea-floor spreading.

The oceanic ridge system is the major topographic feature of the ocean basins, extending 80,000 kilometers as a continuous range throughout all the oceans. Ocean ridges generally rise 2-3 kilometers higher than do the bordering ocean basins. The ridge system in the Atlantic and Indian oceans lies equidistant between the adjacent continents, whereas in the Pacific Ocean, the system is highly asymmetric with respect to the continents. Passing out of the Indian Ocean between Antarctica and Australia, the ridge system continues eastward across the southern Pacific then arcs northward well toward the South American continent. Here known as the East Pacific Rise, it eventually passes into the Gulf of California and presumably under the Basin and Range province of the western United States to reemerge in the Pacific Ocean as the Juan de Fuca Ridge off British Columbia. Although almost entirely submarine, the ridges do rise above sea level at a few places in the Atlantic and Pacific oceans where recently active volcanos have formed islands (for example, Iceland, Tristan da Cunha, and the Galápagos Islands). The ridges are also seismically active, with frequent tensional

473

earthquakes of intermediate strength. Such earthquakes are generally restricted to the oceanic crust within a few kilometers of the ridge crest. The crest of the ridges in the Atlantic and Indian oceans are characterized by a central rift valley that is commonly 2-3 kilometers deep and 20-30 kilometers wide. Volcanism occurs along the centerline of the rift valley, which is also the site of most of the seismic activity. In the Pacific Ocean, earthquakes are confined to a similar narrow zone, but volcanism appears to have been much greater. Here, lava flows have filled the central rift valley to a large part so that the ridge crest appears smooth.

The ocean ridge system is clearly a continuous feature on a global scale, but when viewed in detail, the ridge crests are frequently offset by fracture zones. For example, the Mid-Atlantic Ridge has no less than forty such zones. These fracture zones are steeply cliffed features which vary in width from a few to 50 kilometers. They are mainly confined to the oceanic crust and only rarely approach the continental margins. Fracture zones are only seismically active along that portion of the fault line between the offset ridge crests; the segments extending toward the continents are seismically quiet. Earthquakes between the crests are associated with transverse motion, indicating that the fracture zones are the result of faults in which each side moves horizontally but in opposite directions. These displacements of oceanic crust are known as transform faults.

Throughout the world's oceans, beyond the flanks of the ridge systems, numerous irregularities rise from the sea floor. Small volcanic extrusions that rise less than 1 kilometer from the ocean floor are known as abyssal hills. Larger volcanic features that reach 1 kilometer or more are called seamounts. Seamounts that have flat tops are known as guyots. The flattening is thought to have resulted when seamounts were near sea level and subjected to wave attack and erosion. The composition of seamounts is closely related to their proximity to oceanic or continental crust. For example, near the center of the Pacific Ocean, the seamounts (including those that broach the sea surface to become islands) are composed of basaltic-type rock characteristic of oceanic crust, whereas along the margin of the Pacific Ocean the islands are composed of the granitic-type rocks that are found on the continents. The boundary between these two regions has been named the Andesite Line for the type of rocks found in the volcanic mountains of South America. In tropical areas, seamounts often support coralline reefs in the form of atolls. Atolls are generally circular in plan, consisting of a central lagoon surrounded by a narrow carbonate reef dotted with elongated islands. Presumably, atolls form as volcanic islands subside at a rate that is matched by the upward growth of the encircling reef.

The relatively flat surfaces of the ocean floor, which extend from the mid-ocean ridges to either the marginal trenches or the continental slopes, are known as abyssal plains. Excluding the trenches, these plains are the deepest portion of the ocean. Abyssal plains account for nearly 30 percent of the earth's surface, comprising 75 percent of the Pacific Ocean basin and

33 percent of the Atlantic and Indian ocean basins. Oceanic rises are areas of the ocean floor that are elevated above the abyssal plain, distinctly separated from a continental mass, and of greater areal extent than are typical seamounts or abyssal hills. General oceanic rises lie at least 300 meters above the surrounding ocean floor. Rises are not seismically active and are thought to result from the uplifting of oceanic crust associated with volcanic hot spots (source areas for magma in the upper mantle). Examples are the Bermuda Rise in the North Atlantic Ocean and the Chatham Rise in the southwestern Pacific Ocean.

The oceanic trenches are the deepest parts of the oceans. They are elongate, narrow, and commonly arcuate in shape with the convex side facing to sea. With few exceptions, they occur at the margins of ocean basins. By convention, an ocean deep must be at least 6,000 meters below sea level to be considered a trench. Trenches are found in the Atlantic and Indian oceans but are most common in the Pacific. The deepest are found in the western Pacific, where the Mariana trench plunges to a depth of 11,033 meters. The largest, however, is the Peru-Chile trench adjacent to South America; it is 5,900 kilometers long, averages 100 kilometers wide, and extends to depths below 8,000 meters. Most trenches are associated with island arc systems or with volcanic ranges adjacent continents. Examples include the Guam and Saipan islands west of the Mariana trench and the Andes east of the Peru-Chile trench. Island arcs are volcanic belts that parallel the trench on the continental side. The profiles of the trenches are asymmetrical, with steep sides toward the island arcs. Trenches are also areas of high earthquake activity, low gravitational pull, and low heat flow from the earth.

The concepts of plate tectonics and sea-floor spreading provide the mechanisms necessary for creating the features of the ocean floor. Plate tectonics proposes that the earth's lithosphere is composed of several plates of differing shapes and areas that glide over the plastic asthenosphere. Convection cells caused by radioactive decay of isotopes in the molten rocks of the mantle are the driving mechanism for this motion. These cells circulate the heat upward, causing upwelling in the mantle. The movement of the plates results in areas of separation where magma flows to the surface, creating the volcanic mountains of the mid-ocean ridges. Thus, the ocean floor spreads outward from the ridge crest expanding the dimensions of the ocean. This process occurs in the Atlantic Ocean at the expense of the Pacific Ocean. Where plates collide, such as off the coast of Southeast Asia and South America, deep trenches are formed as oceanic crust is forced under (or subducts) the lighter continental crust. The process is associated with volcanism as the oceanic crust is remelted and island arcs are formed by the accompanying submarine eruptions. The rate of spreading affects the form of the ridge system. Rapid spreading (up to 5 centimeters per year) produces a broad, relatively low ridge without a deep central valley, such as that found in the East Pacific Rise, west of South America. Slow spreading

(1-3 centimeters per year), on the other hand, results in a high-relief ridge with a deep central rift valley such as that of the Mid-Atlantic Ridge.

Hot waters are discharged by hydrothermal vents at active mid-ocean ridges. The chemical composition of ocean water and deep-ocean sediments is influenced by seawater circulating through hot oceanic crust, formed by volcanic eruptions. Some seawater enters the oceanic crust through faults and eventually reaches the vicinity of the magma chambers below the spreading center where molten rock collects before eruption. Reactions with hot basalt charge the seawater with metallic sulfides and remove magnesium and other elements. The hot water then flows into the ocean through irregular, chimneylike vents up to 10 meters high. The vent mounds are made of silica, native sulfur, and metallic sulfide minerals. The bright-colored chimneys and their surrounding deposits resemble valuable ore deposits of copper, zinc, and other metals which are found on the continents. This phenomenon may also be an important process in regulating the chemical composition of seawater, as well as providing a chemical base for a deep-sea biological community that uses chemical energy rather than sunlight to produce organic compounds (chemosynthesis instead of photosynthesis).

Methods of Study

The oceanic crust covers about 70 percent of the earth's surface, yet it has received relatively little attention. For example, deep-sea drilling and sampling of the crust has been completed at only one site for every 500,000 square kilometers of ocean floor. The great depth of the oceans, the tremendous logistical problems of working on and in the sea, and the high cost of oceanic research have all acted to limit the amount of scientific information that is available. Technological advances in the second half of the twentieth century have permitted researchers to explore the oceanic crust with remote sensing techniques as well as through diving excursions to the ocean floor.

The structure of the earth and particularly the oceanic crust have been investigated indirectly by seismic refraction and reflection methods. Studies in the early 1950's showed that the crust was composed of several layers on the basis of the velocity of sound within each layer. Ocean sediment transmits sound at 2 kilometers per second, basalt transmits at 5.1, gabbro transmits at 6.7, and peridotite transmits at 8.1. Precise measurements of the length of time required for a seismic shock wave to penetrate these layers have permitted the thickness of each layer to be calculated.

Accurate and detailed maps of the ocean floor (bathymetric charts) have been compiled from enormous collections of sounding data. By the early 1900's, electronic devices called precision depth recorders (PDRs) became available for oceanographic surveys. These early surveys gave the first realistic view of the ocean's major surface features. Side-scanning sonar, a later development, has provided three-dimensional illustrations of small-scale features of the sea floor. This type of bathymetric data can be recorded continuously as the research ship is under way, and locations can be determined precisely from navigational satellites.

Ocean crustal rock is an average of 7 kilometers thick. This great thickness, plus the hundreds of meters of sediment and thousands of meters of seawater overlying the crust, have made the direct sampling of this rock very difficult. Deep-ocean drilling has, however, permitted samples to be collected from considerable depths below the ocean floor. Beginning in 1968, the Deep Sea Drilling Project (DSDP) has extensively explored the oceanic crust from the ship *Glomar Challenger*, operated by Global Marine, Incorporated. Sponsored by the National Science Foundation and the Office of Naval Research, this project has drilled a total of 160 kilometers of cores. The deepest penetration into the ocean floor was 1.7 kilometers; the deepest water in which drilling took place was 7,000 meters. More than 840 sites were drilled in all parts of the world oceans. Scientists from all over the world participated in this project, which confirmed much of the theory of how the earth's crust moves. DSDP also provided significant data on the age of the ocean basins and the rates of sea-floor spreading.

Other important deep-sea data have been gathered by research submersibles and airborne remote sensors. Starting in the 1960's, submersibles such as Woods Hole Oceanographic Institution's *Alvin* have made some remarkable oceanographic discoveries, especially along the mid-ocean ridges. Much of the knowledge of hydrothermal vents has been obtained through submersible observations. A submersible possesses numerous advantages over surface vessels, including direct observation and sampling. They are, however, dependent on surface ships for support and transport to the dive sites. Aircraft magnetometer surveys have yielded valuable information on paleomagnetism and the earth's gravitational field. These data showed that the polarity of the earth's magnetic field is recorded in the crustal rocks as the ocean floor is formed. Thus, a record was revealed that demonstrated sea-floor spreading by mirror-image of polar reversal patterns on the east and west sides of the Mid-Atlantic Ridge.

Satellites have also played a part in the exploration of the oceanic crust. The short-lived Seasat satellite carried a sophisticated altimeter that could measure the precise distances between the satellite and the ocean surface. Slight differences in sea level were observed which correspond to ocean deeps and density anomalies (such as accumulations of dense rock). For example, the sea stands higher over mid-ocean ridges and lower over trenches. Satellite images are now being used to map previously unknown topographic features of the ocean floor.

Context

The oceanic crust is not a static feature; rather, it is a dynamic layer that profoundly affects the basic processes of the earth. The fundamental aspect of sea-floor spreading is that new sea floor forms along mid-ocean ridges and slowly moves away from the ridge (a few centimeters per year) to be consumed or subducted in trenches at the edge of continents. This process results in many natural hazards, such as earthquakes and volcanoes, that are catastrophic to human populations. On the positive side, this process also results in the formation of valuable mineral deposits as well as regions of spectacular beauty.

Of all the geological hazards, earthquakes are among the most frequent and most destructive. A plot of worldwide seismic activity shows that most earthquakes are confined to rather narrow but continuous zones that surround large stable plates. Most of these boundaries occur in oceanic crust. The earthquakes are associated with the relative movements of these plates, particularly collisions, rifting, and dragging at the plate boundaries. Knowledge of the mechanisms and rates of sea-floor spreading will permit scientists to predict better when and where a potential catastrophe will occur. One of the ultimate goals of crustal dynamics research is to develop a reliable early warning system in order to allow orderly evacuation of earthquake-prone areas.

For the purpose of hazard assessment, volcanoes can be placed into one of two groups, according to the composition of their magmas. Basaltic volcanoes, associated with oceanic crusts, produce large quantities of lava with minimal explosive activity. The volcanoes of the Mid-Atlantic Ridge and the Hawaiian Islands are examples of this type. In contrast, more acidic volcanoes, normally associated with subductive zones, tend to be more violent and unpredictable. The viscosity of the magmas can lead to a buildup of gas pressure, resulting in massive explosions such as Krakatoa in the East Indies and the volcanoes of the west coast of South America. As a human hazard, basaltic volcanoes are relatively benign and may even serve as a safe tourist attraction. Volcanoes are clearly beyond human control, but prediction and risk assessment can reduce the hazards.

Sea-floor spreading can result in mineralization from rising magma on ocean ridges. Pockets of metal-enriched brines and muds have been found along many of the rift valleys of ocean ridges. These deposits result from interaction of seawater and hot magma. In the Red Sea rift, oceanographers have discovered a single pocket which contains an estimated 200 million tons of metal-rich mud. This mud is a multimetallic resource (including copper, zinc, iron, silver, gold, nickel, vanadium, lead, chromium, cobalt, and manganese), the mining of which could provide several salable products. Much research and development work needs to be undertaken in order to build an economically feasible deep-sea mineral extraction device.

Another metallic resource of the deep-ocean floors is a brownish-black, potatosized object known as a manganese nodule. Cross-sectional cuts of these nodules show concentric growth rings that require a million years for 1 millimeter to form. Though this may seem exceedingly slow, the nodules are so numerous, particularly in the Pacific Ocean, that the annual growth totals many millions of tons. Magnesium nodules contain many other valuable metals including copper, nickel, and cobalt. They form by precipitation around a small grain from solutions emanating from the oceanic crust and its cover of sediments.

The deep-sea floor is receiving more and more consideration as a place to store nuclear waste. There are certainly many environmental risks involved, but the deep sea option has some advantages over land-based storage. For example, the deep sea is one of the least valuable and most plentiful pieces of real estate on earth. It has no significant fishing or petroleum use; manganese mining is a possibility, but the deep-sea region is so immense that numerous sites are available. Perhaps the most compelling reason favoring the deep sea is that, except at the ridges and trenches, the ocean floor is extremely stable. Earlier suggestions that radioactive material be placed in the ocean trenches have now been disregarded. The trenches are very unstable areas and zones of subduction with a high incidence of faulting and earthquake activity that could damage storage containers, releasing radioactive material into the environment. Many legal and technological questions need to be answered before deep-sea disposal can begin.

Bibliography

Bonati, Enrico. "The Origin of Metal Deposits in the Oceanic Lithosphere." *Scientific American* 238 (February, 1978): 54-61. Describes the geochemical processes occurring at the spreading centers of ocean ridges that produce metal-rich minerals and the implications for such deposits found throughout the earth's crust.

Edmond, John M., and Karen Von Damm. "Hot Springs on the Ocean Floor." *Scientific American* 248 (April, 1983): 78-84. A discussion of the role of deep-ocean hydrothermal springs in depositing metallic ores and sustaining life in the absence of sunlight.

Gross, M. G. "Deep-Sea Hot Springs and Cold Seeps." *Oceanus* 27 (Fall, 1984). An introduction to a special issue devoted to the exploration, geochemistry, chemosynthesis, and biology associated with water discharge on the ocean floor.

Heirtzler, James R., and W. B. Bryan. "The Floor of the Mid-Atlantic Rift." *Scientific American* 233 (August, 1975): 78-90. Data gathered by FAMOUS (French-American Mid-Ocean Undersea Study) are summarized, particularly investigations of the Mid-Atlantic Rift south of the Azore Islands, involving the use of the submersibles *Alvin, Archimede,* and *Cyana.*

Kennett, James P. *Marine Geology.* Englewood Cliffs, N.J.: Prentice-Hall, 1982. A comprehensive treatment of the geology of the sea floor, including ocean morphology, geophysics, plate tectonics, oceanic crust, and marine sediments.

Menard, H. W. "The Deep-Ocean Floor." *Scientific American* 221 (September, 1969): 126. A summary of the dynamic effects of sea-floor spreading is presented with description of related topographic features of the ocean bottom.

Ross, D. A. *Introduction to Oceanography*. 4th ed. Englewood Cliffs, N.J.: Prentice-Hall, 1983. Representative of a number of informative introductory textbooks on ocean science, this text provides an overview of plate tectonics and geophysics as these subjects relate to the features and dynamics of the oceanic crust as well as oceanography study methods including deep-sea drilling.

Shepard, F. P. *Geological Oceanography*. New York: Crane, Russak, 1977. Examines the concept of sea-floor spreading, features of the ocean bottom, and findings of deep-sea drilling programs.

Tarling, D. H., and M. Tarling. *Continental Drift: A Study of the Earth's Moving Surface*. Garden City, N.Y.: Doubleday, 1971. A comprehensive discussion of the theory of plate tectonics and its early development.

Tokosoz, M. Nafi. "The Subduction of the Lithosphere." *Scientific American* 233 (November, 1975): 88-98. A discussion of how the subduction of oceanic crust is related to ocean trenches, island arcs, volcanism, and earthquakes.

Charles E. Herdendorf

Cross-References

Earth's Core, 171; Earth's Crust, 179; Earth's Mantle, 195; Ocean Basins, 456; Plate Tectonics, 505.

Oil and Gas: Distribution

Oil and gas have been discovered on every principal landmass on earth. Their potential in any particular series of rocks is determined by analyses of the original environment of deposition and its subsequent geologic history and not by present-day climate or location.

Field of study: Petroleum geology and engineering

Principal terms

BASIN: a depressed area of the crust of the earth in which sedimentary rocks have accumulated

FRONTIER: a region of potential hydrocarbon production; little is known of its rock character or geologic history

GEOPHYSICS: the quantitative evaluation of rocks by electrical, gravitational, magnetic, radioactive, and elastic wave transmission and heat-flow techniques

HYDROCARBONS: naturally occurring organic compounds that in the gaseous state are termed natural gas and in the liquid state are termed crude oil or petroleum

PERMEABILITY: the measure of the ability of a porous rock to transmit liquids and gases

POROSITY: the presence of pore space in a sedimentary rock in which hydrocarbons collect

SEDIMENTARY ROCK: rock formed by the deposition and compaction of loose sediment created by the erosion of preexisting rock

STRUCTURE: a physical rearrangement of sedimentary rocks into geometric forms that favor the accumulation and entrapment of hydrocarbons

Summary

Deposits of petroleum and natural gas are found in sedimentary rocks on every principal continent and within the shallow portions of the ocean basins. Sedimentary rocks cover 75 percent of continental land area and extend into the subsurface to maximum depths of 15,000 meters. The occurrence of hydrocarbons varies greatly among differing continents, countries, and individual sedimentary rock basins. Hydrocarbon quantities vary from oil-stained rock and insignificant surface seeps to subsurface reservoirs containing billions of barrels (1 barrel equals 42 U.S. gallons, or about 159 liters) of petroleum or trillions of cubic meters of natural gas. While hydrocarbons have been discovered in rocks ranging in age from 1 million to 1 billion years, generally the younger the sedimentary rock, the greater the chance of commercial hydrocarbon quantities. Approximately 85 percent of crude oil found occurs within the Gulf-Caribbean and the Persian Gulf regions and is contained in rock deposited less than 250 million years of age.

Edwin L. Drake is credited with drilling, in 1859, the first oil well in the United States, near Titusville, Pennsylvania. Wells were first drilled in the search for oil or gas in Europe (France) in 1813, South America (Peru) in 1863, the Far East (Burma) in 1889, Africa (Egypt) in 1908, and the Middle East (Iran) in 1908. By the late 1980's, hydrocarbons had been discovered in eighty-six countries, and oil was being produced from more than 900,000 wells in eighty-two of these countries. Estimated worldwide proven reserves of petroleum exceed 905 billion barrels; gas reserves approximate 100 trillion cubic meters. Indeed, nearly all countries, including members of the Third World, have become dependent on oil and gas as primary sources of energy.

The presence of hydrocarbons is related to the environment prevailing at the time localized columns of sedimentary rock were deposited and to the geologic history to which those rocks have been subjected after being deposited. Hydrocarbons are formed by the chemical alteration of the remains of marine life that collect in the pore space (porosity) of sedimentary rocks, most commonly siltstone, sandstone, and porous varieties of limestone and dolostone. An undeterminable amount of oil and gas created and trapped over millions of years of geologic history has been lost to forces of heat and pressure. Such forces, created through the processes of mountain building and volcanic activity, either destroy hydrocarbons or chemically reduce them to an uneconomic solid state. Large regions of the earth's land area that have been subjected in the geologic past to extended periods of mountain building and volcanism are not considered target areas for hydrocarbon exploration. These "shield" regions, composed of rock generally in excess of 600 million years of age, constitute the geologic nucleus of every continent. An excellent example is the Canadian Shield, centering on the Hudson Bay region of North America.

Regions surrounding the centrally positioned shields are commonly composed of downwarpings, or depressions, in the crust of the earth. Within these "sedimentary basins" are accumulations of sedimentary rocks thousands of meters thick. In such basins, of which approximately 750 have been identified throughout the world, oil and gas are found in quantities depending on conditions of hydrocarbon source rock, geologic structure, and geologic history. Assuming that rocks infilling any basin contain source units, plus porous and permeable rocks, hydrocarbons will generate in the course of geologic time in the central and deeper portions of the basin. With increasing overburden pressure and temperature, oil and gas will slowly migrate outward and upward toward the basin periphery, where either entrapment will take place within a structure or surface seeps will occur.

Each basin must be individually evaluated as to its potential for containing economic deposits of oil or gas. Basins that have been infilled with sedimentary rock deposited under marine (saltwater) conditions offer the best potential, while those infilled under terrestrial environments offer the least. Should the sedimentary rock column be thin, overburden pressure would be insufficient to generate crude oil; if the column is excessively thick, the resultant high temperatures will alter the oil to natural gas or even burn off the gas. Hydrocarbons are destroyed when a basin is subjected to mountain-building forces and escape to the surface to form uneconomic tar sand deposits.

In North America, the principal hydrocarbon-producing regions include the east coast of Mexico, the plains of Alberta and Saskatchewan, and the coastal areas of Alaska. In the lower forty-eight states, the most important producing areas are the coastal states of Alabama, Louisiana, and Texas, the Great Plains and the lower Great Lakes regions, and California. Approximately 10 percent of world oil reserves and 9 percent of world gas reserves are found in North America. Important frontier regions on this continent include the Arctic waters of Alaska and Canada and the Atlantic coastal waters of the United States. The development of both these regions will, however, be constrained by economic and environmental conditions.

Venezuela alone accounts for 84 percent of the oil and 63 percent of the natural gas reserves of South America. The Lake Maracaibo oil fields have historically accounted for a majority of this production. The only other important hydrocarbon-producing country on this continent is Argentina. With large sections of the middle latitudes of South America composed of the Guyana and the Brazilian shields, most of the oil discovered in South America is found in rock less than 250 million years in age and in structures associated with the creation of the Andean mountain ranges. Recent major discoveries in water depths of 900 meters in the Campos basin, offshore Brazil, indicates that significant reserves are yet to be discovered along the Atlantic seaboard of South America.

Africa contains 6 percent of the oil and 16 percent of the gas reserves of the world. North of the Sahara desert, those countries with the greatest hydrocarbon reserves are Libya, Algeria, and Egypt; all border on the prolific Mediterranean basin. In southern Africa, Nigeria controls the majority of production and reserves, while Angola, Congo, and Gabon possess reserves of secondary value. Many sections of this continent have been minimally evaluated for hydrocarbon potential and are considered frontier status. As the African Shield covers much of the 30-million-square-kilometer area of this continent, the best prospects for new oil and gas discoveries are confined to peripheral sedimentary basins, some of which extend into the offshore Atlantic area. These peripheral basins are characterized by thick sedimentary sequences and the common presence of salt deposits; the latter is a positive indicator of hydrocarbon potential.

In the Middle East and Asia, the Soviet Union and the Saudi Arabia peninsula dominate the distribution pattern of oil and gas. Of the proven world reserves of oil, 63 percent are controlled by the countries of Abu Dhabi, Iran, Iraq, Kuwait, and Saudi Arabia. Saudi Arabia alone possesses 170 billion barrels of proven reserves: 19 percent of the world total. With an additional estimated 70 billion probable reserves, Saudi Arabia has the largest total reserves of any country. These reservoirs are being depleted by only 700 producing wells, with the average well flowing approximately 13,000 barrels of oil per day. In contrast, in the United States, more than 500,000 wells produce 9 million barrels of oil, for an average production rate of 18 barrels per day. All hydrocarbons in Saudi Arabia occur within calcium carbonate sedimentary rocks approximately 155 million years in age. The majority of accumulations are contained within anticline structures, convex upward rock folds formed by compressional forces. These flexures, or folds, are located mainly in eastern Arabia, and the longest (including Ghawar, the largest oil field in the world) extends for more than 400 kilometers.

The Soviet Union contains 6 percent of world reserves of oil and 38 percent of world reserves of natural gas. These reserves are principally concentrated in the West Siberian basin, which accounts for two-thirds of Soviet oil production and more than half of its gas production. The West Siberian basin has a cumulative oil production of more than 25 billion barrels. The old hydrocarbon regions of the Soviet Union, the Volga-Ural and North Caucasus regions west of the Ural Mountains and Caspian Sea, contain hydrocarbon reserves of secondary importance. As many sectors of these regions have reached a mature state of exploration, drilling and exploration efforts are being concentrated in the Siberian sector and basins in the Turkmen and Uzbek republics, bordering on Iran and Afghanistan.

West European hydrocarbon reserves are concentrated in the offshore, North Sea extensions of Norway, Denmark, and the United Kingdom. Western Europe contains only 2 percent of world reserves of oil and an insignificant amount of world reserves of natural gas. The increases in European production over the past several decades have been derived mainly from basins located offshore; onshore basins, historically, have never been important producers. Australasia, extending from New Zealand to Malaysia, contains approximately 1 percent of world hydrocarbon reserves. Historically, the principal producing areas have been Sumatra, Java, and Borneo, but exploration programs are assessing the offshore waters of this vast region.

In addition to thickness of sedimentary rock column, type of sedimentary rock, and geologic structure, the geologic age of hydrocarbon source and reservoir rock is significant in evaluating worldwide distribution of oil or gas. The economic presence of hydrocarbons is the result of deposition of an organic shale that undergoes maturation, thus producing the oil or gas (the "source"), followed by migration into a porous and permeable reservoir rock structure. Because source rock maturation occurs within a short period of geologic time, usually measured in millions of years, hydrocarbon distribution is dependent on basin age. In general, the younger the basin and its contained sediments, the greater the volume of hydrocarbons. Approximately 85 percent of known hydrocarbons are contained within rocks younger than 250 million years. These reserves are principally located in California, the Rocky Mountains, the Gulf Coast of Louisiana, Venezuela, the southern Soviet Union, Mexico, Saudi Arabia, Iraq, and Iran. Fourteen percent of world reserves are associated with rocks 250-600 million years in age. These reserves are mainly located in Canada, North Africa, the Soviet Union, and the states of Texas, Oklahoma, Kansas, Illinois, Michigan, Ohio, and Pennsylvania. Small volumes of oil and gas have been found in rocks older than 600 million years, in the Soviet Union, China, Australia, and the United States.

Methods of Study

In the evaluation of the oil and gas potential of a sedimentary basin, whether it be a frontier region or the deeper zones of a mature production area, a routine of analyses has been established. As the geographic distribution of oil and gas accumulations is controlled by regional as well as local factors, such factors must be initially evaluated. The most important of regional factors are those of depositional environment and deformational history.

As a result of the commonly accepted belief that hydrocarbons are derived from oceanic organic matter deposited within sedimentary rocks, basins that have been developed under conditions of marine deposition are given favorable initial evaluation. Minority opinion suggests that hydrocarbons are inorganically (or "abiogenically") derived and, thus, may be found in any type of rock that possesses porosity and permeability. In addition to sedimentary types, such rock might include fractured lava and other igneous rock derived from crystallization of magma (molten rock). This inorganic theory has been tested by the drilling of a deep well in the Scandinavian Shield region of Sweden; an area dominated by heavily fractured granite. As geologic reviews indicated only traces of natural gas and little porosity, the organic theory of hydrocarbon origin appears to be vindicated.

If a world geologic map is superimposed on a hydrocarbon distribution map, relationships dependent on deformational history become apparent. Regionally, oil and gas fields are concentrated in areas of the earth's crust

485

that are downwarped, such as continental margins, coastal plains, and inland plateaus that are low in elevation in comparison to surrounding areas. Such areas generally contain adequate thicknesses of sedimentary rock possessing the organic material, porosity, and permeability necessary to function as source and reservoir rocks.

After a frontier region is evaluated from the above considerations and found to possess potential for oil or gas accumulation, geophysical technology is used to further localize possible hydrocarbon-bearing structures. The most commonly used geophysical techniques are those that employ seismology (the analysis of basin structure by artificially generated elastic wave reflections), gravitational technology (the association of density distributions with rock types), and magnetic technology (the measurement of the earth's magnetic field at different locations). Geophysics technology is used to define fundamental basin architecture and geologic history.

As the surface area of a typical basin is on the order of many thousands of square kilometers, each basin must be studied from the viewpoint of preferred habitats of known oil and gas accumulations. Studies of oil accumulations indicate that more than half (54 percent) of world reserves occur along a basin hinge belt, that zone separating intense downwarping of a basin center from the modest downwarping of the basin periphery. Within the deep basin, only 11 percent of the oil is found, while the peripheral shelf contains the remaining 35 percent.

In the final evaluation of oil and gas distribution patterns, it must be emphasized that the majority of known hydrocarbon deposits are concentrated in three intercontinental depressions in the crust of the earth. In the Western Hemisphere, between North and South America, lies the depression forming the Caribbean Sea and the Gulf of Mexico. Here are found the great oil and gas deposits of Venezuela, Colombia, Mexico, and the Gulf Coast. The second of these depressions contains the hydrocarbon reserves bordering the Red, Mediterranean, Caspian, and Black seas and the Persian Gulf and is formed within the corners of Africa, Europe, and Asia. The last depression is found in the southwest Pacific, between the continents of Asia and Australia, and includes the great fields found in Borneo, Sumatra, and Java. A fourth intercontinental depression, in the early stages of evaluation, occupies the northern limits of North America, Europe, and Asia, surrounding the North Pole. While this region offers great promise and is the site of the largest oil field discovered to date in the United States—the Prudhoe Bay field of Alaska—expanded exploration will be constrained by environmental considerations.

Even with the wide array of available geological and geophysical research procedures, worldwide drilling for new oil and gas reserves is a high-risk endeavor. In spite of continued introduction of new exploration methodologies, approximately 30 percent of all wells drilled since 1859 have been declared dry, and more than 70 percent of the wells drilled annually in frontier areas are unsuccessful.

Context

The worldwide hydrocarbon industry was created in the middle of the nineteenth century to fuel the Industrial Revolution. The recognition of oil and gas as a cheap and efficient energy resource marked the end of wood as an important source of fuel and ultimately reduced "King Coal" to secondary status. By the final decade of the twentieth century, however, the burning of oil and gas was being associated with atmospheric pollution and the greenhouse effect, among other environmental problems. Yet, the promises of solar power have not been realized, and nuclear energy, while economically viable, has been generally considered unsafe.

Oil and gas continue to function in the role that they rapidly assumed after the completion of Edwin Drake's well in 1859; they are the wonder fuels, lubricants, and chemical sources of the modern scientific world. It was popular in the 1960's to predict that the end of the hydrocarbon age would be closely associated with the year 2000, but such predictions have proved premature. The continued exploration for new reserves of oil and gas is partially controlled by economics. Actions taken by the Organization of Petroleum Exporting Countries (OPEC) in the early 1970's caused the value of a barrel of petroleum in the United States to increase from $3 to $35 by 1981. Spurred by this incentive, the American hydrocarbon industry in 1974 created the first annual increase in proven oil reserves in the United States since 1960. In that same year, the world consumption of petroleum products approximated 7 billion barrels annually. Thirty years later, the United States alone was consuming 6 billion barrels of crude oil every year, with some 45 percent of this supply being imported. Worldwide, the demand for oil and gas, in terms of crude oil equivalent energy, is close to 90 million barrels per day.

Oil and gas will probably continue for decades to be the world's primary energy source, and new reserves will continue to be sought worldwide. The distribution of oil and gas in the future will be determined by the same factors governing its distribution today. Geologic factors will continue to dictate the best locations for exploration, with local political stability and economic considerations entering into the decision when and whether to explore.

Bibliography

Baker, Ron. *Oil and Gas: The Production Story*. Austin: University of Texas Petroleum Extension Service, 1983. The first four sections of this primer are easy-to-read introductions to the origin and accumulation of hydrocarbons, exploration and testing technologies, and production stimulation procedures. Each section contains full-color diagrams.

Ball, Max W. *This Fascinating Oil Business*. New York: Bobbs-Merrill, 1940. Although an older reference, this 444-page book contains a wealth of information found in no other review of the business. It is especially valuable for its history of early global exploration in remote regions ranging from Afghanistan to Manchukuo. Written for general readers.

Nawwab, Ismail I., Peter C. Speers, and Paulk F. Hoye, eds. *Aramco and Its World: Arabia and the Middle East.* Washington, D.C.: Arabian American Oil, 1980. A beautifully illustrated volume that explores the history of the discovery and development of oil in a region that possesses 63 percent of the world's petroleum reserves.

Pratt, Wallace E., and Dorothy Good, eds. *World Geography of Petroleum.* Princeton, N.J.: Princeton University Press, 1950. A classic comparative study of the petroleum-producing regions of the world. Coedited by an internationally recognized geologist. A college-level text.

Schackne, Stewart, and N. D'Arcy Drake. *Oil for the World.* 2d ed. New York: Harper & Brothers, 1960. A brief introduction to all aspects of an integrated oil and gas company, from discovery through drilling operations, refining, and marketing.

Tiratsoo, Eric N. *Natural Gas.* 3d ed. Houston: Gulf, 1980. A college-level review of natural gas drilling, production reserves, and economics. Eight chapters discuss principal geographic regions in detail. Includes a valuable compilation of natural gas statistics for the Soviet Union and Eastern Europe.

_____. *Oilfields of the World.* 3d ed. Houston: Gulf, 1985. This easy-to-read companion reference to the above entry describes oilfields in approximately seventy countries. The geology of each field is presented at a college freshman level.

Albert B. Dickas

Cross-References

Oil and Gas: Origins, 489.

Oil and Gas: Origins

Oil and gas are two of the most important fossil fuels. The formation of oil and gas is dependent on the preservation of organic matter and its subsequent chemical transformation into kerogen and other organic molecules deep within the earth at high temperatures over long periods of time. As oil and gas are generated from these organic materials, they migrate upward, where they may accumulate in hydrocarbon traps.

Field of study: Petroleum geology and engineering

Principal terms

FOSSIL FUEL: a general term used to refer to petroleum, natural gas, and coal

HYDROCARBONS: solid, liquid, or gaseous chemical compounds containing only carbon and hydrogen; oil and natural gas are complex mixtures of hydrocarbons

KEROGEN: fossilized organic material in sedimentary rocks that is insoluble and that generates oil and gas when heated; as a form of organic carbon, it is one thousand times more abundant than coal and petroleum in reservoirs, combined

METHANE: a colorless, odorless gaseous hydrocarbon with the formula CH_4; also called marsh gas

NATURAL GAS: a mixture of several gases used for fuel purposes and consisting primarily of methane, with additional light hydrocarbon gases such as butane, propane, and ethane, with associated carbon dioxide, hydrogen sulfide, and nitrogen

PETROLEUM: crude oil; a naturally occurring complex liquid hydrocarbon, which after distillation yields a range of combustible fuels, petrochemicals, and lubricants

RESERVOIR: a porous and permeable unit of rock below the surface of the earth that contains oil and gas; common reservoir rocks are sandstones and some carbonate rocks

Summary

The origin of oil and gas begins with the production of organic matter by plants and plantlike organisms, through a process called photosynthesis.

Photosynthesis converts light energy from the sun into chemical energy and produces organic matter (or carbohydrates, which include sugars, starches, and cellulose) and oxygen. Carbohydrates burn easily and release considerable amounts of energy in the process. It is this property that makes them ideal fuels.

To become a fossil fuel, the organic matter in an organism must be preserved after the organism dies. The preservation of organic matter is a rare event because most of the carbon in organic matter is oxidized and recycled to the atmosphere through the action of aerobic bacteria. (Oxidation is a process through which organic matter combines with oxygen to produce carbon dioxide gas and water.) Less than 1 percent of the organic matter that is produced by photosynthesis escapes from this cycle and is preserved. To be preserved, the organic matter must be protected from oxidation, which can occur in one of two ways: The organic matter in the dead organism is rapidly buried by sediment, shielding it from oxygen in the environment, or the dead organism is transported into an aquatic environment in which there is no oxygen (that is, an anoxic or anaerobic environment). Most aquatic environments have oxygen in the water, because it diffuses into the water from the atmosphere and is produced by photosynthetic organisms, such as plants and algae. Oxygen is removed from the water by the respiration of aerobic organisms and through the oxidation of decaying organic matter. Oxygen consumption is so high in some aquatic environments that anoxic water is present below the nearsurface oxygenated zone. Environments that lack oxygen include such places as deep, isolated bodies of stagnant water (such as the bottom waters of some lakes), some swamps, and the oxygen-minimum zone in the ocean (below the maximum depth to which light penetrates, where no photosynthesis can occur). Large quantities of organic matter may be preserved in these environments.

Some bacteria, called anaerobic bacteria, can survive in water without oxygen and in sediments with anoxic pore water. These bacteria partially decompose organic matter through processes such as fermentation, sulfate-reduction, nitrate-reduction, and methanogenesis (or methane production). Methane is formed as sediment is buried to relatively shallow depths (centimeters to meters), because of the activity of methanogenic bacteria living in this sediment. Bubbles of methane can be seen rising from stagnant water overlying sediment and are often referred to as "marsh gas."

The major types of organic matter preserved in sediments include plant fragments, algae, and microbial tissue formed by bacteria. Animals (and single-celled animal-like organisms) contribute relatively little organic matter to sediments. The amount of organic matter contained in a sediment or in sedimentary rock is referred to as its total organic content (TOC), and it is typically expressed as a percentage of the weight of the rock. To be able to produce oil, a sediment typically must have a TOC of at least 1 percent by weight. Sediments that are capable of producing oil and gas are referred to as "hydrocarbon source rocks." In general, fine-grained rocks such as shales (which have clay-sized grains) tend to have higher TOC than do coarser-grained rocks.

The organic matter trapped in sediment must undergo a series of changes to form oil and gas. These changes take place as the sediment is buried to great depths as a result of the deposition of more and more sediment in the environment over long periods of time. Temperature and pressure increase as depth of burial increases, and the organic materials are altered by these high temperatures and pressures.

As sediment is gradually buried to depths reaching hundreds of meters, it undergoes a series of physical and chemical changes, called diagenesis. Diagenesis transforms sediment into sedimentary rock by compaction, cementation, and removal of water. Methane gas commonly forms during the early stages of diagenesis as a result of the activity of methanogenic bacteria. At depths of a few meters to tens of meters, organic compounds such as proteins and carbohydrates are partially or completely broken down, and the individual component parts are converted into carbon dioxide and water or are used to construct geopolymers, or large, complex organic molecules of irregular structure (such as fulvic acid and humic acid and larger geopolymers called humins). During diagenesis, the geopolymers become larger and more complex, and nitrogen and oxygen content decreases. With increasing depth of burial over long periods of time (burial to tens or hundreds of meters over a million or several million years), continued enlargement of the organic molecules alters the humin into kerogen, an insoluble form of organic matter, that yields oil and gas when heated.

As sediment is buried to depths of several kilometers, it undergoes a process called catagenesis. At these depths, the temperature may range from 50 to 150 degrees Celsius, and the pressure may range from 300 to 1,500 bars. The organic matter in the sediment, while in a process called maturation, becomes stable under these conditions. During maturation, a number of small organic molecules are broken off the large kerogen molecules, a phenomenon known as thermal cracking. These small molecules are more mobile than are the kerogen molecules. Sometimes called bitumen, they are the direct precursors of oil and gas. As maturation proceeds and oil and gas generation continues, the kerogen residue remaining in the source rock gradually becomes depleted in hydrogen and oxygen. In a later stage, wet gas and condensate are formed. ("Condensate" is a term given to hydrocarbons that exist as gas under the high pressures existing deep beneath the surface of the earth but condense to liquid at the earth's surface.) Oil is typically generated at temperatures between 60 and 120 degrees Celsius, and gas is generated at somewhat higher temperatures, between about 120 and 220 degrees Celsius. Large quantities of methane are formed during catagenesis and during the subsequent phase, which is called metagenesis.

When sediment is buried to depths of tens of kilometers, it undergoes the processes of metagenesis and metamorphism. Temperatures and pressures are extremely high. Under these conditions, all organic matter and oil are destroyed, being transformed into methane and a carbon residue, or graphite. Temperatures and pressures are so intense at these great depths that some of the minerals in the sedimentary rocks are altered and recrystallized, and metamorphic rocks are formed.

Accumulations of oil and gas are typically found in relatively coarse-grained, porous, permeable rocks, such as sandstones and some carbonate rocks. These oil- and gas-bearing rocks are called reservoirs. Reservoir rocks, however, generally lack the kerogen from which the oil and gas are generated. Instead, kerogen is typically found in abundance only in fine-grained sedimentary rocks such as shales. From these observations, it can be concluded that the place where oil and gas originate is not usually the same as the place where oil and gas are found. Oil and gas migrate or move from the source rocks (their place of origin) into the reservoir rocks, where they accumulate.

Oil and gas that form in organic-rich rocks tend to migrate upward from their place of origin, toward the surface of the earth. This upward movement of oil occurs because pore spaces in the rocks are filled with water, and oil floats on water because of its lower density. Gas is even less dense than oil and also migrates upward through pore spaces in the rocks. The first phase of the migration process, called primary migration, involves expulsion of hydrocarbons from fine-grained source rocks into adjacent, more porous and permeable layers of sediment. Secondary migration is the movement of oil and gas within the more permeable rocks. Oil and gas may eventually reach the surface of the earth and be lost to the atmosphere through a seep. Under some circumstances, however, the rising oil and gas may become trapped in the subsurface by an impermeable barrier, called a cap rock. These hydrocarbon traps are extremely important because they provide a place for subsurface concentration and accumulation of oil and gas, which can be tapped for energy sources.

There are a variety of settings in which oil and gas may become trapped in the subsurface. Generally, each of these traps involves an upward projection of porous, permeable reservoir rock in combination with an overlying impermeable cap rock that encloses the reservoir to form a sort of inverted container. Examples of hydrocarbon traps include anticline traps, salt dome traps, fault traps, and stratigraphic traps (see figure page 494). There are many types of stratigraphic traps, including porous reef rocks enclosed by dense limestones and shales, sandstone-filled channels, sand bars, or lenses surrounded by shale, or porous, permeable rocks beneath an unconformity. The goal of the exploration geologist is to locate these subsurface hydrocarbon traps. Enormous amounts of geologic information must be obtained, and often many wells must be drilled before accumulations of oil and gas can be located.

Methods of Study

The origin of oil and gas can be determined using physical and chemical analyses. Petroleum contains compounds that serve as biological markers to demonstrate the origin of petroleum from organic matter. Oil can be analyzed chemically to determine its composition, which can be compared to that of hydrocarbons extracted from source rocks in the lab. Generally, oil is associated with natural gas, most of which probably originated from the alteration of organic material during diagenesis, catagenesis, or metagenesis. In some cases, gas may be of abiogenic (nonorganic) origin. Samples of natural gas can be analyzed using gas chromatography or mass spectrometry and isotope measurements.

Commonly, rocks are analyzed to determine their potential for producing hydrocarbons. It is important to distinguish between various types of kerogen in the rocks because different types of organic matter have different potentials for producing hydrocarbons. In addition, it is important to determine the thermal maturity or evolutionary state of the kerogen to determine whether the rock has the capacity to generate hydrocarbons or hydrocarbons have already been generated.

The quantity of organic matter in a rock, referred to as its TOC, can be measured with a combustion apparatus, such as a Leco carbon analyzer. To analyze for TOC, a rock must be crushed and ground to a powder and its carbonate minerals removed by dissolution in acid. During combustion, the organic carbon is converted into carbon dioxide by heating to high temperatures in the presence of oxygen. The amount of carbon dioxide produced is proportional to the TOC of the rock. The minimum amount of TOC considered adequate for hydrocarbon production is generally considered to be between 0.5 and 1 percent TOC by weight.

The type of organic matter in a rock can be determined indirectly through study of the physical and chemical characteristics of the kerogen or directly by using pyrolysis (heating) techniques. The indirect methods of analysis include examination of kerogen with a microscope and chemical analysis of kerogen. Microscopic examination can identify different types of kerogen, such as spores, pollen, leaf cuticles, resin globules, and single-celled algae. Kerogen that has been highly altered and is amorphous can be examined using fluorescence techniques to determine whether it is oil-prone (fluorescent) or inert or gas-prone (nonfluorescent). Chemical analysis of kerogen provides data on the proportions of chemical elements, such as carbon, hydrogen, sulfur, oxygen, and nitrogen. A graph of the ratios of hydrogen/carbon (H/C) versus oxygen/carbon (O/C) is used to classify kerogen by origin and is called a van Krevelen diagram. There are three curves on a van Krevelen diagram, labeled I, II, and III, corresponding to three basic types of kerogen. Type I is rich in hydrogen, with high H/C and low O/C ratios, as in some algal deposits; this type of kerogen generally yields the most oil. Type II has relatively high H/C and low O/C ratios and is usually related to marine sediments containing a mixture of phytoplankton, zooplankton, and bacteria; this type of kerogen yields less oil than Type I, but it is the source material for a great number of commercial oil and gas fields. Type III is rich in oxygen, with low H/C and high O/C ratios (aromatic hydrocarbons), as in terrestrial or land plants; this type of kerogen is comparatively less favorable for oil generation but tends to generate large amounts of gas when buried to great depths. As burial depth and temperature increase, the amount of oxygen and hydrogen in the kerogen decreases, and the kerogen approaches 100 percent carbon. Hence, a van Krevelen diagram can be used to determine both the origin of the organic matter and its relative thermal maturity.

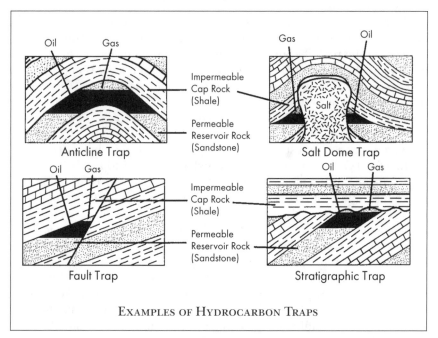

EXAMPLES OF HYDROCARBON TRAPS

The potential that a rock has for producing hydrocarbons can be evaluated through a pyrolysis, or heating, technique, commonly called Rock-Eval. Rock-Eval yields information on the quantity, type, and thermal maturity of organic matter in the rock. The procedure involves the gradual heating (to about 550 degrees Celsius) of a crushed rock sample in an inert atmosphere (nitrogen, helium) in the absence of oxygen. At temperatures approaching 300 degrees Celsius, heating releases free hydrocarbons already present in the rock; the quantity of free hydrocarbons is referred to as S_1. At higher temperatures (300-550 degrees Celsius), additional hydrocarbons and related compounds are generated from thermal cracking of kerogen in the rock; the quantity of these hydrocarbons is referred to as S_2. The temperature at which the maximum amount of S_2 hydrocarbons is generated is called T_{max} and can be used to evaluate the thermal maturity of the organic matter in the rock. In addition, carbon dioxide is generated as the kerogen in the rock is heated; the quantity of CO_2 generated as the rock is heated to 390 degrees Celsius is referred to as S_3. (The temperature is limited to 390 degrees Celsius because at higher temperatures, CO_2 is also formed from the breakdown of inorganic materials, such as carbonate minerals.) These data can be used to determine the hydrocarbon-generating potential of the rock, the quantity and type of organic matter, and the thermal maturity. For example, $S_1 + S_2$, called the genetic potential, is a measure of the total amount of hydrocarbons that can be generated from the rock, expressed in kilograms per ton. If $S_1 + S_2$ is less than 2 kilograms per ton, the rock has little or no potential for oil production, although it has some potential for gas production. If $S_1 + S_2$ is between 2 and 6 kilograms per ton, the rock has moderate potential for oil production. If $S_1 + S_2$ is greater than 6 kilograms per ton, the rock has good potential for oil production. The ratio, $S_1/(S_1 + S_2)$, called the production index, indicates the maturation of the organic matter. Pyrolysis data can also be used to determine the type of organic matter present. The oxygen index is S_3/TOC, and the hydrogen index is S_2/TOC. These two indices can be plotted against each other on a graph, comparable to a van Krevelen diagram.

Context

Oil and gas are derived from the alteration of kerogen, an insoluble organic material, under conditions of high temperatures (50-150 degrees Celsius) and pressures (300-1,500 bars). After oil and gas are generated, they migrate upward out of organic-rich source rocks and come to be trapped and accumulate in specific types of geologic settings. The search for oil and gas deposits trapped in the subsurface can be expensive and time-consuming, and it requires trained exploration geologists. Once a promising geologic setting has been located, the only way to determine whether oil and gas deposits are actually present in the subsurface is to drill a well. Oil and gas are two of the earth's most important fossil fuels. It is important to understand that a finite amount of these hydrocarbons is present within the earth. They cannot be manufactured when known reserves are depleted.

Bibliography

Durand, Bernard, ed. *Kerogen: Insoluble Organic Matter from Sedimentary Rocks*. Paris: Éditions Technip, 1980. This book consists of a series of papers on various aspects of kerogen, ranging from its origin and appearance under the microscope to its chemical composition and structure as determined by a variety of analytical means. The articles are written by specialists; most of them are in English, but a few are in French. Technical but does contain a number of beautiful color plates illustrating the appearance of kerogen-rich rocks and organic microfossils (pollen, spores, acritarchs, dinoflagellates) as seen through the microscope.

North, F. K. *Petroleum Geology*. Boston: Allen & Unwin, 1985. A long book (607 pages) that covers a wide variety of topics related to petroleum geology, it includes five main parts: introduction; the nature and origin of petroleum; where and how oil and gas accumulate; exploration, exploitation, and forecasting; and distribution of oil and gas. Designed as a college textbook to introduce students to many topics with practical application to exploration and drilling in addition to the basics on the origin of oil and gas. Well illustrated with maps and geologic cross sections representing many oil-producing areas around the world. Suitable for geologists and college students.

Peters, K. E. "Guidelines for Evaluating Petroleum Source Rock Using Programmed Pyrolysis." *The American Association of Petroleum Geologists Bulletin* 70 (March, 1986): 318-329. Although rather technical in nature, this article provides information on Rock-Eval pyrolysis, one of the major analytical techniques for analyzing rocks to determine their hydrocarbon potential. Provides a brief summary of the technique and goes into detail using numerous examples, discussing some of the problems encountered in interpreting samples. Suitable for geologists and advanced college students.

Selley, Richard C. *Elements of Petroleum Geology*. New York: W. H. Freeman, 1985. This book is designed as a college textbook for students near the end of their coursework in geology or for geologists beginning careers in the petroleum industry. Fairly technical, it requires basic understanding of geological concepts.

Tissot, Bernard P., Bernard Durand, J. Espitalié, and A. Combaz. "Influence of Nature and Diagenesis of Organic Matter in Formation of Petroleum." *The American Association of Petroleum Geologists Bulletin* 58 (March, 1974): 499-506. This article discusses the generation of hydrocarbons and changes in kerogen that occur during burial. Somewhat technical but well illustrated with graphs. Provides a concise summary of the types of kerogen and depths at which oil and gas are generated.

Tissot, Bernard P., and D. H. Welte. *Petroleum Formation and Occurrence*. 2d ed. New York: Springer-Verlag, 1984. This book is one of the most comprehensive guides to the origin of petroleum and natural gas and should be considered one of the leading references in the field. The book is divided into five parts: the production and accumulation of geologic matter (a geological perspective); the fate of organic matter in sedimentary basins (generation of oil and gas); the migration and accumulation of oil and gas; the composition and classification of crude oils and the influence of geological factors; and oil and gas exploration (application of the principles of petroleum generation and migration). Each part is divided into chapters, which are well written and well illustrated with line drawings and graphs. Easy to read;

up-to-date coverage of the field is provided. An indispensable reference for geologists that is suitable for college-level students.

Waples, Douglas W. *Geochemistry in Petroleum Exploration.* Boston: International Human Resources Development Corporation, 1985. This book provides an overview of the origin of oil and gas and should be considered a leading reference in the field. Concise and well illustrated with line drawings and graphs. Easy to read. A good reference for geologists, it is also suitable for college-level students.

Pamela J. W. Gore

Cross-References

Oil and Gas: Distribution, 481.

Permafrost

Permafrost is a thermal condition existing in any type of earth material that is perennially frozen. It occurs in ground in which the temperature remains below 0 degrees Celsius for at least two consecutive years. Affecting about 25 percent of the earth's surface, the condition hampers construction in the far north in regions such as Siberia and Arctic Canada.

Field of study: Geomorphology

Principal terms

PERIGLACIAL: originally restricted to regions modified by frost weathering such as tundra

PINGO: a growth of ice in a permafrost body associated with frost heave; the result is an upthrusted dome of sediments forming a conical landform several tens of meters high

PLEISTOCENE: the geologic era between two million and sixteen thousand years ago, characterized by extensive continental glaciation and a colder climate

STONE POLYGON: an assemblage of boulders in a roughly polygonal shape ranging from a few centimeters to several meters across; occurring in clusters, they form honeycomb or netlike patterns covering several square kilometers

SUBSIDENCE: a downward vertical movement of the ground, commonly caused by land-cover changes of the surface

TALIK: a layer of unfrozen ground in permafrost; water under pressure frequently occurs in talik layers

THERMOKARST: a group of landforms in flat areas of degrading permafrost and melting ice

Summary

Permafrost, or permanently frozen ground, is a thickness of earth material such as soil, peat, or even bedrock, at a variable depth beneath the ground surface, in which a temperature below 0 degrees Celsius exists continuously for a number of years (from at least two years to several thousands of years). Permafrost is thus not a process or an effect but a temperature condition of earth material on land or beneath continental shelves of the Arctic Ocean. The occurrence of permafrost is conditioned exclusively by temperature, irrespective of the presence of water or of the composition, lithology, and texture of the ground. However, unfrozen water horizons may exist within a body of permafrost as the result of various conditions such as pressure or impurities in the water.

With regard to the climatic history of the earth, cold conditions and glaciers are the exception, not the rule. Because they are so rare, they are referred to as "climatic accidents." During the Pleistocene, advances of continental glaciers covered vast areas of Europe and North America. Glaciers continue to be significant landscape features, since they lock up as ice some 2.15 percent of the earth's waters. Permafrost is a relatively new condition introduced with the onset of the Pleistocene glacial events, and it will be maintained as long as the climate remains cold.

As the earth's climate warmed during the past twenty thousand years and the continental ice receded, the location, distribution, and thickness of permafrost have changed. Today, the area underlain by permafrost is vast; it has been estimated that about 25 percent of the earth's surface is underlain by permafrost. One reason why vast areas of the earth possess permafrost is that the continents widen toward higher latitudes in North America and Eurasia. In the Northern Hemisphere, Mongolia, Russia, and China contain about 12.2 million square kilometers of land underlain with permafrost. North America and Greenland have 8.8 million square kilometers of permafrost. In the Southern Hemisphere, Antarctica accounts for 13.5 million square kilometers of permafrost. During the Pleistocene, perennially frozen ground had a different distribution. Permafrost produces soil and sediment structures that remain after the climate warms. Relict features such as ancient ice wedges can be identified, and former permafrost areas can be mapped. Evidence of preexisting permafrost conditions, and thus a record of climate change, has been extensively documented in Britain, southern Canada, and the central United States.

In terms of aerial distribution and continuity, permafrost may be classified as either continuous or discontinuous. At higher-latitude land circling the Arctic Ocean, the lateral extent of permafrost is uninterrupted nearly everywhere, except in areas of recent deposition and under large lakes that do not freeze to the bottom. Discontinuous permafrost includes areas of frozen ground separated by non-permafrost areas. A transect from the Arctic Ocean southward, into Canada or Russia, reveals more and more permafrost-free areas. Further south, in central Canada and central Russia, only patchy or sporadic zones of perennially frozen ground are to be found.

Permafrost is a unique condition because it is maintained very close to its melting point. Several thousand square kilometers are warmer than −3 degrees Celsius. If global warming continues, as many scientists predict, most discontinuous permafrost would degrade. Russian scientists, in fact, have documented the northward retreat of permafrost near Archangel at a rate of 400 meters per year since 1837. In Canada, permafrost has retreated northward more than 100 kilometers since 1945, as ground temperatures have increased by 2 degrees Celsius in the northern prairie provinces. Changes in temperature may result from a local microclimate change, alteration of land cover, or changes in the atmosphere. Conversely, permafrost is not only degraded by global warming but may indeed contribute to atmospheric change. Scientists suggest that thawing of Arctic terrain, which is rich in peat accumulation, may affect the chemistry of the atmosphere. It has been determined that one-quarter of the earth's terrestrial carbon is stored in organic matter in the permafrost and active layer. Long-term warming would release enormous quantities of greenhouse gases, methane and carbon dioxide in particular, accelerating global warming.

A cross section of excavated earth in the high Arctic reveals the vertical characteristics of permafrost. The thickness of permafrost is variable because of differences in air temperatures, occurrence of water, composition and texture of earth materials, and many other factors. The permafrost in Siberia may be at least 5,000 meters thick, whereas in Alaska thicknesses of 740 meters have been recorded. Further south, thicknesses decrease as discontinuous permafrost becomes more common. The thickness of permafrost is in part determined by a balance between the increase of internal heat with depth and the heat loss from the earth's surface. If the thickness and depth of permafrost remains unchanged for many years, the heat loss at the surface and the heat at the base of the frozen ground are in equilibrium, and a steady state is maintained. At the surface, the ground seasonally thaws and freezes; this section of the permafrost is known as an "active layer." Its thickness is controlled by numerous variables in addition to seasonal temperature variations, and it exhibits great variability in thickness. Vegetation, snow cover, albedo, water distribution, and human alteration of the surface can all act as insulators and affect the thickness of the active layer. However, in areas not disturbed by people, the ground will typically thaw to a depth of from 1 to 12 meters during the summer.

The deeper boundary separating the active layer from the permafrost is known as the "permafrost table" or "zero curtain." Within the permafrost, unfrozen lenses known by the Russian term *talik* occur. These unfrozen bodies frequently contain water and function as protected water sources or aquifers in the Arctic environment. Talik layers may be isolated bodies completely enclosed in impermeable permafrost and result from a change in the thermal regime in permafrost. Conversely, talik may be laterally extensive even in continuous permafrost and is frequently a valuable source of groundwater. Talik lenses are larger in discontinuous permafrost, and the active layer thickens at the expense of the permafrost.

Below the permafrost table, the perennially frozen ground has low permeability. Maps and photographs of the Arctic reveal many lakes, swamps, and marshes, suggesting high precipitation. However, evaporation rates are low because of the cold air temperatures, and standing water occurs because the ground is frozen at depth. Throughout the region, precipitation is probably less than 35 centimeters annually, equivalent to that of marginal deserts. The abundant water and poor drainage result from the occurrence of permafrost at depth, not to high atmospheric precipitation. As seasonal warming and cooling occur, the land surface expands and contracts, particularly in continuous permafrost areas. In essence, the lowering of the ground temperature causes thermal contraction of the ground, and vertical fissures or frost cracks occur. Water filling and freezing in these cracks creates ice wedges in the permafrost that widen with seasonal expansion and contraction. Ice wedges are often 3 to 4 meters wide and extend 5 to 10 meters into the perennially frozen ground below the active layer. In areas of degradation, ice wedges in the thawing permafrost begin to melt, creating thermokarst features characterized by numerous depressions similar to karst plains. As temperatures rise above freezing, the active layer extends deeper and deeper and eventually may intercept an ice wedge. If large ice wedges begin to thaw, the ground subsides, creating steep-sided conical depressions. The perimeter of a depression, often composed of very moist sediment, will slump into the depression, thus making it wider. Water then fills the depression, creating a thermokarst, or thaw lake. Averaging about 3 meters across, the lakes occur in clusters, are elliptical in shape, and are all oriented at right angles to the prevailing wind direction.

Methods of Study

To examine and obtain data on the characteristics of permafrost is difficult because of the isolation and hostile climatic setting of Arctic regions. Fieldwork is usually done in spring and summer, when conditions are warmer. However, wet conditions frequently exist; the thawed active layer above the permafrost is often poorly drained and boggy, making trekking difficult. Geophysical methods, various types of thermometers, and aerial images have helped to unravel the complexities of permafrost.

The application of electrical resistivity is an example of a geophysical method; this technique is most effective in continuous permafrost extending to depths of about 60 meters and with a temperature not lower than 15 degrees Celsius. Ice acts as an insulator, whereas unfrozen ground regions have a smaller specific resistivity. If a hole is bored into the ground, a probe may be lowered to record the electrical resistivity at depth, and this information can then be interpreted as a temperature profile on a chart. More recently, airborne geophysical methods, which are more suitable for reconnaissance surveys of large regions, have been put to use. Although less detailed than ground surveys, they do provide information on the thickness of taliks in continuous permafrost areas and the general distribution of discontinuous permafrost.

Shallow ground temperatures yield significant data. They establish, for example, the thickness of the active layer, the presence of taliks, the depth of seasonal fluctuation of ground temperature, and the temperature profile of permafrost. These and related data are important for engineers to have prior to the construction of buildings, airstrips, or oil and gas platforms.

Maximum and minimum thermometers positioned in a bore hole record permafrost temperature at a specific depth. Maximum temperature data are used in winter, when ground temperatures exceed air temperatures, and minimum data can be used in summer, when the permafrost is colder than the air.

Thermocouple thermometers work on electrical concepts and are especially suitable for permanent installation. As many as twenty thermocouples can be installed in a pit or a drilled hole and joined so that readings may be taken quickly at different locations.

Aerial photography and images are used to make maps of Arctic environments and to determine spatial relationships and land covers of vast and less accessible areas. The occurrence and thickness of the active layer—and, to some degree, whether permafrost is continuous or discontinuous—is revealed in the distribution of water, landforms, and vegetation types. In general, landforms in cold climates can be divided into those associated with the aggradation of permafrost and those related to the degradation of permafrost. "Paisas" are low conical mounds 1 to 50 meters in height that occur in discontinuous permafrost, whereas stone polygons are common landforms in continuous permafrost. Thermokarst lakes are transitional features sited on the boundary between continuous and discontinuous permafrost. By mapping these features, the aerial extent and type of permafrost in a landscape can be determined.

Context

Following World War II, two events stimulated the study of permafrost environments. With the onset of the Cold War, both Eastern and Western governments quickly realized that permafrost phenomena materially impeded the settlement and development of northern regions, particularly with regard to military early-warning systems and airbases. The lack of knowledge of appropriate construction techniques in the cold environment resulted in the deformation, in many cases beyond repair, of roads, bridges, and other structures that were not built to withstand frost heaving or subsidence. In addition, issues emerged regarding the resources of the north that required a clear understanding of the behavior and alteration of this thermally controlled environment, especially the exploration for oil and gas in the Arctic coastal zone, the construction of pipelines for oil transport, and rising environmental concerns.

Essentially, construction in permafrost areas requires that the sensitive permafrost table and active layer be considered so that the thermal conditions remain unchanged if possible. Construction of concrete runways and roads alters drainage and vegetation, and these structures absorb heat during long summer days. The heat is conducted to the active layer, which may become thicker, making the permafrost thinner. Such environmental disturbances can cause a structure to subside unevenly. If a house or building is constructed on a concrete slab that is placed directly on the active layer, the structure may subside as the active layer thickens at the expense of the underlying thawing permafrost. Should the thicker active layer encounter a water-charged talik lens, seepage into the building is likely to occur. To avoid the problem, structures are built above ground level and placed on pilings.

In 1977, the 1,285-kilometer Trans-Alaskan Pipeline was built from Prudhoe Bay southward to the Pacific port of Valdez, Alaska. Originally, a buried pipeline was proposed. However, since the oil is transported at a temperature of from 70 to 80 degrees Celsius, a buried pipeline would thaw the permafrost, causing the pipe to sag and possibly break. To avoid potential damage to the pipeline and to the environment, one-half of the structure was suspended above the ground. A thorough understanding of the sensitivity and the mechanisms of permafrost must thus be considered so that similar problems affecting structures and the environment can be avoided.

Thanks to an environment that is very cold or frozen most of the time, organic matter derived from the scanty Arctic vegetation does not decay readily. Peat deposits make Canada a top exporter of this resource. To scientists, the peat provides clues to the past environmental conditions of the Arctic. Seeds and pollen preserved in these organic layers or in lake sediments help researchers to reconstruct past climatic conditions. Radiocarbon analyses reveal some peat beds to be as much as fifty thousand years old. This information, however, does not tell scientists how long the ground was frozen. In Siberia and in Alaska, numerous frozen mammoth carcasses have been discovered. The fossil remains include not only the bones but also the flesh of these elephant-like mammals. Radiocarbon tests indicate that the mammoths fall into one of two distinct age groups, one dating from about forty-five thousand to thirty thousand years ago, and a younger cluster ranging in age from fourteen thousand to eleven thousand years ago. The animals were frozen and were not eaten by scavengers. This suggests that they must have been rapidly covered, perhaps by mud flows, and frozen quickly. The stomach-content remains of examined mammoths reflect the past vegetative diversity of the Arctic. Included were larch needles, willow, tree bark, grasses, sedges, and even rare poppies. Since the animals were not decomposed, scientists were able to infer that the ground must have been continually frozen for thousands of years.

Bibliography

Ballantyne, C. F., and C. Harris. *The Periglaciation of Great Britain.* New York: Cambridge University Press, 1994. A good upper-level description of permafrost in Britain. The discussion ranges from basic freeze-and-thaw processes to permafrost conditions and distribution in the geologic past.

Clark, M. J., ed. *Advances in Periglacial Geomorphology.* New York: Wiley InterScience, 1988. A series of state-of-the-art reviews of numerous processes and landforms in cold climates. Emphasizes processes of formation; very technical. Suitable for advanced earth science students as a reference and literature source.

Davis, J. L. *Landforms of Cold Climates.* Cambridge, Mass.: MIT Press, 1969. A basic primer on permafrost and related topics. Permafrost conditions and distribution are well described, and the discussion is jargon-free.

French, H. M. *The Periglacial Environment.* New York: Longman, 1978. A descriptive account of permafrost and its environment in North America and Eurasia. Somewhat technical, yet readable and supplemented with numerous photographs and other graphics.

Harris, S. A. *The Permafrost Environment.* Totowa, N.J.: Barnes & Noble, 1986. A primer on the consequences of construction in permafrost regions. Topics include the soil properties and methods of permafrost analysis. Written for contractors and environmental specialists.

Muller, S. W. *Permafrost or Permanently Frozen Ground and Related Engineering Problems.* Ann Arbor, Mich.: Edwards Brothers, 1947. A comprehensive account of permafrost by the man who coined the term. Emphasizes the problems of building in permafrost areas. Gives excellent examples of poor planning in a poorly understood environment.

Simpson, Sherry. "Permafrost." *Alaska* 63, no. 10 (December, 1997/ January, 1998): 28-34.

Tricart, Jean. *Geomorphology of Cold Climates.* New York: St. Martin's Press, 1970. A translation of a classic work on cold climate processes and landforms. Includes glacial and periglacial phenomena. Somewhat advanced, but not overwhelmingly technical.

Williams, P. J., and M. W. Smith. *The Frozen Earth: Fundamentals of Geology.* New York: Cambridge University Press, 1989. A college-level textbook for science majors, emphasizing the physics of cold regions. Equations are used to illustrate significant physical ideas.

C. Nicholas Raphael

Cross-References

Continental Glaciers, 93; Karst Topography, 348; Weathering and Erosion, 701.

Plate Tectonics

Plate tectonics is the theory that the earth's surface is composed of major and minor plates that are being created at one edge by the formation of new igneous rocks and consumed at another edge as one plate is thrust, or subducted, below another. This elegant theory accounts for the formation of earthquakes, volcanoes, and mountain belts, the growth and fracturing of continents, and many types of ore deposits.

Field of study: Tectonics

Principal terms

ANDESITE: a volcanic rock that occurs in abundance only along subduction zones

BASALT: a dark-colored, fine-grained igneous rock

CONTINENTAL RIFT: a divergent plate boundary at which continental masses are being pulled apart

CONVERGENT PLATE BOUNDARY: a compressional plate boundary at which an oceanic plate is subducted or two continental plates collide

DIVERGENT PLATE BOUNDARY: a tensional plate boundary where volcanic rocks are being formed

EARTHQUAKE FOCUS: the area below the surface of the earth where active movement occurs to produce an earthquake

OCEANIC RISE: a type of divergent plate boundary that forms long, sinuous mountain chains in the oceans

SUBDUCTION ZONE: a convergent plate boundary where an oceanic plate is being thrust below another plate

TRANSFORM FAULT: a large fracture transverse to a plate boundary that results in displacement of oceanic rises or subduction zones

Summary

Plate tectonics is the theory that the earth's crust is composed of six major rigid plates and numerous minor plates with three types of boundaries. The divergent plate boundary is a tensional boundary in which basaltic magma (molten rock material that will crystallize to become calcium-rich plagioclase, pyroxene, and olivine-rich rock) is formed so that the plate grows larger along this boundary. The rigid plate, or lithosphere, moves in conveyer-belt fashion in both directions away from a divergent boundary across the ocean floor at rates of 0-18 centimeters per year. The lithosphere consists

505

of the crust and part of the upper mantle and averages about 100 kilometers thick; it is thicker over continental than over oceanic crust. The lithosphere seems to slide over an underlying plastic layer of rock and magma called the asthenosphere. Eventually, the lithosphere meets a second type of plate boundary, called a convergent plate margin. If lithosphere-containing oceanic crust collides with another lithospheric plate containing either oceanic or continental crust, then the oceanic lithospheric plate is thrust or subducted below the second plate. If both intersecting lithospheric plates contain continental crust, they crumple and form large mountain ranges, such as the Himalaya or the Alps. Much magma is also produced along convergent boundaries. A third type of boundary, called a transform fault, may develop along divergent or compressional plate margins. Transform faults develop as fractures transverse to the sinuous margins of plates, in which they move horizontally so that the plate margins may be displaced many tens or even hundreds of kilometers.

Divergent plate margins in ocean basins occur as long, sinuous mountain chains called oceanic rises that are many thousands of kilometers long. The rises are often discontinuous, as they are displaced long distances by transform faults. The two longest oceanic rises are the East Pacific Rise, running from the Gulf of California south and west into the Antarctic, and the Mid-Atlantic Ridge, running more or less north-south across the middle of the Atlantic Ocean. The oceanic rises are deep-sea mountain ranges, and there is a rift valley that runs down the middle of the highest part of the mountain chain. The rift valley apparently forms along the ocean rises as the plates move outward from the rises in both directions and pull apart the lithosphere. The oceanic floor descends from a maximum elevation at the oceanic rises to a minimum in the deepest trenches along subduction zones. Thus, the lithosphere moves downhill from the oceanic rises to the convergent plate margins. It is thought that the lithosphere gradually cools and contracts as it moves from the oceanic rises to the convergent margins.

The oceanic rises are composed of piles of basalts forming gentle extrusions. There is high heat flow out of oceanic rises because of the large volume of magma carried up toward the surface. The magnetic minerals in the lavas are frozen into alignment with the earth's magnetic field. Half the magnetized lavas move out from the oceanic rises in one direction, and the other half move out in the opposite direction. The magnetic field of the earth appears to reverse itself periodically over geologic time. The last magnetic reversal occurred about 730,000 years ago. This last reversal can now be observed at the same distance in both directions away from the oceanic rises. A series of such magnetic reversals can be traced back across the Pacific ocean floor for a period of about 165 million years. Many shallow-focus earthquakes occur at depths of up to 100 kilometers below the surface, along the rises and transform faults. Presumably they result from periodic movement that releases tension in the lithosphere.

A second type of divergent plate margin, called a continental rift zone, occurs in continents. Examples are the Rio Grande Rift, occurring as a sinuous north-south belt in central New Mexico and southern Colorado, and the East African Rift, occurring as a sinuous north-south belt across eastern Africa. These rift zones occur as down-dropped blocks forming narrow, elongate valleys that fill with sediment. The rift valleys often contain rivers or elongate lakes. They are characterized by abundant basalts with high potassium contents and, often, smaller amounts of more silica-rich rocks called rhyolites. Rhyolites are light-colored volcanic rocks containing the minerals alkali feldspar (potassium, sodium, and aluminum silicate), quartz (silica), sodium-rich plagioclase, and often minor dark-colored minerals. Shallow-focus earthquakes result in these areas from the tension produced as the continental crust is stretched apart, much as taffy is pulled.

Many rift valleys never become very large. Others grow and may actually rip apart the continents to expose the underlying oceanic crust and rise, as is occurring in the Red Sea. There the oceanic crust is near enough to the continents that it is covered with sediment. Eventually, the continents on both sides of the Red Sea may be pulled apart so far that the underlying oceanic floor will be exposed, with no sediment cover. About 240 million years ago, the continents of North and South America, Europe, and Africa were joined in an ancient landmass called Pangaea. They slowly broke apart along the north-south Mid-Atlantic Ridge from about 240 to 70 million years ago. At first, only a rift valley similar to the East African Rift was formed.

Major Tectonic Plates and Mid-Ocean Ridges

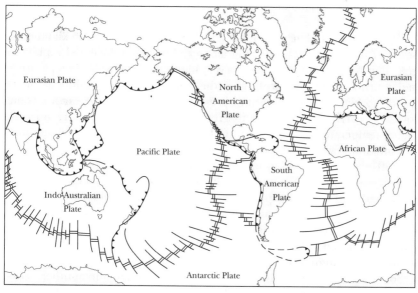

Types of Boundaries: Divergent ⫽ Convergent ⤳ Transform ╱

Later it opened, much like the area of the Red Sea today. Finally, the continents drifted far enough apart during the last 70 million years to form a full-fledged ocean basin, the Atlantic Ocean.

As the lithosphere moves slowly across the ocean floor, minor volcanic activity is generated over hot spots on the ocean floor. The Hawaiian Islands are situated over one of these hot spots. The basalts produced there are much richer in potassium than are those formed over oceanic rises. The Hawaiian Islands are part of a linear, northwest-trending chain of islands, about 2,000 kilometers long, that extends to the island of Midway. The volcanic rocks become progressively older from the Hawaiian Islands to Midway Island. Presumably, Midway Island formed first as the plate slid over the hot spot. As the plate moved to the northwest, the source of magma was removed from Midway, and newer volcanoes began progressively to form over the same hot spot.

Eventually, the lithospheric plate with oceanic crust reaches a compressional plate boundary and may be subducted below other oceanic crust; one result is the island arcs in the western Pacific Ocean, such as Japan. Or the lithospheric plate may be pulled below continental crust, often at angles of 20-60 degrees to the horizontal (the Andes in western South America are the result of such movement). The intersection of the two colliding plates is marked by a sinuous trench forming the deepest portions of the ocean floors. Sediment collects along the slopes of the trench, carried down from the topographic highs of the upper plate. Mountain belts are built up on the nonsubducted plate, as a result of the tremendous amounts of igneous rock that form and of the compressional forces of the plate collision, which throw much sediment and metamorphic rock in the nonsubducted plate to higher elevations.

The subducted plate can be traced to depths as great as 700 kilometers. Some of the sediments collecting along the trench are carried rapidly to great depths, where they undergo a very high-pressure and low-temperature metamorphism. (Metamorphism is the transformation of minerals in response to high temperatures and pressures deep within the earth.) Some rocks are carried more slowly to great depths and have a more normal, higher-temperature metamorphism. During metamorphism, many minerals containing water along the subducted plate gradually break down and give off water vapor, which moves up into the overlying plate. The water vapor is believed to lower the melting point of these rocks within the subducted and overlying plates so that widespread melting takes place, producing the abundant basalts and andesites that build up island arcs or continental masses above the subducted plate. In addition, much rhyolitic magma is formed in the continental crust, presumably through the melting of some of the higher-silica rocks in the continents.

Sometimes a continent is carried by an oceanic plate into another continent at a subduction zone, which happened when India collided with the Asian continent. Such a collision crumples the continents into very high mountains; the Himalayas were formed in this way. This process produces an earthquake zone that is more diffuse (with foci to depths up to 300 kilometers) than are those along subducted plates. No volcanic rocks are produced in these continental-continental plate collisions. Instead, abundant granites crystallize below the surface. Granites contain the same minerals as do rhyolites. Rhyolites form small crystals by quick cooling when they crystallize rapidly in volcanic rocks; granites form larger crystals from magma of the same composition by slow cooling below the earth's surface.

Methods of Study

Plate tectonics is a major, unifying theory that clarifies many large-scale processes on the earth. The major concepts to support the theory were put together only in the late 1950's and the 1960's. Yet, many of the keys to developing the theory had been known for many years. Beginning in the seventeenth century, a number of people noticed the remarkable "fit" in the shape of the continents on opposing sides of the Atlantic Ocean and suggested that the continents could have been joined at one time. It was not until the early twentieth century that Alfred Wegener put many pieces of this puzzle together. Wegener noticed the remarkable similarity of geological structures, rocks, and especially fossils that were currently located on opposite sides of the Atlantic Ocean. Most notably, land plants and animals that predated the hypothesized time of the breakup of the continents, at about 200 million years before the present, were remarkably similar on all continents. Subsequently, their evolution in North and South America was quite different from their development in Europe and Africa. Climates could also be matched across the continents. For example, when the maps of the continents were reassembled into their predrift positions, the glacial deposits in southern Africa, southern South America, Antarctica, and Australia could be explained as having originated as one large continental glacier in the southern polar region.

One of the biggest problems with the concept of continental drift at that time was the lack of understanding of a driving force to explain how the continents could have drifted away from one another. Then, in 1928, Arthur Holmes proposed a mechanism that foreshadowed the explanation geologists later adopted. He suggested that the mantle material upwelled under the continents and pulled them apart as it spread out laterally and produced tension. The basaltic oceanic crust would then carry the continents out away from one another much like rafts. When the mantle material cooled, Holmes believed, it descended back into the mantle and produced belts along these areas. From the 1920's to the early 1960's, however, continental drift theories had no currency, for there was no real evidence for driving forces that might move the continents. It was not until the ocean floors began to be mapped that evidence was found to support a plate tectonic model. The topography of the ocean floor was surveyed, and large mountain ranges, such as the Mid-Atlantic Ridge with its rift valleys, and the deep ocean trenches were discovered. Harry Hess suggested in the early 1960's that the oceanic ridges were areas where mantle material upwelled, melted, and spread laterally. Evidence for this sea-floor spreading hypothesis came from the mirror-image pattern of the periodically reversed magnetic bands found in basalts on either side of the ridges. The symmetrical magnetic bands could be explained only by the theory that they were originally produced at the ridges, as the earth's magnetic field periodically reversed, and then were spread laterally in both directions at the same rate.

Supporting evidence for plate tectonics began to accumulate during the 1960's. Further magnetic pattern surveys on ocean floors confirmed that the symmetrical pattern of matching magnetic bands could be found everywhere around ridges. Also, earthquake, volcanic rock, and heat-flow patterns were discovered to be consistent with the concept of magma upwelling along rises and sea-floor material being subducted along oceanic trenches. Oceanic and lithospheric plates could then be defined, and the details of the interaction of the plate boundaries could be understood. With this overwhelming evidence, most geologists became convinced that the plate tectonic model was valid.

Context

The plate tectonic theory is highly significant in that it accounts for a wide variety of phenomena: the formation of volcanoes, earthquakes, mountain belts, and many types of ore deposits, as well as the growth, drift, and fracturing of continents. For scientists, it is a rich and fascinating theory with many implications.

Plate tectonics is important economically because of the theory's usefulness in predicting and explaining the occurrence of ore deposits. Plate boundaries such as the mid-oceanic rises are areas of high temperature in which hot waters are driven up toward the surface. These hot waters are enriched in copper, iron, zinc, and sulfur, so sulfide minerals such as pyrite (iron sulfide), chalcopyrite (copper and iron sulfide), and sphalerite (zinc sulfide) form along oceanic rises. One such deposit in Cyprus has been mined for many centuries. Tensional zones sometimes formed in basins behind subduction zones may form deposits similar to those at oceanic rises. In addition, ferromanganese nodules form in abundance in some places by chemical precipitation from seawater. These nodules are enriched in cobalt and nickel, as well as in iron and manganese as complex oxides and hydroxides. They could potentially be mined from ocean floors.

Deposits enriched in chromium occur in folded and faulted rocks on the nonsubducted plate next to the oceanic trench in subduction zones. This deposit is found in some peridotites (olivine, pyroxene, and garnet rocks) or dunites (olivine rock) that have been ripped out of the upper mantle and thrust up into these areas. The ore mineral chromite (magnesium and chromium oxide) is found in pods and lenses that range in size from quite small to massive. Many intrusions of silica-rich magma above subduction zones contain water-rich fluids that have moved through the granite after it solidified. The water-rich fluids deposit elements such as copper, gold, silver, tin, mercury, molybdenum, tungsten, and bismuth throughout a large volume of the granite in low concentrations. Hundreds of these deposits have been found around subduction zones in the Pacific Ocean.

Bibliography

Motz, Lloyd M., ed. *The Rediscovery of the Earth*. New York: Van Nostrand Reinhold, 1979. An unusual book, as it is written by many of the experts who developed the plate tectonic model. Begins at an elementary level so that someone without much background in plate tectonics should be able to understand the discussion; the discussion progresses, however, to an advanced level. Beautifully illustrated with photographs and diagrams.

Press, Frank, and Raymond Siever. *Earth*. 4th ed. New York: W. H. Freeman, 1986. A well-written introductory textbook on geology. Chapters 19 and 20 deal with plate tectonics. Well illustrated; contains a glossary. Appropriate for general readers.

Seyfert, Carl K., and L. A. Sirkin. *Earth History and Plate Tectonics*. New York: Harper & Row, 1973. This book integrates the plate tectonic concept with the evolution of plants and animals through geologic time. Some understanding of plate tectonics, rocks, and minerals would be helpful before using this source. Written as an introductory text in historical geology, so important concepts are reviewed. Good illustrations.

Skinner, Brian J., et al. *Resources of the Earth*. Englewood Cliffs, N.J.: Prentice-Hall, 1988. A good book for the layperson who is interested in the history, use, production, environmental impact, and geological occurrence of ore deposits. Technical terms are kept to a minimum. Well illustrated; contains a glossary. Suitable for someone who is taking a course in geology.

Wessel, Paul, and Loren Kroenke. "A Geometric Technique for Relocating Hotspots and Refining Absolute Plate Motions." *Nature* 387, no. 6631 (May 22, 1997): 365-370.

Windley, Brian F. *The Evolving Continents*. 2d ed. New York: John Wiley & Sons, 1984. A more advanced source than the others listed. Summarizes how the continents have evolved through geologic time. The reader should understand plate tectonic processes well before attempting to read this book.

Robert L. Cullers

Cross-References

Continental Growth, 102; The Lithosphere, 375; Magmas, 383; Ocean Basins, 456; The Oceanic Crust, 471; Subduction and Orogeny, 615; Transform Faults, 638.

Pyroclastic Rocks

Pyroclastic rocks form from the accumulation of fragmental debris ejected during explosive volcanic eruptions. Pyroclastic debris may accumulate either on land or under water. Volcanic eruptions that generate pyroclastic debris are extremely high-energy events and are potentially dangerous if they occur near populated areas.

Field of study: Volcanology

Principal terms

ASH: fine-grained pyroclastic material less than 2 millimeters in diameter

IGNIMBRITE: pyroclastic rock formed from the consolidation of pyroclastic-flow deposits

LAPILLI: pyroclastic fragments between 2 and 64 millimeters in diameter

PUMICE: a vesicular glassy rock commonly having the composition of rhyolite; a common constituent of silica-rich explosive volcanic eruptions

PYROCLASTIC FALL: the settling of debris under the influence of gravity from an explosively produced plume of material

PYROCLASTIC FLOW: a highly heated mixture of volcanic gases and ash that travels down the flanks of a volcano; the relative concentration of particles is high

PYROCLASTIC SURGE: a turbulent, low-particle-concentration mixture of volcanic gases and ash that travels down the flanks of a volcano

STRATOVOLCANO: a volcanic cone consisting of both lava and pyroclastic rocks

TEPHRA: fragmentary volcanic rock materials ejected into the air during an eruption; also called pyroclasts

TUFF: a general term for all consolidated pyroclastic rocks

VOLATILES: fluid components, either liquid or gas, dissolved in a magma that, upon rapid expansion, may contribute to explosive fragmentation

Summary

Pyroclastic rocks form as a result of violent volcanic eruptions such as that of Mount St. Helens in 1980 or Mount Vesuvius in A.D. 79. Molten rock, or magma, within the earth sometimes makes its way to the surface in the form of volcanic eruptions. These eruptions may produce lava or, if the eruption is highly explosive, fragmental debris called tephra or pyroclasts. The term "pyroclastic" is from the Greek roots pyros (fire) and klastos (broken). Dissolved water and gases (volatiles) are the source of energy for these explosive eruptions. All molten rock contains dissolved fluids such as water and carbon dioxide. When the molten rock is still deep within the earth, the confining pressure of the overlying rock keeps these volatiles from being released. When the magma rises to the surface during an eruption, the pressure is lowered, and the gases and water may be violently released, causing fragmentation of the molten rock and some of the rock surrounding the magma. This type of explosive eruption is more common in rocks rich in silica, which are more viscous (flow more thickly) than those that are silica-poor. External sources of water, such as a lake or groundwater reservoir, may also provide the necessary volatiles for an explosive eruption.

Pyroclastic debris can be produced from any of three different types of volcanic eruptions: magmatic explosions; phreatic, or steam, explosions; and phreatomagmatic explosions. Magmatic explosions occur when magma rich in dissolved volatiles undergoes a decrease in pressure such that the volatiles are rapidly released or exsolved. The solubility of volatiles in magma is partially controlled by confining pressure, which is a function of depth. Solubility decreases as the magma rises toward the surface. At a certain depth, carbon dioxide and water begin exsolving and become separate fluid phases. At this point, the magma may undergo explosive fragmentation either through an open vent or by destroying the overlying rock in a major eruptive event.

As a magma rises toward the surface, it may encounter a groundwater reservoir or, in a subaqueous vent, interact with surface water. In both cases, the superheating and boiling of water followed by its explosive expansion to gas may fragment the magma and the surrounding country rock. The ratio between the mass of water and the mass of magma controls the type of eruption. If there is little water in relation to magma, the explosive activity may be confined to the eruption of steam and is called a phreatic explosion. If the magma contains significant quantities of dissolved volatiles and encounters a large amount of water, the resulting explosion is termed phreatomagmatic.

Pyroclastic deposits are composed of tephra, or pyroclasts. These fragments can have a wide range of sizes. Particles less than 2 millimeters in diameter are termed ash, those between 2 and 64 millimeters are called lapilli, and those greater than 64 millimeters are called blocks, or bombs. There are three principal components that make up pyroclastic debris: lithic fragments, crystals, and vitric, or juvenile, fragments. Lithic fragments can be subdivided into pieces of the surrounding rock explosively fragmented during an eruption (accessory lithics), pieces of already solidified magma (cognate lithics), and particles picked up during transport of eruptive clouds down the flanks of a volcano (accidental lithics). Crystal fragments are whole or fragmented crystals that had solidified in the magma before eruption. Vitric or juvenile fragments represent samples of the erupting, still molten, magma. They may be either partly crystallized or uncrystallized (glass). Pumice is a type of juvenile fragment that contains many vesicles, or holes, as a result of the rapid exsolution of gases during eruption. Small, very angular glass fragments are called shards.

Three types of pyroclastic deposit can be distinguished based on the type of process that forms the deposit: pyroclastic-fall deposits, pyroclastic-flow deposits, and pyroclastic-surge deposits. These types can all be formed by any of the previously described different types of volcanic eruption. Any of these deposits may be termed a tuff if the grain size is predominantly less than 2 millimeters.

Pyroclastic-fall deposits form from the settling, under the influence of gravity, of particles out of a plume of volcanic ash and gases erupted into the atmosphere forming an eruption column. Tuffs formed in this way are coarsest near the eruption center and become progressively finer farther away. Ash falls can also be derived from the tops of more dense pyroclastic flows, as the finer-grained material is turbulently removed from the upper portion of the pyroclastic flow and then settles to the ground.

Two types of pyroclastic deposits result from the formation of dense clouds of ash during an eruption and the subsequent transport of debris in the form of a hot cloud of ash, lapilli, and gases. Pyroclastic flows have a relatively high particle concentration and, in some areas, the western United States for example, form enormous deposits with volumes as large as 3,000 cubic kilometers. Pyroclastic-flow deposits rich in pumice are termed ignimbrites. Pyroclastic surges are expanded, low-particle-concentration density currents that are generally very turbulent. Surge deposits are volumetrically less important than those of pyroclastic flows but can be very destructive. Both flows and surges may have emplacement temperatures of up to 800 degrees Celsius. Pyroclastic surges, however, because of their lower density and turbulent nature, may attain velocities up to 700 kilometers per hour. It is this combination of speed and temperature that makes pyroclastic surges so dangerous.

Most pyroclastic rocks are associated with stratovolcanoes, also called composite volcanoes. These volcanic edifices are built by a combination of extrusive and explosive processes and thus are formed of both pyroclastic debris and lava. Well-known examples of stratovolcanoes include Mount St. Helens, Mount Fuji, and Mount Vesuvius. Volcanoes such as those found on the Hawaiian Islands are of a less energetic variety called shield volcanoes and produce insignificant quantities of pyroclastic material. Stratovolcanoes are located around the world and are associated with the global process of plate tectonics, which explains the movement of the continents, the generation of new "plates," which include oceanic crust, and the destruction of old plates. It is at the zone of plate destruction that rock is melted and makes its way to the surface, sometimes producing pyroclastic eruptions.

A deposit of pyroclastic flow emitted by the eruption of Mount St. Helens. *(Cascades Volcano Observatory)*

Methods of Study

Geologists study pyroclastic rocks using field techniques, laboratory analyses, and theoretical considerations of eruption processes. Observations of deposits in the field remain the cornerstone of much of geologic interpretation. Pyroclastic deposits form essentially as sedimentary material, that is, as fragments or clasts moved in air or water and deposited in layers. As such, many of the techniques used by sedimentologists (geologists interested in the formation and history of sedimentary rocks) are employed in the study of pyroclastic deposits. Careful examination of a variety of different sedimentary features within pyroclastic deposits can aid in the interpretation of the processes of transport and deposition. This information, studied over as wide a geographic area as possible to ascertain systematic changes in the deposits, will assist in understanding the geologic history of the region.

Much can be learned through analysis of the composition of pyroclastic rocks, which is generally done using a variety of laboratory techniques. The use of specialized microscopes allows geologists to examine very thin sections of rock in order to observe textures and to discern mineral composition. Geochemical techniques have become very popular and powerful in the study of all kinds of rocks, including pyroclastic rocks. By looking at the amounts of certain elements that occur in extremely low abundances and also at relative proportions of certain types of isotopes (naturally occurring forms of the same element that differ only in the number of neutrons in the nucleus), scientists can understand more about the processes taking place

515

deep within the earth that lead to the formation of magma and eventually to the eruption of pyroclastic debris. Scanning electron microscopes have been used to study in detail the surface features and textures of fine volcanic ash particles. This information can lead to better understanding of eruptive and transport processes that formed and deposited the pyroclastic particles.

Theoretical studies associated with pyroclastic rocks revolve primarily around considerations of the mechanics of high-temperature, high-velocity eruption clouds and their transport and deposition. This type of reasoning allows a geologist to infer certain conditions of eruption from an analysis of the deposits. The geologic rock record has abundant pyroclastic deposits, and it is through this type of inference that geologists interpret the geologic history of a region. A comprehensive understanding of pyroclastic deposits must include a thorough understanding of the processes by which the deposit forms.

Context

Violent volcanic eruptions that may produce pyroclastic deposits are among the most powerful events occurring on the earth. Historically, many of the most destructive volcanic eruptions have involved pyroclastic surges. The eruption in A.D. 79 of Mount Vesuvius generated pyroclastic debris that buried the towns of Pompeii and Herculaneum, killing most of the inhabitants. In 1902, on the island of Martinique in the Caribbean, the violent eruption of Mount Pelée produced a pyroclastic surge that swept down on the city of St. Pierre, killing all but a handful of a population of about thirty thousand. The eruption of Mount St. Helens in 1980 and of El Chichón in Mexico in 1982 both produced pyroclastic surges. Two thousand people were killed as a result of the El Chichón eruption. Pyroclastic surges and flows generally do not present a hazard beyond about a 20-kilometer radius.

Pyroclastic deposits form a major portion of some volcanic terrains. Some of these deposits are enormously extensive, indicating that the eruptions that produced them were much larger than any witnessed in modern times. It is not clear whether these deposits reflect an overall increase in volcanic activity in the earth's past or whether this type of titanic eruption occurs sporadically throughout geologic time. Titanic eruptions inject so much debris into the upper atmosphere that global weather can be affected. The earth experienced brilliant red sunsets and lowered temperatures because dust blocked the sun for several years following the 1883 eruption of Krakatoa in the strait between Java and Sumatra, which completely destroyed an island and discharged nearly 20 cubic kilometers of debris into the air. The explosion was heard nearly 5,000 kilometers away in Australia, and darkness fell over Jakarta, 150 miles away. Some geologists speculate that enormous volcanic eruptions in the earth's past have even led to extinctions, including that of the dinosaurs, by producing so much ash that the amount of sunlight received on the earth's surface is reduced, and the entire food chain is disrupted.

Bibliography

Blong, R. J. *Volcanic Hazards: A Sourcebook on the Effects of Eruptions.* Sydney: Academic Press, 1984. Discusses the nature of volcanic hazards with case histories. Suitable for college-level students.

Cas, R. A. F., and J. V. Wright. *Volcanic Successions: Modern and Ancient.* Winchester, Mass.: Unwin Hyman, 1987. Eleven of the fifteen chapters in this excellent book deal wholly or in part with pyroclastic rocks. Takes a sedimentological approach to the study and interpretation of pyroclastic deposits. Indispensable for geologists interested in pyroclastic rocks. Suitable for college-level students.

Decker, Robert, and Barbara Decker. *Volcanoes.* San Francisco: W. H. Freeman, 1981. This book provides the reader with a good overview of different types of volcanoes and volcanic processes, including those that produce pyroclastic debris. Suitable for high school students.

Fisher, R. V., and H. U. Schmincke. *Pyroclastic Rocks.* New York: Springer-Verlag, 1984. This book, along with that of Cas and Wright, provides the most comprehensive treatment of pyroclastic deposits and the processes which form them. Suitable for college-level students.

Simkin, Tom, L. Siebert, L. McClelland, D. Bridge, C. Newhall, and J. H. Latter. *Volcanoes of the World: A Regional Directory, Gazeteer, and Chronology of Volcanism During the Last Ten Thousand Years.* Stroudsbourg, Pa.: Hutchinson & Ross, 1981. Suitable for all readers.

Bruce W. Nocita

Cross-References

Igneous Rocks, 315; Plate Tectonics, 505; Stratovolcanoes, 599; Volcanoes: Recent Eruptions, 675; Volcanoes: Types of Eruption, 685.

Reefs

Reefs are among the oldest known communities, existing at least 2 billion years ago. They exert considerable control on the surrounding physical environment, influencing turbulence levels and patterns of sedimentation. Ancient reefs are often important hydrocarbon reservoirs.

Field of study: Sedimentology

Principal terms

CALCAREOUS ALGAE: green algae that secrete needles or plates of aragonite as an internal skeleton; very important contributors to reef sediment

CARBONATE ROCKS: sedimentary rocks such as limestone, which is composed of the minerals calcite or aragonite, or dolostone, which is composed of the mineral dolomite

CORALLINE ALGAE: red algae that secrete crusts or branching skeletons of high-magnesium calcite; important sediment contributors and binders on reefs

REEF: a biogenically (organically) produced carbonate structure including an internal framework that traps sediment and confers resistance to wave action

RUGOSE CORALS: a Paleozoic coral group also known as "tetracorals"; sometimes colonial, but more often solitary and horn-shaped

SCLERACTINIAN CORALS: modern corals or "hexacorals," different from their more ancient counterparts in details of the skeleton and the presence of a symbiosis with unicellular algae in most shallow-water species

STROMATOLITES: layered columnar or flattened structures in sedimentary rocks, produced by the binding of sediment by blue-green algal (cyanobacterial) mats

STROMATOPOROIDS: spongelike organisms that produced layered, mound-shaped, calcareous skeletons and were important reef builders during the Paleozoic era

TABULATE "CORALS": colonial organisms with calcareous skeletons that were important Paleozoic reef builders; considered to be more closely related to sponges than to corals

Summary

Reefs or reeflike structures are among the oldest known communities, extending back more than 2 billion years into the earth's history. These earliest reefs were vastly different in their biotic composition and physical structure from modern reefs, which are among the most diverse of biotic communities and display amazingly high rates of biotic productivity (carbon fixation) and calcium carbonate deposition, despite their existence in a virtual nutrient "desert." Reefs are among the few communities to rival the power of humankind as a shaper of the planet. The Great Barrier Reef of Australia, for example, forms a structure some 2,000 kilometers in length and up to 150 kilometers in width.

It is necessary to distinguish between "true," or structural, reefs and reeflike structures or banks. Reefs are carbonate structures that possess an internal framework. The framework traps sediment and provides resistance to wave action; thus, reefs can exist in very shallow water and may grow to the surface of the oceans. Banks are also biogenically produced but lack an internal framework. Thus, banks are often restricted to low-energy, deep-water settings. "Bioherm" refers to mound-like carbonate buildups, either reefs or banks, and "biostrome" to low, lens-shaped buildups.

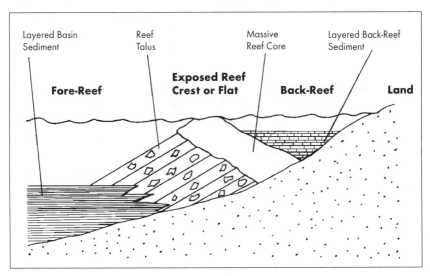

Modern reefs are classified into several geomorphic types: atoll, barrier, fringing, and patch. Many of these may be further subdivided into reef crest or flat, back-reef or lagoon, and fore-reef zones (see figure). Atoll reefs are circular structures with a central lagoon, thought to form on subsiding volcanic islands. Barrier reefs are elongate structures that parallel coastlines and possess a significant lagoon between the exposed reef crest and shore. These often occur on the edges of shelves that are uplifted by faulting.

Fringing reefs are elongate structures paralleling and extending seaward from the coastline that lack a lagoon between shore and exposed reef crest. Patch reefs are typically small, moundlike structures, occurring isolated on shelves or in lagoons. The majority of fossil reefs would be classified as patch reefs, although many examples of extensive, linear, shelf-edge trends are also known from the geologic record.

Reefs form one of the most distinctive and easily recognized sedimentary facies (or environments). In addition to possessing a characteristic fauna consisting of corals, various algae, and stromatoporoids, they are distinguished by a massive (nonlayered) core that has abrupt contacts with adjacent facies. Associated facies include flat-lying lagoon and steeply inclined fore-reef talus, the latter often consisting of large angular blocks derived from the core. The reef core is typically a thick unit relative to adjacent deposits. The core also consists of relatively pure calcium carbonate with little contained terrigenous material.

Knowledge of modern reef communities is much greater than that of their ancient counterparts. Information gained from the study of modern reefs will enhance scientists' knowledge of ancient reefs as well. Modern reefs are restricted to certain environments. They occur abundantly only between 23 degrees north and south latitudes and tend to be restricted to the western side of ocean basins, which lack upwelling of cold bottom waters. This restriction is based on temperature, as reefs do not flourish where temperatures frequently reach below 18 degrees Celsius. Reef growth is largely restricted to depths greater than 60 meters, as there is insufficient penetration of sunlight below this depth for symbiont-bearing corals to flourish. Reefs also require clear waters lacking suspended terrigenous materials, as these interfere with the feeding activity of many reef organisms and also reduce the penetration of sunlight. Finally, most reef organisms require salinities that are in the normal oceanic range. It appears that many fossil reefs were similarly limited in their environmental requirements.

Some of the most striking features of modern reefs include their pronounced zonation, great diversity, and high productivity and growth rates. Reefs demonstrate a strong bathymetric (depth-related) zonation. This zonation is largely mediated through depth-related changes in turbulence intensity and in the quantity and spectral characteristics (reds are absorbed first, blues last) of available light. Shallow (1- to 5-meter) fore-reef environments are characterized by strong turbulence and high light intensity and possess low-diversity assemblages of wave-resistant corals, such as the elkhorn coral, *Acropora palmata*, and crustose red algae. With increasing depth (10-20 meters), turbulence levels decrease and coral species diversity increases, with mound and delicate branching colonies occurring. At greater depths (30-60 meters), corals assume a flattened, platelike form in an attempt to maximize surface area for exposure to ambient light. Sponges and many green algae are also very important over this range. Finally, corals possessing zooxanthellae, which live in the coral tissues and provide food for

520

the coral host, are rare or absent below 60 meters because of insufficient light. Surprisingly, green and red calcareous algae extend to much greater depths (100-200 meters), despite the very low light intensity (much less than 1 percent of surface irradiance). Sponges are also important members of these deep reef communities.

Coral reefs are among the most diverse of the earth's communities; however, there is no consensus on the mechanism(s) behind the maintenance of this great diversity. At one time, it was believed that reefs existed in a low-disturbance, highly stable environment, which allowed very fine subdivision of food and habitat resources and thus permitted the coexistence of a great number of different species. Upon closer inspection, however, many reef organisms appear to overlap greatly in food and habitat requirements. Also, it has become increasingly apparent that disturbance, in the form of disease, extreme temperatures, and hurricanes, is no stranger to reef communities.

Coral reefs exhibit very high rates of productivity (carbon fixation), which is a result of extremely tight recycling of existing nutrients. This is necessary, as coral reefs exist in virtual nutrient "deserts." Modern corals exhibit high skeletal growth rates, up to 10 centimeters per year for some branching species. Such high rates of skeletal production are intimately related to the symbiosis existing between the hermatypic or reef-building scleractinian corals (also gorgonians and many sponges) and unicellular algae or zooxanthellae. Corals that, for some reason, have lost their zooxanthellae or that are kept in dark rooms exhibit greatly reduced rates of skeleton production.

In addition to high individual growth rates for component taxa, the carbonate mass of the reefs may grow at a rate of some 2 meters per 1,000 years, a rate that is much higher than that of most other sedimentary deposits. This reflects the high productivity or growth rates of the component organisms and the efficient trapping of derived sediment by the reef frame. Although the framework organisms, most notably corals, are perhaps the most striking components of the reef system, the framework represents only 10-20 percent of most fossil reef masses. The remainder of the reef mass consists of sedimentary fill derived from the reef community through a combination of biosynthesis (secretion) and bioerosion (breaking down) of calcium carbonate. An example of the relative contributions of reef organisms to sediment can be found in Jamaica, where shallow-water, back-reef sediment consists of 41 percent coral, 24 percent green calcareous algae, 13 percent red calcareous algae, 6 percent foraminifera, 4 percent mollusks, and 12 percent other grains. The most important bioeroders are boring sponges, bivalves, and various "worms," which excavate living spaces within reef rock or skeletons, and parrot fish and sea urchins, which remove calcium carbonate as they feed upon surface films of algae.

A diversity of organisms has produced reef and reeflike structures throughout earth history. Several distinct reef community types have been noted, as well as four major "collapses" of reef communities. The oldest reefs or reeflike structures existed more than 2 billion years ago during the Precambrian eon. These consisted of low-diversity communities dominated by soft, blue-green algae, which trapped sediment to produce layered, often columnar structures known as stromatolites. During the Early Cambrian period, blue-green algae were joined by calcareous, conical, spongelike organisms known as archaeocyathids, which persisted until the end of the Middle Cambrian. Following the extinction of the archaeocyathids, reefs again consisted only of blue-green algae until the advent of more modern reef communities in the Middle Ordovician period. These reefs consisted of corals (predominantly tabulate and, to a much lesser extent, rugose corals), red calcareous algae, bryozoans (moss animals), and the spongelike stromatoporoids. This community type persisted through the Devonian period, at which time a global collapse of reef communities occurred. The succeeding Carboniferous period largely lacked reefs, although algal and crinoidal (sealily) mounds are common. Reefs again occurred in the Permian period, consisting mainly of red and green calcareous algae, stromatolites, bryozoans, and chambered calcareous sponges known as sphinctozoans, which resembled strings of beads. These reefs were very different from those of the earlier Paleozoic era; in particular, the tabulates and stromatoporoids no longer played an important role. The famous El Capitan reef complex of West Texas formed during this interval. The Paleozoic era ended with a sweeping extinction event that involved not only reef inhabitants but also other marine organisms.

After the Paleozoic extinctions, reefs were largely absent during the early part of the Mesozoic era. The advent of modern-type reefs consisting of scleractinian corals and red and green algae occurred in the Late Triassic period. Stromatoporoids once again occurred abundantly on reefs during this interval; however, the role of the previously ubiquitous blue-green algal stromatolites in reefs declined. Late Cretaceous reefs were often dominated by conical, rudistid bivalves that developed the ability to form frameworks and may have possessed symbiotic relationships with algae, as do many modern corals. Rudists, however, became extinct during the sweeping extinctions that occurred at the end of the Cretaceous period. The reefs that were reestablished in the Cenozoic era lacked stromatoporoids and rudists and consisted of scleractinian corals and red and green calcareous algae. This reef type has persisted, with fluctuations, until the present.

Methods of Study

Modern reefs are typically studied by scuba (self-contained underwater breathing apparatus) diving, which enables observation and sampling to a depth of approximately 50 meters. Deeper environments have been made accessible through the availability of manned submersibles and unmanned, remotely operated vehicles that carry mechanical samplers and still and video cameras. The biological compositions of reef communities are determined by census (counting) methods commonly employed by plant ecologists. Studies of symbioses, such as that between corals and their zooxanthellae, employ radioactive tracers to determine the transfer of products between symbiont and host. Growth rates are measured by staining the calcareous skeletons of living organisms with a dye, such as Alizarin red, and then later collecting and sectioning the specimen and measuring the amount of skeleton added since the time of staining. Another method for determining growth is to X-ray a thin slice of skeleton and then measure and count the yearly growth bands that are revealed on the X-radiograph. Variations in growth banding reflect, among other factors, fluctuations in ocean temperature.

Reef sediments, which will potentially be transformed into reef limestones, are examined through sieving, X-ray diffraction, and epoxy impregnation and thin-sectioning. Sieving enables the determination of sediment texture, the relationships of grain sizes and abundance (which will reflect environmental energy and the production), and erosion of grains through biotic processes. X-ray diffraction produces a pattern that is determined by the internal crystalline structure of the sediment grains. As each mineral possesses a unique structure, the mineralogical identity of the sediment may be determined. Thin sections of embedded sediment or lithified rock are examined with petrographic microscopes, which reveal the characteristic microstructures of the individual grains. Thus, even highly abraded fragments of coral or algae may be identified and their contributions to the reef sediment determined.

Because of their typically massive nature, fossil reefs are usually studied by thin-sectioning of lithified rock samples collected either from surface exposures or well cores. Reef limestones that have not undergone extensive alteration may be dated through carbon 14 dating, if relatively young, or through uranium-series radiometric dating methods.

Context

Modern reefs serve as natural laboratories, enabling the geoscientist to witness and study phenomena, such as carbonate sediment production, bioerosion, and early cementation, that have been responsible for forming major carbonate rock bodies in the past. The study of cores extracted from centuries-old coral colonies shows promise for deciphering past climates and perhaps predicting future trends. This is made possible by the fact that the coral skeleton records variations in growth that are related to ocean temperature fluctuations. The highly diverse modern reefs also serve as ecological laboratories for testing models on the control of community structure. For example, the relative importance of stability versus disturbance and recruitment versus predation in determining community structure is being studied within the reef setting.

Modern reefs are economically significant resources, particularly for many developing nations in the tropics. Reefs and the associated lagoonal seagrass beds serve as important nurseries and habitats for many fish and invertebrates. The standing crop of fish immediately over reefs is much higher than that of adjacent open shelf areas. Reef organisms may one day provide an important source of pharmaceutical compounds, such as prostaglandins, which may be extracted from gorgonians (octocorals). In addition, research has focused upon the antifouling properties exhibited by certain reef encrusters. Reefs also provide recreational opportunities for snorkelers and for scuba divers, a fact that many developing countries are utilizing to promote their tourist industries. Finally, reefs serve to protect shorelines from wave erosion.

Because of the highly restricted environmental tolerances of reef organisms, the occurrence of reefs in ancient strata enables fairly confident estimation of paleolatitude, temperature, depth, salinity, and water clarity. In addition, depth- or turbulence-related variation in growth form (mounds in very shallow water, branches at intermediate depths, and plates at greater depths) enables even more precise estimation of paleobathymetry or turbulence levels. Finally, buried ancient reefs are often important reservoir rocks for hydrocarbons and thus are important economic resources.

Bibliography

Bathurst, Robin G. C. *Carbonate Sediments and Their Diagenesis.* 2d ed. New York: Elsevier, 1975. Provides an excellent general reference on carbonate sediments, from reef and other environments, and their diagenesis.

Darwin, Charles. *The Structure and Distribution of Coral Reefs.* Berkeley: University of California Press, 1962. Darwin's book, originally published in 1851, is replete with observations on coral reefs from around the world. In addition, the theories presented on the formation of reef types such as atolls have withstood the test of time.

DeVantier, L. M., and G. De'Ath, et al. "Ecological Assessment of a Complex Natural System: A Case Study from the Great Barrier Reef.: *Ecological Applications* 8, no. 2, (May, 1998): 480-497.

Frost, S. H., M. P. Weiss, and J. B. Sanders, eds. *Reefs and Related Carbonates: Ecology and Sedimentology.* Tulsa, Okla.: American Association of Petroleum Geologists, 1977. Includes a broad array of papers covering such aspects of the ecology and geology of modern and ancient reefs as bioerosion, diagenesis, paleoecology of ancient reefs, the role of sponges on reefs, and sedimentology of basins adjacent to modern reefs.

Goreau, Thomas F., et al. "Corals and Coral Reefs." *Scientific American* 241 (August, 1979): 16, 124-136. A good overview of the ecology of modern coral reefs. The discussion of coral physiology and the symbiotic relationship with zooxanthellae is particularly valuable.

Jones, O. A., and R. Endean, eds. *Biology and Geology of Coral Reefs.* 4 vols. New York: Academic Press, 1973-1977. This series of volumes encompasses both biological and geological aspects of coral reefs. Of particular value are review chapters covering, for example, the reefs of the western Atlantic.

Kaplan, Eugene H. *A Field Guide to Coral Reefs of the Caribbean and Florida, Including Bermuda and the Bahamas.* Boston: Houghton Mifflin, 1988. In addition to providing descriptions and illustrations (many in color) of common reef organisms, this book provides an excellent overview of modern reef community structure, zonation, and environments.

Laporte, Leo F., ed. *Reefs in Time and Space: Selected Examples from the Recent and Ancient.* Tulsa, Okla.: Society of Economic Paleontologists and Mineralogists, 1974. Includes papers on reef diagenesis, reef geomorphology, and the distribution of carbonate buildups in the geologic record.

Newell, Norman D. "The Evolution of Reefs." *Scientific American* 226 (June, 1972): 12, 54-65. Provides an overview of the composition of reef communities throughout the earth's history, including the various collapses and rejuvenations.

Smith, F. G. *Atlantic Reef Corals: A Handbook of the Common Reef and Shallow-Water Corals of Bermuda, the Bahamas.* Rev. ed. Baltimore, Md.: University of Miami Press, 1971. Smith provides taxonomic keys, descriptions, illustrations, and zoogeographic distributions for the Atlantic reef corals. In addition, he provides much general information on the distribution of coral reefs and their ecology.

Stoddart, D. R. "Ecology and Morphology of Recent Coral Reefs." *Biological Reviews* 44 (1969): 433-498. This article provides a review of the global distribution of coral reefs with emphasis on their ecology and geomorphology.

W. David Liddell

Cross-References

Continental Growth, 102; The Fossil Record, 257.

River Flow

Worldwide, rivers are the most important sources of water for cities and major industries. Hydroelectric power is a major source of electrical power, and transport of heavy, bulky goods by river barge is a vital link in most transportation systems. Understanding and predicting low, high, and average flows of rivers are therefore important to the people and industries that depend on them.

Field of study: Sedimentology

Principal terms

DISCHARGE: the total amount of water passing a point on a river per unit of time

EVAPOTRANSPIRATION: all water that is converted to water vapor by direct evaporation or passage through vegetation

HYDRAULIC GEOMETRY: a set of equations that relate river width, depth, and velocity to discharge

HYDROGRAPH: a plot recording the variation of stream discharge over time

HYDROLOGIC CYCLE: the circulation of water as a liquid and vapor from the oceans to the atmosphere and back to the oceans

HYDROLOGY: broadly, the science of water; the term is often used in the more restricted sense of flow in channels

RATING CURVE: a plot of river discharge in relation to elevation of the water surface; permits estimation of discharge from the water elevation

TURBULENT FLOW: the swirling flow that is typical of rivers, as opposed to smooth, laminar flow

Summary

There are two very different fundamental types of flow of water. Laminar flow is a smooth flow, in which two particles suspended in the water will follow parallel, nearly straight paths. Turbulent flow is a complex, swirling flow in which the paths of two suspended particles have no necessary relation to each other. Turbulent flow may have velocity components that are up or down from, sideways or upstream of the average flow direction, although the average flow direction is always in the downstream, or downslope, direction. Flow near the bottom and sides of a river or stream is always laminar, but the zone of laminar flow is very narrow. Turbulent flow dominates throughout the cross section of stream flow.

Because flowing water exerts a shear stress, or viscous shear, along the bottom and sides of the river, average flow velocity is least at the bottom and increases upward into the body of the river. Because of energy losses to surface waves, the average turbulent downstream velocity (hereafter referred to simply as velocity) at the river's surface is slightly below the maximum velocity. The maximum velocity occurs at about 0.6 of the river depth above the bottom of the river.

Because of viscous shear between flowing water and the bottom materials, the small (and sometimes large) sediment particles that make up most river bottoms are moved by the flowing water. Once in motion, small particles whose settling velocity is less than the upward component of turbulent flow move downstream with the water, settling out only in backwaters, such as the still water behind a dam, where flow velocities are very low. This continuously moving sediment is referred to as suspended load and is restricted, in most cases, to clay- and silt-sized material. Sand-sized and larger sediment grains are moved during periods of high flow velocity and dropped where, or when, flow velocity decreases. This coarse material is the "bed load." Bed-load transport normally occurs only near the river bed. It is the sediment in transport that does the work of the river: erosion and transport. Erosional features of river valleys—canyons, potholes, and many others—are produced by the "wet sandblast" of the suspended bed load, a process that is much more effective during periods of high flow velocity.

During periods of rainfall or snowmelt, direct precipitation into streams and runoff from adjacent land provide stream flow. Between periods of rainfall or snowmelt, stream flow comes from the slow seepage of groundwater into surface streams. Small streams, the smallest of which flow only during wet periods, join to form larger streams, which join to form rivers. Because rainfall or snowmelt occurs only occasionally in an area drained by a river (the river's drainage basin), the quantity of water flowing past any point on a river, or any point on any of its tributary streams, varies with time.

The quantity of water passing a point on the river is measured in cubic meters per second (or, in common North American and British practice, cubic feet per second) and is called the discharge. For a period of time after a rainfall or snowmelt event—for example, a flood—discharge increases at all points in the drainage basin. If the flood event occurs in only one part of the drainage basin, say the higher part of the basin with the smaller streams, the increase in discharge will occur first in the higher part of the basin and will occur in the larger streams lower in the basin at some later time.

If the discharge at a given point on a stream is measured continuously over a period of days and the discharge is plotted against time, with discharge on the vertical axis of the graph and time in days on the horizontal axis, the increase in discharge related to a storm or snowmelt (a flood) will appear as a hump in the discharge curve. A graph of discharge with time is called a hydrograph, and a hydrograph that shows a flood-related hump is a flood hydrograph. Flood hydrographs tend to be more pronounced—that is, the curve is higher and has steeper sides—on streams near the source of the flood water and longer and less pronounced—that is, lower and with more gently sloping sides—farther downstream. Putting it another way, the flood hydrograph attenuates, or dies out, downstream. The low, flat portion of the hydrograph that measures flow between flood events is called baseflow. Experience with flood hydrographs for a river enables hydrologists to predict the effect of future rainstorms and snowmelts of varying intensity and to predict low flow during prolonged dry periods.

An increase in discharge is accompanied by an increase in velocity. For rivers with sandy bottoms (sandy streambeds), the higher water velocity causes erosion, or scour, of the bottom; that is, the sand begins to move along the bottom with the water, and the river channel becomes deeper. The elevation of the water surface relative to some fixed point on the river bank also increases. As a rule, the banks are not vertical but sloping, and the increase in elevation of the water surface causes an increase in width. In summary, an increase in discharge is accompanied by increases in flow velocity, stream width, water surface elevation, and depth of channel (the last two items add up to an overall increase in depth). A reduction in discharge has just the opposite effect, including the deposition of new sand, arriving from upstream, in the channel as the velocity decreases.

In 1953, Luna Bergere Leopold and Thomas Maddock, Jr., introduced a concept that helps explain the relationships among the variables that change when discharge changes. The concept is called hydraulic geometry, and it is embodied in the following three equations: $w = aQ^b$, $d = bQ^f$, and $v = kQ^m$, where w is stream width, d is average depth of the stream, v is the flow velocity, and Q is discharge. The coefficients a, b, and k and the exponents b, f, and m are determined empirically—that is, by measurement in the field. Since the discharge is equal to the cross-sectional area of the stream (wd) times the flow velocity (that is, $wdv = Q$), $wdv = aQ^b \times bQ^f \times kQ^m = Q$. For this to be true, the product of a, b, and k must be 1 ($abk = 1$), and the sum of the exponents b, f, and m must also be 1 ($b + f + m = 1$). Many field studies have shown these equations to be good approximations of actual variation in width, depth, and velocity with variation in discharge.

The coefficients of the hydraulic geometry equations have little effect, relative to the powerful exponents, and are usually ignored. The exponents of the hydraulic geometry equations depend on the physical characteristics of the drainage basins and stream channels involved, but for a given point on a given stream, average values are $b = 0.1$, $f = 0.45$, and $m = 0.45$, which means that the increase in discharge during a flood event is expressed primarily in increases in depth and flow velocity.

The discharge of rivers increases downstream because of the larger length of stream receiving baseflow and the contribution from many tributaries, and hydraulic geometry equations may also be applied to downstream changes in discharge. Average values for the exponents in the downstream hydraulic geometry equations for width, depth, and velocity are $b = 0.5$, $f = 0.4$, and $m = 0.1$. This phenomenon is a paradox. Casual observation would suggest that small streams in the higher parts of the drainage basins flow faster than the large rivers in the lower parts of the drainage basins. Yet, appearances are deceptive: Water in the wider, deeper channels of the larger streams actually flows faster than does water in the smaller headwater streams.

When a flood flow exceeds the capacity of the river channel, the surface elevation of the river rises above the elevation of the river banks, and flooding of the surface adjacent to the stream occurs. This is what has generally been called a flood, or an overbank flood. Overbank floods do serious damage to homes, businesses, industrial facilities, and crops on the flooded areas, and much time and money are devoted to flood prevention. The principal method of flood prevention is the construction of levees, large earth embankments or concrete walls along the stream bank, at locations where the potential economic loss because of flooding justifies the expense of their construction and maintenance.

Methods of Study

Research on flow in rivers, or hydrology, almost invariably involves the determination of flow velocity and discharge. Because discharge (Q) equals stream width (w) times average depth (d) times average flow velocity (v), velocity and discharge are closely related. As implied by the world "average" in the preceding sentence, depth and velocity vary across the width of a river or stream. Velocity is lowest in the shallower parts of the river. Therefore, the cross section of the stream is divided into sections, and the discharge is taken as the sum of discharges of all the sections.

For a relatively small river, a rope or strong cord is stretched across the river and marked at regular intervals, typically 2 meters. The depth is measured at each marked point, and a current meter is used to measure the velocity at each point. The flow in rivers and streams is turbulent, and the current meter actually measures the average downstream component of velocity. Average downstream velocity varies with depth of the channel and also through the vertical section at any point on the stream, the lowest velocities occurring at the bottom and at the surface. Average velocity at any given vertical section of flow occurs at about 0.6 d above the bottom, which is where the current meter is positioned. Alternatively, two velocity measurements may be taken, at 0.2 and 0.8 d above the bottom, and averaged. The equation $Q = wdv$, where w is the interval at which measurements were taken, is computed for each section, or interval, across the stream; all the resultant discharges are summed to obtain the total discharge at the point on the stream, which is now called a station.

The process described above is laborious. If discharge at a station must be more or less continuously monitored, a quicker method is desirable,

The erosional effects of river flow can carve such enormous geological features as the Grand Canyon in Arizona. *(Lynn Abigail)*

which is accomplished by developing a rating curve for the station. Discharge is measured in the manner described above at several different times, with as wide a range of discharges as practical. A staff gauge (a board mounted vertically in the stream and marked in units of length, usually feet in American and British practice) is erected at the station, and the level of the water surface, called the stage of the river, is determined each time the discharge is measured. The rating curve consists of a plot of stage against discharge. Once the rating curve has been determined, discharge is estimated from river stage by use of the rating curve. Because of the scour and fill of the river bottom that occur with each increase and decrease in discharge and velocity, the rating curve is not a straight line. Moreover, because the scour and fill may, over time, change the character of the river channel at the station, it is necessary actually to measure the discharge periodically to check the validity of the rating curve. If large changes in the channel occur, it is necessary to establish a new rating curve.

The U.S. Geological Survey maintains a large number of rating stations, or gauging stations, and periodically reports stream discharges. At most stations operated by the agency, the staff gauge is replaced by a vertical pipe driven in the streambed and perforated near the bed. Water is free to flow in and out of the pipe as river stage changes, but the water surface inside the pipe is not disturbed by surface waves. A cable with a weight at the end is attached to the float, and the cable is passed over a wheel near the top of the pipe. The float is free to move with the water surface inside the pipe and, as it moves, it turns the wheel. Sensitive instruments monitor the position of the wheel and therefore the river stage as well. In this way, the stage is periodically and automatically reported to a central office by telephone line or radio.

Context

All aspects of river flow—including the nature of flow (laminar and turbulent), flow velocity, quantity of flow, transport and deposition of sediment by flowing water, shaping of river channels by erosion and deposition, and their variation in time and space—constitute the scientific discipline of fluvial hydrology (river hydrology).

From a practical standpoint, the most important aspects of river flow are the quantity of flow, or discharge, and the velocity. Planning for water supply, treated waste disposal, flood control, river transportation, irrigation, and hydroelectric power requires knowledge of long- and short-term variation in discharge and flow velocity. As an example, the areas adjacent to rivers, the floodplains, are desirable sites for heavy industry because they are level, close to major supplies of water, and convenient to river transport. The drawback is that overbank flows, which flood these areas, occur every three years on average, with major floods occurring two or three times per century. One solution is to construct levees that prevent flooding of specific areas of the floodplain. Knowledge of river flow enables the levee designers to predict how high the levees must be and how they must be constructed to withstand the high flow velocities associated with high discharges.

Minimum flows also are of interest. Typically, low river flows occur at some season of the year, in the late summer and autumn in most of North America. If the low flow is inadequate to supply water for a major city, for example, reservoirs must be constructed. A knowledge of the expected degree and duration of low flow is required to determine what the capacity of the reservoirs must be.

Low flows are also critical in wastewater treatment. Although treated sewage is biologically clean, it has a high content of dissolved solids and, undiluted, is not suitable for reuse at downstream locations. The treated wastewater must be released into a river with a minimum discharge large enough to dilute it to an acceptable dissolved-solids content. If the expected minimum discharge of a given river is not adequate, the treatment plant must be relocated.

A very challenging problem arises in connection with the design of irrigation canals. Flow velocity in the canal must be high enough to prevent the settling, or deposition, of silt and clay in the irrigation water because this would block the canal, but the velocity must not be so high that the water will erode the banks of the canal. Methods of designing stable and efficient irrigation canals have been developed from study of erosion and deposition in natural rivers.

Every aspect of the management of water resources, a vital element in a complex industrial society, is dependent on knowledge of river flow. Intelligent planning for the multiple uses of rivers mentioned above requires knowledge of expected low, average, and high discharges and the expected timing of these events. River hydrology provides the basis for this planning.

Bibliography

Dingman, S. L. *Fluvial Hydrology*. New York: W. H. Freeman, 1984. A very clear development of all the equations that are essential to fluvial hydrology. For full understanding, a year of college calculus is necessary, but about 80 percent of the material can be mastered by a student with knowledge only of algebra and trigonometry. An excellent treatment of the subject, with many answered problems and a helpful annotated bibliography. The introductory chapters on units of measurement (dimensions), fundamentals of physical equations, and physical properties of water should be required reading for all science students.

Dunne, Thomas, and Luna B. Leopold. *Water in Environmental Planning*. San Francisco: W. H. Freeman, 1978. A thorough (818-page), essentially nonmathematical examination of the subject. Emphasis is on design to avoid environmental problems associated with water in all its manifestations. Many worked problems, most involving only basic mathematics.

Kindsvatter, C. E. *Selected Topics of Fluid Mechanics*. U.S. Geological Survey Water Supply Paper 1369-A. Washington, D.C.: Government Printing Office, 1958. Develops in an exceptionally clear way the fundamental concepts of fluid mechanics that underlie fluvial hydrology. Less than 10 percent of the material requires a knowledge of calculus. Many aspects of the approach are much superior to more recent texts.

Leopold, Luna B. *Water: A Primer*. San Francisco: W. H. Freeman, 1974. A brief (172-page), very readable introduction to the subject. Highly recommended for all newcomers to the field and most professionals.

Leopold, Luna B., M. G. Woolman, and J. P. Miller. *Fluvial Processes in Geomorphology*. San Francisco: W. H. Freeman, 1964. A lucid explanation of the fundamentals of the field by the originator of hydraulic geometry, written ten years after the first paper on the subject. Highly recommended for both beginners and professionals.

Manning, J. C. *Applied Principles of Hydrology*. Westerville, Ohio: Charles E. Merrill, 1987. An excellent short introduction to the principles of surface water and groundwater hydrology. Incorporates very little of the mathematics involved in the two fields, but the descriptions and illustrations of methods of making field measurements are clear and informative.

Richards, Keith. *Rivers, Form, and Process in Alluvial Channels.* London: Methuen, 1982. A college-level text requiring some familiarity with calculus for complete understanding. Its great strength is the very large number of research papers cited in the text. Conveys a sense of the quantity and type of research that has been done.

Robert E. Carver

Cross-References

Alluvial Systems, 1; Deltas, 126; Drainage Basins, 141; Floodplains, 241; River Valleys, 534.

River Valleys

The valleys in which streams flow are produced by those streams through long-term erosion and deposition. The landforms produced by fluvial action are quite diverse, ranging from spectacular canyons to wide, gently sloping valleys. The patterns formed by stream networks are complex and generally reflect the bedrock geology and terrain characteristics.

Field of study: Geomorphology

Principal terms

AGGRADATION: the process by which a stream elevates its bed through deposition of sediment

BASE LEVEL: the theoretical vertical limit below which streams cannot cut their beds

FLUVIAL: of or related to streams and their actions

INTERFLUVE: an upland area between valleys

STREAM EQUILIBRIUM: a state in which a stream's erosive energy is balanced by its sediment load such that it is neither eroding nor building up its channel

UNDERFIT STREAM: a stream that is significantly smaller in proportion than the valley through which it flows

VALLEY: that part of the earth's surface where stream systems are established; it includes streams and adjacent slopes

Summary

River valleys consist of valley bottoms and the adjacent valley sides. Between valleys are undissected uplands known as interfluves. Valley floors may be quite narrow, as in the case of the Black Canyon of the Gunnison River, or they may be quite wide, as in the case of the Hwang Ho or the Brahmaputra. Similarly, valley sides may have very gentle to rolling slopes, or they may be nearly sheer, as in the case of the Arkansas River's Royal Gorge. In many areas, the interfluves are simply divides between adjacent valleys, but on tablelands such as the Colorado Plateau, they may be tens of kilometers wide in places.

River channels and river valleys are products of the streams that flow through them. As a stream erodes a channel for itself in newly uplifted terrain, it eventually carves a valley whose form is determined by the erosive power of the stream, by the structural integrity of the rock and debris of the valley walls, by the length of time that the stream has been operating on its surroundings, and by past environmental conditions. These past environmental conditions are attested by stream channel and valley profiles that have not entirely erased landforms produced during the most recent episodes of glaciation and climate change. The valley of the Mississippi River, for example, is formed in the complex deposits of what was a much larger, more heavily laden glacial meltwater stream that existed only 15,000 years ago. River channels and valleys may be referred to as palimpsests, a term originally used to describe parchment manuscripts that had been partly scraped clean, then reused. On the fluvial (river-carved or river-deposited) landscape, previous landscape elements are seen, just as old words show through on a recycled piece of parchment.

In many parts of the world, streams are found flowing through valleys that appear to have been formed by a far larger stream. The valley width, amplitude of meanders, and caliber of coarse sediment are proportional to far larger stream courses. Such streams are said to be underfit, and the valleys are largely remnant features from times of wetter and cooler climates that accompanied glaciation, or the streams were glacial meltwater channels during deglaciation.

The fact that stream channels and valleys are not chance features on the landscape was first noted by British geologist John Playfair in 1802. Playfair suggested that instead of being isolated features on the landscape, streams are part of well-integrated networks. More important, these networks are finely adjusted to the landscape and to one another in such a way that tributaries almost always join the main trunk stream at the same level as that stream. Streams that must plunge over waterfalls to join a larger stream are quite rare. Such discordant streams are found primarily in recently glaciated terrain, where the stream-valley system has not yet become fully adjusted, such as in Yosemite Valley. This remarkable consistency of stream accordance is the strongest evidence indicating that streams carve their own valleys.

Streams develop into network patterns that are strongly influenced by bedrock structure. Where the bedrock is relatively uniform and without strong joints and faults, a dendritic pattern of drainage develops. In this pattern, a stream system is branched like a tree. On inclined plains such as the Atlantic coastal plain, the stream pattern is often parallel, with major streams flowing directly down the topographic slope to the sea. Inclined mountain systems such as the Sierra Nevada also produce parallel drainage. Where structural folding of the terrain has produced linear ridge and valley topography, main trunk streams occupy the linear valleys and are quite long, with short, steep tributaries feeding them off the flanks of the hills or mountains. In areas such as the Great Valley of Virginia, this type of pattern has developed and is known as trellis drainage. Volcanoes often produce a radial pattern of drainage; Mount Egmont, New Zealand, is often cited as an example.

The depth of a river valley is a function of the height of the land above base level, the length of time that has passed since the stream began to erode, the resistance of the bedrock to erosion, the load of sediment that the stream is carrying, and the spacing of adjacent streams. "Base level" refers to the theoretical lower limit to which a stream can cut. Ultimate base level for all streams is sea level, but local base level is significantly higher for many streams. Streams require a minimum slope to transport their sediment to the sea; this limits the depth to which a stream may cut. The upper section of the Hwang Ho near Tibet flows at elevations of more than 3,000 meters, but the river must still flow more than 4,000 kilometers to the sea. In order to carry its heavy load of sediment over that distance, the river must maintain its channel at a rather high elevation.

Valleys that are deep and relatively narrow are called canyons; those valleys that are especially narrow are called gorges. Streams require time to erode great canyons, although the rate of cutting can be quite rapid compared to most geologic processes. The process of downwearing of interfluves is far slower, so young (recently uplifted) landscapes produce the deepest canyons; as downcutting by the main stream ceases, upland weathering and erosion lowers the local relief. The deepest, narrowest canyons are eroded in strong, homogeneous rock such as granite and quartzite. The Royal Gorge of the Arkansas River and the Colca Gorge of Peru are excellent examples; both are cut in recently uplifted masses of resistant rock formations. The Colca Gorge is far narrower than is the Grand Canyon and, at more than 3,000 meters, it is nearly twice as deep.

As a mountain mass or tableland is uplifted, streams often develop that flow directly down the initial slope of the land; such streams are called consequent streams because their course is a direct consequence of the terrain slope. Because streams seek the path of least resistance, their courses often follow the outcrop pattern of weak rocks such as shale, producing what is known as a subsequent stream pattern. In many cases, however, streams seem to ignore the terrain and structural slope of the land entirely, flowing through mountains of quite resistant rock. An excellent example is the Black Canyon of the Gunnison River in Colorado, where the river carves a deep canyon through a high plateau, with its channel cut in resistant gneisses and igneous intrusives. What makes this so surprising is that much lower terrain, underlain by thick sequences of weak shale, lies only 3 kilometers to the west of the head of the gorge, which would seem to provide a much easier path to the sea. The reason for this course—and many other anomalous stream courses—is that the course of the Gunnison was established prior to the uplift of the plateau. As the land rose beneath the river, it maintained its position, carving an ever-deepening gorge in the rocks as they rose. Far older and more extensive is the anomalous course of the New River, which cuts directly across the structure of the Appalachians, flowing a great distance to the Ohio River rather than the shorter, more direct route to the Atlantic. Once again, this river is one that was established prior to the uplift of the land in which it is now entrenched, in this case the ancient Appalachians. Thus, the river is rather ironically named, considering that its course is older than the Appalachians.

At any given time, a stream may either erode or aggrade (build up) its bed. This vertical change in stream profile is determined by many factors

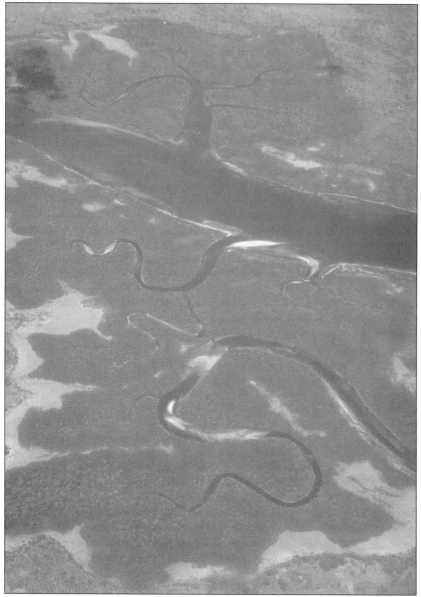

The patterns carved by rivers and streams are strongly influenced by the underlying bedrock. *(R. Kent Rasmussen)*

internal to the stream, as well as outside environmental factors. Over geological time, the tendency of streams is to erode their beds deeper and deeper. Over shorter time intervals, however, a stream may reach a state of equilibrium between its erosive energy and the load of sediment that the stream is carrying. A delicate balance is maintained within the stream channel: If stream energy is increased by an increase in flood volume or frequency, for example, the stream is likely to respond by eroding its bed. Conversely, if stream power remains constant and the sediment load is significantly increased, the stream will respond by aggrading. This channel aggradation increases the channel slope, giving the stream more energy to transport the sediment load, and a new equilibrium is reached.

Methods of Study

River valleys are studied by means of several different techniques. The earliest studies were based entirely on field observations. Among the ancient natural scientists who speculated upon the origins of fluvial landforms were Aristotle and the Arab philosopher and physician Avicenna (c. A.D. 1000), who promoted the idea that river valleys could have been carved by running water. Near the end of the fifteenth century, Leonardo da Vinci advanced the understanding of landforms through his insightful observations and interpretations of the topography of southern Europe. Deductive studies advanced significantly during the eighteenth and nineteenth centuries as the concept of uniformitarianism began to take hold and the concept of deep geologic time began to gain acceptance. Uniformitarianism holds that the present is the key to the past, which means that processes occurring on the earth today are the same processes that have always occurred on the earth. Therefore, the erosion of the Grand Canyon or Tiger Leap Gorge on the Yangtze occurred as the result of the same fluvial processes that operate there currently. The most important corollary of the law of uniformitarianism is that vast amounts of time are required for great mountains to be uplifted, for valleys to be carved, and for uplands to be leveled.

As detailed topographic and geologic maps became available in the early twentieth century, deductive analysis of river valleys became much more precise. By comparing precise topographic cross sections of river valleys to the underlying geology, geomorphologists came to understand the importance of bedrock lithology in controlling valley wall slope and the importance of lithology and zones of rock weakness (joints and faults) in determining stream courses, network patterns, and stream spacing.

Stream gauges, which indicate the discharge of rivers, have been operating over large areas since the beginning of the twentieth century. Thus, a great wealth of information has become available concerning the manner in which streams erode and transport sediment. With gauge data concerning rates of runoff and sediment load, researchers are able to project over time the rates of regional denudation (lowering of the landscape by erosion). Results of such studies show that mountainous areas are being lowered at much faster rates than are piedmont and lowland areas, but the global average rate of denudation is on the order of a few centimeters per thousand years. From these same data, much has been learned about the relative importance of catastrophic floods and more frequent, low-magnitude runoff events.

Together with the compilation of up-to-date maps, the advent of low-altitude aerial photography has allowed researchers to study stream channel and valley alterations over relatively short time intervals. By comparing sequential high-resolution photographs, which are obtained about every ten years in the United States, it is possible to gain detailed knowledge of the magnitude and locations of such changes. In some cases, the changes are great enough to be detected from commercial satellite images, which provide repeat coverage as often as every fourteen days.

Laboratory studies are of increasing importance in studying river valleys. From simple stream tables to elaborate scaled physical models of river channels, such studies provide fluvial geomorphologists with an understanding of how streams operate and the nature of changes that will occur with environmental changes such as an increase in runoff, an increase in the caliber of sediment delivered to the stream, or the construction of a dam upstream. The results not only help in understanding basic fluvial processes but can aid greatly in management decisions as well.

River network formation and slope retreat are sometimes studied by observing the development of stream channels on recently denuded terrain such as construction sites. Although it is not possible to eliminate entirely the important role of geologic time in such an experiment, it is possible to gain insight into the nature of such processes by observing these fast-changing environments.

Computer simulation of stream network development and river valley processes has been utilized by a number of researchers. Computer modeling is based upon known physical properties of rock and soil materials, plus the laws of physics that govern flows of energy and materials. As models become more sophisticated, their ability to simulate and predict slow geologic processes in a short period of time will improve.

Context

Perhaps the greatest significance of river valleys proceeds from the importance of streams in sculpting the face of the earth. Fluvial erosion and valley formation are the most important processes by which uplifted masses of land are lowered over long periods of time. In order to gain an understanding of, and a true appreciation for, the "lay of the land," one must first understand the processes by which streams carve their valleys. It is no wonder, then, that students of landforms—from Aristotle to modern fluvial geomorphologists—have generally focused on streams and valleys.

Stream network patterns are almost always indicative of underlying geology, so reconnaissance geology often takes first aim at the stream channel network of an area. By understanding where and why stream channels have carved their valleys, one can learn much about the underlying geologic structure.

As future changes in climate—caused by carbon dioxide pollution and an attendant global greenhouse warming—are debated, substantial changes in stream flow will occur. One study suggested that a 10 percent reduction in precipitation in the Colorado River basin could result in a 50 percent reduction in stream flow of the Colorado River. Such a change would be catastrophic in and of itself, but there would be significant changes in the behavior of the stream as well. Given a constant supply of sediment, the river would likely aggrade its bed, threatening streamside developments, irrigation diversions, and vegetation. Researchers believe that minor changes in discharge-sediment load relationships in Chaco Arroyo, Arizona, led to stream entrenchment that destroyed the advanced irrigation system built by the Anasazi people, leading to the abandonment of Chaco Canyon. An understanding of fluvial dynamics is essential to predicting such changes.

The oldest direct application of fluvial dynamics is the design of stable irrigation canals. The earliest civilizations using irrigation, such as the early Mesopotamian culture, discovered that canals had a tendency either to erode or to fill themselves with sediment if they were not carefully adjusted to take into account the volume of water flowing through the canal, the slope of the canal, the materials through which the canal was cut, and the sediment load of the water. More recently, it has become necessary to predict the changes in stream character below large man-made dams. Below Glen Canyon Dam, the Colorado River has been relieved of its heavy sediment load and has compensated by eroding the beaches along the river that are used for recreation.

Bibliography

Bloom, A. L. *Geomorphology: A Systematic Analysis of Late Cenozoic Landforms.* Englewood Cliffs, N.J.: Prentice-Hall, 1978. This book is intermediate in difficulty between the Butzer and Chorley geomorphology texts listed below. Chapters 8 through 12 cover the processes of fluvial landscape development thoroughly and evenly. The author does a good job of explaining the interrelations between valley form, structural control, and fluvial processes. Well illustrated and supported with numerous citations. Suitable for college-level readers.

Butzer, K. W. *Geomorphology from the Earth.* New York: Harper & Row, 1976. This geomorphology text is aimed at an introductory audience and does a good job of avoiding unnecessary jargon. Although not as thoroughly illustrated as some books on the subject, its level of presentation makes it accessible to a far larger audience than most. Suitable for advanced high school students and beyond.

Chorley, R. J., S. A. Schumm, and D. E. Sugden. *Geomorphology.* New York: Methuen, 1985. An advanced text in geomorphology that is best used as a reference. Provides detailed citations of the important scholarly works on each subject addressed and is quite thorough. Many diagrams, graphs, and line drawings supplement the material, but there are few photographs. Suitable for college students who have already gained some background in geomorphology.

Hunt, C. B. *Natural Regions of the United States and Canada.* San Francisco: W. H. Freeman, 1974. This book is primarily a physiography text, aimed at describing and explaining the pattern of landforms across North America. The author produced it in the light of "the general public's interest in the natural environment and the need for an authoritative account of it in language as nontechnical as possible." In both regards, the book is quite successful. Well illustrated with diagrams, line drawings, and photographs. Introductory chapters provide an explanation of the general phenomena, and the remaining chapters provide thorough, understandable coverage of the regional expression of landforms. Suitable for high school students and beyond.

Lutgens, F. K., and E. J. Tarbuck. *Essentials of Geology.* 2d ed. Westerville, Ohio: Charles E. Merrill, 1986. This short (310-page) physical geology text provides a brief introduction to all the major topics in the field. Its strengths are its full-color illustrations and its accessibility to the general reader. Suitable for anyone interested in the earth sciences.

McKnight, T. L. *Physical Geography: A Landscape Appreciation.* 2d ed. Englewood Cliffs, N.J.: Prentice-Hall, 1987. This physical geography text provides very thorough but understandable coverage of the concepts of stream networks, stream channel history, and fluvial dynamics. Richly illustrated with color diagrams and photographs. Suitable for high school students and beyond.

Skinner, Brian J., and Stephen C. Porter. *Physical Geology.* New York: John Wiley & Sons, 1987. A consistently excellent text in introductory physical geology, richly and generously illustrated in full color. Very thorough and quite extensive (more than seven hundred pages). Chapter 11, "Streams and Drainage Systems," is one of the best chapters. Suitable for advanced high school students and beyond.

Michael W. Mayfield

Cross-References

Alluvial Systems, 1; Alpine Glaciers, 8; Drainage Basins, 141; Floodplains, 241; River Flow, 526; Rocks: Physical Properties, 550; Weathering and Erosion, 701.

Rock Magnetism

Rock magnetism is the subdiscipline of geophysics that has to do with how rocks record the magnetic field, how reliable the recording process is, and what conditions can alter the recording and therefore raise the possibility of a false interpretation being rendered by geophysicists.

Field of study: Geophysics

Principal terms

BASALT: a very common, dark-colored, fine-grained igneous rock

BLOCKING TEMPERATURE: the temperature at which a magnetic mineral becomes a permanent recorder of a magnetic field

CURIE TEMPERATURE: the temperature above which a permanently magnetized material loses its magnetization

DAUGHTER PRODUCT: an isotope that results from the decay of a radioactive parent isotope

DETRITAL REMANENT MAGNETIZATION: the magnetization that results when magnetic sediment grains in a sedimentary rock align with the magnetic field

FERROMAGNETIC MATERIAL: the type of magnetic material, such as iron or magnetite, that retains a magnetic field; also called permanent magnet

GRANITE: a low-density, light-colored, coarse-grained igneous rock

MAGNETITE: a magnetic iron oxide composed of three iron atoms and four oxygen atoms

RADIOACTIVITY: the spontaneous disintegration of a nucleus into a more stable isotope

THERMAL REMANENT MAGNETIZATION: the magnetization in igneous rock that results as magnetic minerals in a magma cool below their Curie temperature

Summary

The direct study of the earth's magnetic field is only four centuries old. This study involves the measurement of the field with scientific instruments and subsequent analysis of the resulting data. Four centuries is a very small fraction of the 4.6 billion years that the earth has existed; thus, direct study affords scientists very little understanding of the nature of the field over long periods of time. It is useful to know what happened to the earth's magnetic field in those billions of years before the present, because the field can be a source of information about conditions on the earth's surface and its interior. Magnetic minerals in rocks serve as recording devices, giving scientists clues regarding the nature of the ancient magnetic field.

A moving electric charge, such as an electron, produces a magnetic field that is the ultimate source of any larger magnetic field. An atom is composed of a nucleus, with its protons and neutrons, and the electrons that surround the nucleus. The protons do not move within the nucleus, and therefore they do not produce a magnetic field. The electrons, however, orbit the nucleus, and this movement produces a weak magnetic field. In addition, the electrons spin on their axes, and this activity also gives rise to a small magnetic field.

Because all atoms have electrons orbiting and spinning, one might think that all materials should have a permanent magnetic field, but the situation is more complicated. Strictly speaking, every material is magnetic, but there are different types of magnetism. Some materials are paramagnetic: When they are placed in an external magnetic field, the atoms align with the field. The atoms act as small compasses, orienting with the field, and the material is magnetized; the magnetic fields produced by the atom's electrons add to the intensity of the external field. When the external field is removed, however, the atom's orientation becomes randomized because of vibrations caused by heat, and the material is consequently demagnetized. Many materials, such as quartz, are paramagnetic and are not able to record the earth's magnetic field.

A much smaller number of minerals are ferromagnetic. There are various types of ferromagnetism, but the underlying principle is the same. In ferromagnetic materials, an external magnetic field again aligns the atoms parallel to the field, and the material is magnetized. When the field is removed, however, the atoms remain aligned, and the substance retains its magnetization; it is "permanently" magnetized. Actually, the substance can be demagnetized by heating or stress. Dropping a bar magnet on the floor or striking it with a hammer will demagnetize it slightly. The shock randomizes some of the atoms so that they cease to contribute to the overall magnetic field. The heating of a magnet above its Curie temperature also destroys its magnetization by randomizing the atoms and making the material paramagnetic. As the temperature drops below the Curie point, the material becomes slightly remagnetized, because the weak field of the earth aligns some of the atoms.

In ferromagnetic materials, atoms are not all aligned in one direction; rather, they are found in aligned groups, called domains. Under a microscope, the domains are barely visible. Within a particular domain, the atoms are aligned, but all the domains are not aligned in the same direction. A "permanent" magnetic material that is unmagnetized has all the domains randomly aligned, and the overall field cancels to zero. When placed in a magnetic field, some of the domains realign parallel to the direction of the

543

field and stay aligned after the field is removed. It is these domains that give the material its overall magnetization. If a high enough magnetic field is applied, all the domains align with the field, and the magnetization has reached its saturation point; the strength of the material's magnetic field is at a maximum. One of the areas of research for physicists is the quest for materials that have high magnetic field strengths but with less material. Such materials are useful in making small, but powerful, electric motors.

Rocks are classified into three main groups: igneous, formed from crystallized molten rock; sedimentary, formed from weathered rock material; and metamorphic, produced when other rock is modified with heat, pressure, and fluids. Most magnetic minerals occur in igneous and sedimentary rocks.

Materials such as iron, cobalt, and nickel are ferromagnetic; for this reason they are used in making various permanent magnets. These metals are not found naturally on the earth's surface in the uncombined state, so they do not contribute to rocks' recording ability. Most of the minerals that make up rocks, such as quartz and clay, are not ferromagnetic. These minerals are useless as recorders, but many rocks contain magnetite or hematite, which are good recorders. These common magnetic minerals are oxides of iron.

Hematite is Fe_2O_3, which means that there are two iron atoms for every three oxygen atoms. Hematite is red in color, similar to rust on a piece of iron. Most reddish-brown hues in sedimentary rock are caused by hematite. This magnetic mineral is not a very strongly magnetized compound, but it is a very stable recorder in sedimentary rocks. Unfortunately, in many cases, its formation postdates that of the rock in which it occurs, so it does not necessarily record the magnetic field at the time of the rock's formation.

Magnetite (Fe_3O_4) has been known as lodestone for several millennia. It is a strongly magnetized iron compound that makes some igneous rocks very magnetic and supplies some of the recording ability of sedimentary rocks. The magnetite in rocks can record the field direction by one of several methods. In igneous rocks, magnetite crystals form as the magma cools. As the crystals grow, they align themselves with any magnetic field present. This process is called thermal remanent magnetization, or TRM. If the crystals are quite small or quite large, they cannot permanently record the field direction; after a short time the recording fades and becomes unreadable. The magnetism of such small grains is called superparamagnetism: They do align with a magnetic field, but they easily lose their orientation. The larger grains contain many magnetic domains that become misaligned over time so that the recording fades.

Grains the size of fine dust are good recorders. Unfortunately, not all igneous rocks have grains of the proper size. The size of the mineral crystal depends on the rate of cooling: When magma is cooled very slowly, large crystals are produced, while a rapid cooling results in smaller crystals. Granite is coarse-grained and thus is not the best recorder. The best igneous recorder is basalt, a black, fine-grained rock. Basalt can contain so much magnetite that a piece the size of a spool of thread acts like a bar magnet so that paper clips can be suspended from it. Basalt is fairly common on the surface of the earth, particularly in the ocean basins, where nothing but basalt underlies the sediment on the basin floor.

A useful magnetic recorder must provide information about how old it is. Basalt again fills this requirement, for its crystallization can be dated by measuring the amount of radioactive elements and their daughter products it contains. Clearly, basalt is an ideal source of information on the magnetic field. Unfortunately, it does not occur everywhere on the earth; moreover, as a recorder, it covers only times of eruptions of magma. Some other recorder must be used to fill in the blanks.

Sedimentary rock is formed from the products of the rock weathering that accumulate mostly in watery environments, such as rivers, lakes, and oceans. Clastic sedimentary rocks are formed from fragments of rock and mineral grains, such as grains of quartz in sandstone. Chemical sedimentary rock is derived from chemical weathering products, such as calcium carbonate or calcite, which is the major constituent of limestone. Most of the material in sedimentary rocks is not ferromagnetic, but there are a few grains of magnetite and other ferromagnetic compounds. As the grains fall through the water, they align with the magnetic field present at the time. When they hit the bottom, they retain the orientation, for the most part, and are subsequently covered by more sediment. This process is termed detrital remanent magnetization, or DRM.

An interesting aspect of DRM is the role that organisms play in its formation. The grains of magnetic minerals that fall through the water are oval-shaped, and when they strike the surface of the sediment they become misaligned with the field. Organisms such as worms disturb the sediment in a process known as bioturbation, which moves the sediment around and realigns the magnetic grains with the field. In the mid-1980's, it was discovered that certain varieties of bacteria have small grains of magnetite in their bodies. The bacteria use the grains like compasses to find their way down into the sediment on which they feed. The bacteria eventually die, and the magnetite grains become part of the sediment, aligned with the magnetic field; this phenomenon is known as biomagnetism.

The grain-size problem also occurs in DRM, given that sediment particle can be the size of a particle of clay, a boulder, or anything in between. Conglomerate, a rock composed of rounded pebbles and other large particles, is not a good recorder, nor is coarse sandstone. Finer sandstones, shales, siltstones, and mudstones are much better. Most chemical rocks, such as halite (common table salt), are poor recorders; limestone may or may not be good, depending on the conditions of formation.

The magnetization in sedimentary rocks is generally between one thousand and ten thousand times weaker than is the magnetization in a basalt. Very sensitive magnetometers are needed to measure the magnetic field in these specimens. To be useful in geomagnetic studies, sedimentary rocks must be dated, but this is a difficult task, as they cannot be dated using radioactive methods. By a complex method of determination, fossils can act as indicators of the age of the rock in which they are found. If igneous rock layers are located above and below the rock layer of interest, and if these igneous rock layers can be dated, an intermediate age can be assigned to the sedimentary layer.

Methods of Study

A magnetometer useful in the study of rock magnetism is the superconducting rock magnetometer, or SCM. Superconductivity is the phenomenon of a material's losing its resistance to electric current at low temperatures. Liquid helium is used to cool a portion of the magnetometer, composed of a cylinder of lead closed at one end. As the lead cools, it becomes superconducting, and if done in a region of low magnetic-field intensity, this low field is "trapped" inside the cylinder. Magnetic field sensors known as SQUIDS, or superconducting quantum interference devices, are very sensitive to low-intensity magnetic fields. The sample is lowered into the device, and its electronic display shows the intensity of the sample's magnetization. Such devices are useful in studying the rock magnetism of low-intensity sedimentary rocks.

The Curie temperature is important for establishing the thermal remanent magnetization for igneous rocks. A sample of a particular ferromagnetic material in a magnetic field is heated and the temperature is measured; the sample's Curie temperature is determined when the pull of the magnetic field on the sample weakens. The Curie point for various ferromagnetic materials is established by this method. Once that is done, the procedure is reversed. A sample of an unknown ferromagnetic material can be heated in a magnetic field to determine its Curie point, which can then be compared with the established table of values to identify the magnetic mineral. This method does not establish the exact composition of the material, but it does narrow down the possibilities, which is of value because other methods for determining composition are more expensive. In addition, it has been discovered that Curie temperature is not the only factor critical to the recording process. At the Curie point, the material is ferromagnetic but the recording ability is weak. The material has to cool through the blocking temperature for recording stability. Thereafter, magnetic minerals are magnetically stable for periods of billions of years.

Another area of study is the determination of the best grain size and shape for magnetic recording. Researchers experiment with different sizes and shapes of magnetic grains in magnetic fields of various strengths and directions and measure their responses to changes. It was found that crystals of magnetic materials such as magnetite develop features known as domains. These are areas where the atoms are aligned in one direction and produce the unified magnetic field for the domain. A small crystal has only one domain that can easily shift to another direction; therefore small crystals are poor recorders. If the crystal is quite large, it has many domains in which it is again easy to shift direction. Crystals with one large domain or several small domains are magnetically "hard" in that it is more difficult to shift the magnetic alignment. For magnetite, these are dust-sized particles, around 0.03 micron in diameter.

Other research reveals that the rock's recording of the field is not as "neat and clean" a process as portrayed in the previous paragraphs. Many events can lead to the alteration of the magnetic alignment. If the rock is heated above the Curie point and then cooled, the magnetic alignment is that of the field present at that time, and the old alignment is erased. The rock may be changed chemically, and old magnetic minerals may be destroyed and new ones produced. This process is referred to as chemical remanent magnetization, or CRM. These secondary magnetizations can be removed in some cases, and they can even provide more information on the rock's history. One method of magnetic cleaning or demagnetization involves subjecting the rock sample to an alternating magnetic field while other magnetic fields are reduced to zero. This "cleaning" will remove that portion of the mineral's magnetization that is magnetically "softer" than is the maximum alternating field. The magnetization above the level is unaffected and should represent the original magnetization. Heating a sample to a certain temperature is another method of demagnetization.

Context

The study of the earth's magnetic field history, and all the inferences about the earth drawn from that study, depends on the ability of rocks to record information about the magnetic field at the time of the rocks' formation. That ability, in turn, is dependent upon the magnetic characteristics of a few permanently magnetized minerals, such as magnetite.

The study of rock magnetism is rather esoteric; only a few individuals worldwide are involved in this subdiscipline of geomagnetism. Yet, such study has shown that rocks can faithfully record the history of the earth's magnetic field. This record is used to infer conditions on the earth hundreds of million years ago. Such studies have lent support to the idea that the continents have actually moved over the surface of the globe, and thus the theory of plate tectonics was born, with all its implications for the formation and location of petroleum and ore deposits, the origin of earthquakes and volcanoes, and the formation of mountain ranges such as the Himalaya. Such is an example of the odd twists and turns that science can take. Seemingly inconsequential findings can lead to a theory with great potential for making the earth and its workings much more understandable.

Bibliography

Cox, Allan, ed. *Plate Tectonics and Geomagnetic Reversals.* San Francisco: W. H. Freeman, 1973. Cox provides fascinating introductions to chap-

ters that are composed of seminal papers concerning magnetic reversals and their contribution to the development of the theory of plate
tectonics. Information on rock magnetism is scattered throughout the
book in discussions on baked sediments, magnetization of basalt,
magnetic intensity, and self-reversals in rocks. The papers are advanced
for the average reader, but there are many graphs, diagrams, and
figures that merit attention.

Glen, William. *The Road to Jaramillo: Critical Years of the Revolution in Earth
Science.* Stanford, Calif.: Stanford University Press, 1982. This book
gives a history of the plate tectonics revolution of the mid-1950's to the
mid-1960's. Rock magnetism is specifically covered on pages 103-109.
Other aspects of rock magnetism are discussed in various portions of
the book, for example, those dealing with the magnetic minerals
associated with rock magnetism and deep-sea core work. Of particular
interest are the sections devoted to instruments used to measure rock
magnetism.

Hargraves, R. B., and S. K. Banerjee. "Theory and Nature of Magnetism
in Rocks." In *Annual Review of Earth and Planetary Sciences,* vol. 1, edited
by F. Donath. Palo Alto, Calif.: Annual Reviews, 1973. The article covers
the theories of the natural remanent magnetization of rocks (NRM).
NRM is the combined magnetization of all magnetic minerals in a rock,
such as those resulting from TRM. The various carriers of remanence
are also covered, with a table that lists each mineral and its composition, crystal structure, magnetic structure, and other pertinent information. The paleomagnetic potential of rocks is also discussed. A few
mathematical equations, but nothing too formidable. Numerous figures and a long list of references.

Lapedes, D. N., ed. *McGraw-Hill Encyclopedia of Geological Sciences.* New
York: McGraw-Hill, 1978. Pages 704-708, under the heading "Rock
Magnetism," provide concise descriptions of how rock magnetization
occurs, the present field, magnetic reversals, secular variation, and
apparent polar wandering. The text is very readable, with no mathematics and a fair number of graphs, tables, and figures.

O'Reilly, W. *Rock and Mineral Magnetism.* New York: Chapman and Hall,
1984. O'Reilly covers the atomic basis for magnetism, the magnetization process, the various remanent magnetizations such as TRM, the
magnetic properties of minerals, and, finally, the applications of rock
and mineral magnetism. Many tables, figures, and photographs of
minerals. Some mathematics and chemistry are also included but
should not be too difficult. References are included at the end of each
chapter.

Stacey, F. D. *Physics of the Earth.* New York: John Wiley & Sons, 1977. Under
section 9.1, "Magnetism of Rocks," the author provides a short, technical description of rock magnetism. Several figures show the various
types of magnetic alignments. The domain structure of ferromagnetic

materials is also discussed. Many other aspects of the earth's magnetic field are also covered at a technical level.

Tarling, D. H. *Paleomagnetism: Principles and Applications in Geology, Geophysics, and Archeology.* New York: Chapman and Hall, 1983. A very good resource on the subject of paleomagnetism, or the ancient magnetic field. Chapter 2 is devoted to the "physical basis" for the magnetization of material, with a discussion of the atomic level and the resulting magnetic domains. The various remanent magnetizations are covered in detail. Chapter 3 deals with the various magnetic minerals and their identification. The magnetization of the various rock types are covered in chapter 4. Chapter 5 discusses instruments used in paleomagnetic work. The remainder of the text deals with mathematical analysis used in paleomagnetic work and, finally, paleomagnetic applications.

Stephen J. Shulik

Cross-References

Earth's Age, 155; Igneous Rocks, 315; Plate Tectonics, 505.

Rocks: Physical Properties

Rocks and rock products are used so widely in everyday life that natural and physical properties of rocks affect everyone. Major properties of rocks fall into two categories: those properties that compose the exterior nature of the rock, such as color and texture, and those properties that make up the rock's internal nature, such as strength, resistance to waves, and toughness.

Field of study: Petrology

Principal terms

COEFFICIENT OF THERMAL EXPANSION: the linear expansion ratio (per unit length) for any particular material as the temperature is increased

COMPRESSIVE STRENGTH: the ability to withstand a pushing stress or pressure, usually given in pounds per square inch

DENSITY: mass per unit volume

ELASTICITY: the maximum stress that can be sustained without suffering permanent deformation

HARDNESS: the resistance to abrasion or surface deformation

SHEAR, OR SHEARING, STRENGTH: the ability to withstand a lateral or tangential stress

TOUGHNESS: the degree of resistance to fragmentation or resistance to plastic deformation

Summary

Physical properties of rocks are important not only to geologists and geophysicists but also to construction engineers, technicians, architects, builders, and highway planners. In fact, rock properties affect everyone. Major physical properties include hardness, toughness, density, coefficient of thermal expansion, compressive strength, shear strength, tensional and bending (transverse) strength, elasticity, absorption rate of fluids, electrical resistivity, ability to propagate waves, rates of weathering, chemical activity, spacing between fractures, response to freeze-thaw tests, color, texture, radioactivity, and melting point.

Hardness, or resistance to abrasion, can be found by using the Mohs scale, which was derived by Friedrich Mohs in 1822 to measure relative resistance to abrasion in minerals and rocks. Any substance will be able to scratch another substance softer than or as hard as itself. Ten standard minerals are used: Talc is the softest and designated as hardness equal to 1, and diamond is the hardest and designated as hardness equal to 10. Rock hardness is of particular importance to those people who use rocks or rock products in building façades, monuments, tombstones, patios, and other structures. A variety of tests are used to determine resistance to abrasion over a time interval.

Toughness differs from hardness. The property is defined as the degree of resistance to brittleness or plastic deformation of a particular substance. Toughness is determined by a test using repeated impacts of a heavy object. A hammer is dropped from a specified height upon a sample, and this height is continually increased. The height of the fall in centimeters upon breakage is then defined as toughness. An example of a very tough material is jade, whereas rock salt is very brittle. The so-called French coefficient of wear, which is 40 divided by percentage wear in a test known as the Deval test, is also used to measure toughness. Toughness is a very important property in rocks that are used in building roads and airstrips, which are subject to repeated stresses.

The density of a rock is its mass or weight divided by gravitational acceleration, divided by unit volume (amount of space). More frequently, the rock is compared with an equal volume of water that has a density of almost exactly 1 gram per cubic centimeter. This number, which is dimensionless, is known as specific gravity. There is a considerable variation in specific gravity among rock types and even within them. Most limestone and dolomite rocks range from about 2.2 to 2.7 in specific gravity, whereas basalt and traprock are considerably heavier, at 2.8 to 3.0. Density is not synonymous with strength and toughness. Some low-density materials are strong, and some very dense materials cut by fractures are weak.

The coefficient of thermal expansion measures expansion along a line through the rock with increase in temperature. It is also termed linear expansion. The property is expressed as a ratio of change in length divided by unit length times change in temperature. Typical values for common rocks used in crushed stone or building stone range from about 4×10^{-6} (four millionths) to 12×10^{-6} (twelve millionths). Knowledge of the coefficient of thermal expansion is very important to engineers who design structures such as highways and bridges. On bridges, expansion-contraction joints are commonly used in consideration of this property.

Compressive strength, or bulk modulus, is measured on the basis of the highest pushing pressure or stress a rock can withstand per square unit, usually measured in pounds per square inch (psi). This strength is generally greater than transverse, tensional, or shearing strength. Limestones average about 15,000 psi, granites about 25,000 psi, quartzite about 30,000 psi, and basalts and traprock about 48,000 psi. This strength will vary considerably from rock to rock of the same composition because of variation in structural properties, as most rocks have fractures and voids and differ in grain size and shape.

THE MOHS HARDNESS SCALE

The Mohs scale measures the hardness of a rock or mineral in terms of the ease with which it can be scratched by other substances. The scale is based on the following minerals, in ascending order of hardness:

1. talc
2. gypsum
3. calcite
4. fluorite
5. apatite
6. orthoclase
7. quartz
8. topaz
9. corundum
10. diamond

For example, a rock or mineral that can be scratched by topaz but not by quartz has a Mohs hardness of 7-8. Supplementary tests can be made with such objects as a fingernail (2+), a copper penny (approximately 3), a knife blade (5+), window glass (5½), or a steel file (6½).

Shearing strength, or modulus of rigidity, is measured on the basis of the highest lateral stress in pounds a rock can withstand per square inch. An example of shearing stress is the stress exerted by a car sideswiping another car. Values for limestones average about 2,000 psi and granites about 3,000 psi.

Bending, or transverse, strength is defined as the strength of a slab loaded at the center and supported only by adjustable knife edges. This strength is determined by the "modulus of rupture," which is a function of the rupture load in pounds, the length of a slab, the width or breadth of a slab, and the thickness of a slab.

The so-called modulus of torsional rigidity, or elasticity (Young's modulus), is measured in pounds per square inch times 10^6 (millions) or dyne-centimeters^2. This property is a measure of how easily a material can return to its original shape when stressed. If stressed beyond the limits of this modulus, permanent changes in shape or deformation occur. This modulus is extremely variable, ranging typically from about 3×10^6 psi to 6×10^6 psi for limestone, to about 6×10^6 psi to 8×10^6 psi for granite. Elasticity refers to the ability of a material to spring back to its original shape without plastic deformation or rupture. Beyond a certain value of stress, rupture or flow will occur.

A measurement important to petroleum geologists, geophysicists, and engineers is electrical resistivity of rocks. The resistivity may be defined as the reciprocal of electrical conductivity. Resistivity is measured in ohm-centimeters; that is, electrical resistance along a centimeter's length. It will vary greatly depending on whether a rock is dry, contains fresh water in its pores or cracks, contains saline water, or contains organic compounds such as petroleum or natural gas. Igneous rocks such as basalts and traprock will vary from about 1×10^4 to 4×10^5 ohm-centimeters, and granites about 10^7 to 10^9 ohm-centimeters. Basalts have less resistivity and greater conductivity than granite because they generally contain more abundant amounts of dark, nonmetallic minerals than do granites. Limestones and sandstones commonly range from 10^3 to 10^5 ohm-centimeters, but those containing much saline water will have values much less, because water with dissolved salts is a good conductor of electricity. Metallic ore deposits will show very low resistivities because of the high conductivity of metals.

Absorption rate of fluids by rocks is another property often measured. A dry rock sample of known weight (or dry aggregate of rock chips) is soaked in water for twenty-four hours, dried under surface conditions, and weighed. The weight of water absorbed is given as a percentage of the dry weight of the rock. This property is extremely variable in rocks. For example, limestones may vary from 0.03 percent to 12 percent absorption rate. Very fine-grained limestones will commonly have a higher absorption rate than coarse-grained limestones.

Ability to transmit waves is an important rock property to construction engineers designing buildings, roads, bridges, and dams in earthquake-prone areas and to petroleum geologists and engineers. This property depends on the density and the strength of the rock, especially its compressibility and its rigidity (shear) modulus. Basically, there are two types of waves that result from earthquakes or man-made explosions: longitudinal and transverse waves. Longitudinal waves, similar to sound waves, create pushing and pulling effects on molecules while traveling along a straight line. Transverse waves move sideways as they travel. Compressive strength (pushing strength) is important in determining the speed of longitudinal waves, whereas modulus of rigidity, or shear, strength is more important in determining the speed of transverse waves. In general, the greater the strength, the greater the wave speed through a material; the greater the density, the less the wave speed through a material or rock under a given temperature and pressure. The density of darker igneous rocks is generally greater than that of lighter-colored igneous and sedimentary rocks. Also, the density of rocks increases toward the earth's center, but velocity is not always less, because strength tends also to increase. This factor may equal or even outweigh the density factor and produce a net increase in velocity or speed.

Soundness, or resistance to weathering, is important because rock materials are exposed to outside conditions, where temperature and moisture conditions may vary considerably. Chemicals may also attack the rock, and in Arctic or temperate climates, freezing and thawing may occur. In a climate where much freezing and thawing occur, the breakage and deterioration of cement and rock materials caused by this phenomenon will be noticed. Engine-driven vehicles and equipment would not last very long in winter if antifreeze were not added to the engine block and radiator. Examples of chemical weathering can be found in an old cemetery, where gravestone inscriptions of the same age on chemically resistant rocks such as quartzite are much clearer than those on chemically reactive rocks such as marble. Several tests may be run to denote relative resistance to chemical and physical weathering.

Color is one of the most interesting physical properties of rocks. It is determined by the colors of the component minerals, arrangement of these minerals, and weathering pigments. Color and patterns of colors are important in assessing the aesthetic qualities of ornamental and building stone. The unweathered colors must be pleasing to the eye, and staining detrimental to the appearance must not occur in appreciable amounts. The quality and beauty of many marbles are renowned when used in ornamental stone and sculpture. Building limestone must have a pleasing gray or tan color and not contain iron or manganese minerals, which would stain the rock brown or black during weathering. The pink-and-black or white-and-black mottling of attractive granites, diorites, and syenites is a familiar sight in business, school, university, and other public buildings. It is not, therefore, necessary or even always desirable that a color be uniform throughout.

Spacing between fractures is another structural property of rocks. For building stone, it is usually desirable to have massive rock or rock separated at intervals of tens of meters by layering or bedding planes or by vertical cracks or joints. Preferably, the distance between planes will be fairly constant and the cracks flat and even. At least some fractures are necessary, because one face of the block must be blasted, and some incipient cracks are needed to drive in the charges. Other faces may be sawed out by means of wire saws and silica grit. For some uses of building stone, such as roofing or facing slates and paving stones, smooth, closely spaced fractures are desirable, provided the stone is relatively strong and free from fractures in between.

Texture relates to grain size and grain shape, as well as fabric, which is the relationship of the grains to one another. Appearance, resistance to weathering, absorption rate characteristics, and strength are all related to texture.

Radioactivity, another physical-chemical property, is demonstrated by those rocks that spontaneously emit radiant energy (which results from the disintegration of unstable atomic nuclei). Scientists use the heat generated by absorption of the radiation emitted by these rocks to establish the temperature of the earth's interior and its thermal evolution. Radioactivity has been of interest since the early 1980's because of the health hazard posed by radon. Radon is a cancer-causing intermediate product of radioactive decay, and it can seep into basements and collect in unventilated homes. Proper designing or repair of basement interiors may be necessary to reduce rates of radon infiltration.

Rock melting point is a rock property that is generally of interest to petrologists, engineers on deep-drilling projects, and engineers designing furnaces or kilns that require rock or brick that has a melting point higher than that of the materials being processed in the kiln. Except in the case of dry, completely one-mineral rocks, melting will occur over a range of temperatures rather than at merely one temperature. This type of melting occurs for three reasons: several types of minerals, each with a different melting point when pure, usually occur in rocks; each mineral mutually affects the melting points of the other minerals, usually lowering the melting points and extending them over a range of temperatures; and water and other fluids may mutually lower the melting point of a rock.

Methods of Study

A great variety of tests are used to determine the degree or types of physical properties of rocks. Hardness may be determined by comparing a rock with known hardness points made from minerals of known hardness or from ceramic cones of various known hardnesses, either on the Mohs ten-point scale or on a more elaborate 1,000-point industrial scale. Another test is called the Dorry hardness test. Rock cones 2.5 centimeters in diameter are loaded with a total weight of about 1,000 or 1,250 grams, then subjected to abrasive action of fine quartz of known size fed upon a rotating cast-iron disk. The loss in weight of the core after 1,000 revolutions is used to compute value of hardness. One of the most widely used machines to test hardness is the Los Angeles Rattler (abrasion testing machine). It has a steel drum 70 centimeters in diameter and 50 centimeters long. The sample is inserted with a certain number of steel balls, and the drum is rotated five hundred revolutions at thirty revolutions per minute. The material is then sized into further grades, and the coarser sets of remaining fragments are subjected to one thousand further revolutions using various numbers of steel balls weighing about 5,000 grams. Another test, the Brinell test, determines hardness by pressing a small ball of hard material on the sample and measuring the size of the depression.

Toughness is tested on samples with a diameter of about 2.5 centimeters. The sample is held in a test cylinder on an anvil and subjected to the fall of a steel hammer or plunger weighing 2 kilograms. The first fall is from a 1-centimeter height and is progressively increased by 1 centimeter until breakage occurs. Height of fall is then expressed as toughness.

Specific gravity is measured by placing the sample in water. It is measured, dried to obtain dry weight, then immersed in water for twenty-four hours, surface dried, and weighed. Finally, it is weighed again in water. Loss of weight from dried weight initially is true specific gravity. The second step, involving the twenty-four-hour immersion, is necessary to ascertain that pore spaces in the sample are filled with water so that a true measure of buoyancy can be taken in the third step.

Compressive strength is measured on test rock cylinders 5 centimeters in diameter and 5.6 centimeters long. The sample is placed in a cylindrical casing in a press, and the press is lowered at a specified rate. Amount of compression in pounds per square inch at failure is the compressive strength.

The so-called soundness test measures resistance to chemical weathering or, alternatively, resistance to a major type of mechanical weathering, called ice or frost wedging. In the first case, the sample (usually about 1,000 grams) is covered with a saturated solution of sodium or magnesium sulfate for eighteen hours. Then it is oven-dried. The test is repeated usually five times. If the sample decomposes, it is called unsound; if it shows signs of decomposition but is still intact, it is called questionable. If it shows no or extremely minor signs of decomposition, it is called sound. In the case of freeze-thaw tests, different-sized materials are tested by freezing in water to an air temperature of –22 degrees Celsius, then thawed at room temperature (about 20 degrees Celsius). A first cycle lasts twenty-four hours and may be repeated. Damage to the sample is then noted.

Wave velocities, or speeds through rocks, are dependent on density and strength. These velocities can be determined theoretically or in the laboratory if density and strength can be measured. Conversely, compressive and shearing strength may be measured knowing rock density and velocity of wave transfer.

Geophysicists and engineering geologists may determine wave speeds through the earth's crust by setting up a seismograph station in the field. Explosives are set off, and travel time and velocity of waves are monitored by a system of geophones that relay the wave energies to the seismograph, which then measures the waves on a recording drum with a stylus. Characteristic wave reflections occur at certain depths. By knowing the times of explosive detonation and the time of arrival of the two kinds of earth body waves (primary and secondary waves) and by understanding that distance traveled for the two waves is the same, scientists can determined the wave speeds. This information is vital when drilling for oil or gas and digging deep foundations, especially in limestone sinkhole areas or in areas prone to earthquakes.

Context

Knowledge of particular rock physical properties is important to geologists, engineers, geophysicists, hydrologists, builders, architects, industrial city and regional planners, and the general public in a variety of ways. Rock properties are especially important in regard to the construction of buildings, dams, roads, airstrips, human-made lakes, and tunnels; drilling for petroleum and natural gas; mining and quarrying; monitoring earthquake hazards; measuring properties of aggregate and construction materials; determining rates of heat flow and mechanical strain in the earth; monitoring radioactive hazards; studying chemical makeup and physical structure of the earth's interior; and evaluating aesthetic properties of ornamental construction materials.

Within earthquake-prone areas (as in California) or areas affected by wind shear (as at the tops of high buildings), the nature and strength of material surrounding foundations, the foundations themselves, groundwater behavior and its effects on building strength, and the ability of building materials to withstand vibrations are of utmost importance. Engineering codes have been enacted within some of these areas so that earthquake-resistant or wind-shear-resistant buildings, dams, tunnels, and highways may be constructed. In the 1970's and 1980's, it was recognized that areas not frequently subjected to earthquakes in the past (as in middle Mississippi Valley) are still at risk. In these areas, earthquake codes must be enacted for further building sites. Siting of dams is a problem that affects geologists and engineers and everyone who lives downstream. Failure of several dams in the western and southeastern United States during the 1970's and 1980's has demonstrated the importance of careful siting. In one case, foundations for one side of a dam were situated in soluble materials because adequate geologic study had not been made.

Radioactive minerals in rocks or soils of certain localities have allowed radon to seep into basements through walls and cracks at potentially dangerous levels. In some cases, improved construction, prediction of settling (sedimentation), and proper design of basement interiors may reduce this hazard. Radon can be measured through the use of widely available kits.

Geologists, geophysicists, planners, and engineers must work closely together to ensure that building materials are durable and safe and that health, aesthetic, and economic needs are satisfied. The geologist and engineer have important functions in locating and quarrying building stones, such as granites and limestone, that are of durable and aesthetic quality for building fronts, trim, and ornamentation. The beauty of many downtown areas, parks, and public buildings is enhanced by suitable use of these materials. Road materials and aggregate must be durable in order to take the impact of heavy traffic, freeze-and-thaw processes, heat, and chemical stresses. Durability is important from the viewpoint of safety, convenience, and economics (durable materials can save taxpayers money in the long run and can reduce potential damage to vehicles).

Bibliography

Birch, Francis, J. F. Schairer, and H. Cecil Spicer. *Handbook of Physical Constants.* Geological Society of America Special Paper 36. Baltimore: Waverly Press, 1942. An excellent, if older, compilation of physical and chemical properties of rocks and minerals. Some prefatory discussion is given with most of the tables. Especially important are discussions and tables concerning rock strength and wave propagation. Contains an extensive section on chemical properties of rocks and minerals. Suitable for college-level students.

Carmichael, Robert S., et al. *Practical Handbook of Physical Properties of Rocks and Minerals*. Boca Raton, Fla.: CRC Press, 1989. A more recent compilation of major physical and chemical properties of rocks and minerals. Of special interest are the sections on radioactive and electrical properties of rocks. Most of the tables are preceded by discussion of the particular property illustrated. Suitable for college-level students, and indispensable for geologists and engineers.

Gillson, Joseph L. *The Carbonate Rocks in Industrial Minerals and Rocks*. New York: American Institute of Mining, Metallurgical, and Petroleum Engineers, 1960. This source is a classic one, discussing physical properties, occurrence, uses, resource amounts, and quarrying methods for limestones, dolomites, and marbles. Includes an excellent discussion of tests for physical properties of these rocks and of other rock types as well. Also includes a few tables of physical properties. Suitable for high school readers but excellent at all levels.

Severinghaus, Nelson. *Crushed Stone in Industrial Minerals and Rocks*. New York: American Institute of Mining, Metallurgical, and Petroleum Engineers, 1960. Discussion in this source centers on the physical properties of good crushed stone, occurrence, economics of use, uses and methods of quarrying and processing, and testing. Some physical properties are listed, as well as tests for determining hardness, toughness, and strength of crushed stone and aggregate. Suitable for high school readers but excellent at all levels.

David F. Hess

Cross-References

Minerals: Physical Properties, 415; Stress and Strain, 607.

Sand Dunes

Sand dunes form when wind or water deposit sand. Unlike many other geological processes that produce no visible change in a single human life span, the deposition of sand as dunes occurs on a short time line. Moving dunes overrun cropland and forests, supply and deny barrier islands and mainland beaches, and act as a barrier to prevent inland floods.

Field of study: Sedimentology

Principal terms

BARCHAN DUNE: a crescent-shaped sand dune of deserts and shorelines that lies transverse to the prevailing wind direction

CROSS-BEDDING: layers of rock or sand that lie at an angle to horizontal bedding or to the ground

ECOSYSTEM: in the environment, the unit of the interactions between living elements and nonliving factors in a sustaining system

GEOMORPHOLOGY: the branch of earth science interpreting surface land-forms

LONGITUDINAL DUNE: a long sand dune parallel to the prevailing wind

SALTATION: the hopping, jerking, and jumping movements of sand grains by wind or water action

SLIP FACE: the downwind or steep leeward front of a sand dune that continually stabilizes itself to the angle of repose of sand grains

STAR DUNE: a starfish-shaped dune with a central peak from which three or more arms radiate

Summary

Perhaps the most familiar sand dunes are those of the great flat deserts such as the Sahara and Kalahari in Africa, the Mojave in North America, and the Gobi in Asia. It is these mounds of sand that come to the minds of most people when they think of deserts, yet less than 12 percent of arid lands are covered by dunes. Moreover, dunes are not solely a product of desert dynamics; they also form on the beaches of lakes, rivers, and oceans, on barrier islands and river floodplains, under water, and on Mars. Wherever there is a ready supply of sand and an agent of transportation, such as wind or water, a mound of sand with a crest—the dune—will form.

Nevertheless, most of the sand dunes of the world are found in the giant deserts. The 900,000-square-mile Arabian Desert bordering the Red Sea and the 280,000-square-mile Rub' al-Khali, the empty Quarter, farther south appear to be ridge after endless ridge of dunes and blowing sand. Other arid regions, such as China's Takla Makan Desert (crossed by Marco Polo); Australia's Simpson Desert; Death Valley and the deserts of Colorado, Chihuahua, and Sonora in North and Central America; and India's Thar Desert contain sand dunes.

Less familiar are coastal dunes, along the Bay of Biscay and the Dutch Frisian Islands in Europe; Bermuda Island; the eastern United States barrier islands from which rise Miami, Florida, and Atlantic City; the Indiana Dunes on the south shore of Lake Michigan; the Oregon dunes and Clatsop Spit on the Pacific coast; the Namib coastal desert in southwestern Africa; and the Atacama Desert in coastal Peru. Nor are dunes limited to ocean shores; the lee side of channels along the Mississippi River and the beaches of other large rivers, such as the Koyukuk and Kobuk in Alaska, are windblown dunes. Shoreline dunes are confined to a belt just inland from the beach. Otherwise they are similar in characteristics and design to those of the desert. Where valleys of inland rivers, such as the Platte, Arkansas, and Missouri rivers, cross the Great Plains, dunes form. The stable dunes of the Sand Hills region of western Nebraska cover about 57,000 square kilometers (22,000 square miles). The Juniper Dunes area of southwestern Washington is a remnant of the largest known flood in earth history, the Spokane flood. Nor are sand dunes confined to the earth. In 1976, the Viking 1 Mars probe discovered crescent-shaped dunes and other desert phenomena on the plains of Memnonia on Mars. Sand dunes also form under water. Beneath tidal inlets, such as Cook Inlet in Alaska, are ripples and dunes formed as water swiftly flows into and out of the basin. Dunes require neither dry land nor hot air to form.

The sand dune is the mature form, the "adult," of the "infant" sand pile. Both shoreline and desert sand dunes begin with an obstacle—a large rock, pebbles, a small sand heap, or vegetation—around which sand builds up. On deserts, loose sand is supplied to the wind by the weathering of sand-rich bedrock. On beaches, longshore currents and waves supply sand. As the stream of windblown sand deflects and separates around an obstacle, a wind shadow zone forms between the forks. The wind shadow is an island of sheltered air in which the wind blows much more slowly than in adjacent streams. Zones of slower wind speed provide a resting place for some of the sand, which piles up into a drift that then presents a larger barrier to additional blowing sand.

With constant wind and sand supply, dunes continue to build in size. Some Sahara dunes may reach 100 meters (330 feet) in height, although dunes of 30 meters (100 feet) are more common. The typical dune shape is distinctive: a long, low-angled windward slope rising to a peak and a steeper leeward slope. Sand grains pushed over the crest eventually settle to a constant 30- to 35-degree angle on the downwind slope. This angle of repose is the maximum slope of the intersection of the leeward dune side and the ground. Sand moves either by surface creep or saltation. Sand grains driven forward by the wind hit resting sand grains that either are forced forward from the impact like a billiard ball (creep) or are launched into the air like a Ping-Pong ball (saltation). A saltating grain of sand can bounce about a meter high and with such force that it can move a grounded grain six times larger. Usually larger sand particles roll along the surface, while finer particles saltate and sandblast anything in their path. Like a ground fog, saltation layers may be dense enough to obscure the ground beneath yet form a sharp edge with clear air above. Saltating sand makes a dune appear to be "smoking."

Dunes are extremely mobile. During the process of dune growth, as the saltating grains spill over and pile up on the leeward side, or slip face, the entire dune moves downwind. The windward slope, then, is constantly eroding while the sand grains are deposited on the leeward slope. Dunes march forward at a barely detectable 3 meters (10 feet) per year, although dunes in areas of strong, constant winds can move more than 120 meters (350 feet) in one year, or about one-third of a meter (one foot) every day. On low-lying coasts with continuous supplies of sand, such as those in the Netherlands and southwestern Africa, dunes migrate inland unless anchored with stabilizing vegetation.

Different methods of classifying sand dunes have been developed. The most commonly used system distinguishes dunes by shape, sand supply, and orientation to the wind. Another classification scheme factors in the pres-

Most of the world's sand dunes are found in large deserts. *(Mojave National Park)*

ence of vegetation as well as crosswinds in determining dune form. Earth scientists usually categorize a sand dune as either a traverse dune, a longitudinal dune, or a star dune.

One type of traverse dune is the crescent-shaped barchan that lies transverse to—that is, across—the predominant wind direction. The tips of the crescent grow forward on the downwind side of the dune. The gently sloping leeward side is concave, and the steeper windward slip face is convex. When sand supply is limited and the wind direction is fairly uniform, a barchan dune will form. The barchan is much broader than it is tall; a maximum-height dune reaching 33 meters (100 feet) would be about 400 meters (1,200 feet) wide. The steady northeast trade winds supply sand to the barchan dunes of the North African desert. When sand supply is more abundant, traverse dunes form—not sharp-edged crescents but more amorphous, wavy ridges still perpendicular to the direction of the wind. Typical shoreline dunes, supplied with beach sand, belong in this category, as do some dunes of the Saudi Arabian desert.

Longitudinal dunes, such as the sinuous line of ridges, called seifs, in the Sahara Desert, lie parallel to the average direction of prevailing winds. When wind direction varies within a 90-degree pie-shaped wedge, the moderate supply of sand is forced from slightly different angles to form a straight-lined dune with a "peak and saddle" profile. An aerial survey of dunes in the world's most famous deserts would find primarily linear sand ridges. The southeast trade winds have built spectacular parallel dunes in the Great Sandy, Simpson, and Victoria deserts of Australia. In Egypt's Western Desert, continuous 70- to 100-meter-tall (200- to 300-foot) ridges, snaking 30-40 kilometers in length, are not uncommon. In southern Iran, 225-meter-tall longitudinal dunes are known.

The star dune, formed in a confined basin by variable winds blowing from radically different directions, is the most complex and the least studied of the three types. Star dunes have been described from the Namib Desert in southwestern Africa and the Mojave Desert in California, although only about 5 percent of documented dunes belong to this category. The typical dune form of this type resembles a starfish with a central peak and three or more radiating arms. Arms of star dunes in the northern Mojave correspond to the dominant wind directions over the seasons. As soon as any one wind direction stabilizes, the star dune grades into a less distinctive composite shape. Less stable, star dunes often are flanked by and grade into both barchan and linear dunes.

As sand dune deposits are buried to become part of the geologic record, their overall shape, such as crescent or linear, is disturbed. What remains in desert-formed dunes is the conspicuous slip face, the downwind slope, which appears as a distinctive cross-bedded wedge. Wedges group together and intersect one another from different directions, with no evidence of normal horizontal layering. The approximately 100-million-year-old Navajo sandstone of the Colorado Plateau shows spectacular cross-bedding. An older dune formation, the Permian Coconino sandstone, and the younger Lower Bunter sandstones near Birmingham, England, are also ancient sand dunes. Since beach dunes are less frequently preserved, few have been unequivocally identified in the geologic record. Sections of the St. Peter sandstone of the Mississippi river valley are thought to be sand dune deposits. Sand deposited in sand dunes has distinct characteristics that remain in the fossil record. Windblown quartz grains, limited to a narrow range of small sizes, are extremely round, with a frosted glass surface resulting from the constant collisions of saltation.

The vastness and inhospitability of sand dunes in the great deserts, such as the Arabian and Egyptian, evoked awed respect from early Arab traders and later European explorers, such as the famous T. E. Lawrence ("Lawrence of Arabia"), Charles Montague Doughty, and Carsten Niebuhr. Adding to the mysterious reputation of sand dunes are the incredible booming or roaring noises heard by travelers. Native tales attributed the drumlike rumbles to cantankerous spirits, such as Rul, jinni of the dunes, or to a legend of tolling bells buried beneath the sands. Crossing the dunes of the Takla Makan Desert in Mongolia, Marco Polo reported similar unexplained sounds: "Often you fancy you are listening to the strains of many instruments, especially drums and the clash of arms." The loud noises are attributed to the endless movement of sand on a dune's slip face. As the sand on the steep forward slope avalanches to readjust to its angle of repose, grains vibrate and produce sounds.

Humankind's primary interaction with the sand dunes of ocean beaches, saltwater estuaries, and, to a lesser extent, the deserts has been an attempt to restrain their constant migration. A line of dunes is a natural barrier against flooding from exceptionally high tides or violent storm-driven waves. Inland migrating dunes may, however, overpower inhabited beachfronts. The usual way to stabilize dunes and reduce migration is to plant vegetation such as sea oats, marram, and other grasses or to erect slatted fences. Vegetation must then be protected from the intense use of dune buggies and excessive visitation. In the oil-abundant Iranian desert, a petroleum product has been sprayed on the shifting dunes to provide a water-holding mulch in which shrubs and trees can root. Another factor important to long-term dune survival is an assured supply of sand. If dunes are cut off from ocean-supplied sand, they eventually wither. Sea walls and jetties erected by engineers attempting to reduce beach erosion also prevent waves from supplying sand to the dunes. Bay Ocean, an unsuccessful resort development on Tillamook Spit of the Oregon coast, erected jetties to harbor ships in Tillamook Bay. The jetty also, however, prevented sand, typically carried by longshore currents up and down the coast, from supplying the dunes. First, the dunes and the beach itself disappeared, followed, a few years later, by the spit, which was breached by a severe storm. One of the most famous programs of dune stabilization was the United States government's attempt to stabilize 360 kilometers (600 miles) of barrier island beaches off the

Atlantic coast. During the Depression of the 1930's, a sand fence was installed to encourage the development of exceptionally large, artificial dunes. Beach grasses, shrubs, and trees were then planted on the offshore slope. Forty years later, the man-made dunes on Hatteras Island continued to rise in height, but the original beach, 220 kilometers (650 feet) wide, had receded to less than 33 meters (100 feet), and the marshes behind the towering dunes were drying up. The unnaturally high dunes had prevented sand deposition on the beach and transport of seawater to the marsh at the back of the island. In 1973, the federal government changed its policy of dune stabilization to allow nature to take its course. Scientists estimate that Cape Hatteras should return to its previous size and condition by the year 2000.

Methods of Study

Specialists from several disciplines of earth science contribute to the interpretation of sand dunes. The geologist in the specialty of geomorphology, or the study of land features, in addition to tedious personal observations on foot, uses remote-sensing technologies, such as aerial photography and satellite imagery, to piece together a cohesive view. The sedimentologist, who studies processes of rock formation from sediments, uses data collected from both field observations and laboratory environments such as a flume or wind tunnel. Engineers have contributed inadvertently to understanding the behavior of sand dunes by a failure to recognize the dune as a phenomenon dependent on at times unidentified but essential processes.

Like other natural landscape phenomena, sand dunes were first studied by local inhabitants, explorers, and adventurers traveling overland. Earliest European explorations, such as the thirteenth century trek across a portion of the Mongolian Gobi and Marco Polo's journey across the high, frigid Talka Makan Desert, yielded general observations about the arid lands. Using the traditional transportation mode, the camel, Europeans touched the vast tracts of the Arabian Desert and the Rub' al-Khali, the Empty Quarter, in the late nineteenth and early twentieth centuries. Later, using automobiles, explorers and scientists such as Roy Chapman Andrews in Mongolia and Ralph Bagnold in Egypt's Western Desert contributed careful observations about dune shapes and sand movement. Dune classifications have been proposed by Bagnold from his studies in the Egyptian Sand Sea, by John T. Hack in his studies of Arizona dunes, by C. T. Madigan describing the sea of dunes in the immense Australian deserts, and by others. Field studies of the relationship between wind regime—that is, variation of wind direction and wind velocity—and dune morphology, or form, continue today. Where geologists can wet down a dune face, they use shallow trenches and take core samples to study airflow, sedimentary processes, and internal dune structure and deposits.

In the 1930's, Bagnold pioneered studies of sand transport by wind with experiments in the wind tunnels of the Imperial College of Science and Technology in London. Controlled laboratory experiments such as these provide data on the relationship between wind speed and sand grain size; sand grains of a specific size do not move until the wind reaches a certain speed. On one type of sand tested by Bagnold, wind speed reached a velocity of about 18 kilometers (11 miles) per hour before the grains began to roll. Models of sand grain transport and saltation are continually being updated by field studies and laboratory experiments. Sedimentologists use laboratory flume studies to provide data on the formation of dunes under water. Dunes in a water environment form and remain stable only at a specific stream velocity; dunes do not form if the velocity is too slow and will disappear at higher velocities. Flow of water and eddies around stream-bed ripples and dunes resembles airflow in wind-formed dunes.

Technological advances in remote sensing, the field of data collection that uses aerial photography and images from satellites, provide insights unavailable to earlier scientists. Geologists use color and infrared aerial photographs taken from multiple, overlapping flights of specialized airborne cameras to map sand dunes. Detailed maps showing airflow and sand-grain flow and ripple migration direction combine data gathered from aerial photographs and field studies. Photographic images of topographic forms such as sand dunes are produced by a type of sensing equipment called side-looking airborne radar. Computer-enhanced photographs taken over the same area on repeating dates by Landsat and National Oceanic and Atmospheric Administration (NOAA) satellites have revealed time-lapse views of dune migration, such as the westward advance of dunes of the African Namib Desert into the Atlantic Ocean. Space exploration has added to the technology of studying sand dunes. Experiments carried by the *Columbia* space shuttle revealed previously unknown characteristics of the Selima sand sheet in the Libyan desert, and the Viking 1 Mars probe provided close-ups of the Martian sand dunes, which greatly resemble those of the Egyptian Desert.

Geological and civil engineers have discovered previously little-understood features of sand dune migration and relationship to beach sand sources by attempting to stabilize dunes with vegetation and fences. The ocean-fronting beaches and continent-facing marshes of barrier islands on the eastern United States seaboard were reduced considerably by forty years of a federally appointed dune stabilization program. Relationships between the continuing existence of the barrier islands and sand dune mobility were understood only after "stabilized" islands such as Hatteras began to disappear. In many cases, the use of vegetation to anchor shoreline dunes and to inhibit their advance has been more successful. If road or dwelling construction or a natural increase of sand supply does not disturb established vegetation, the balance between sand and vegetation will hold the dune in place.

Context

Sand dunes do not exist only in the middle of vast, desolate, uninhabited reaches of deserts. They encroach on valuable cropland and forests in developing countries, and they march inland from the ocean's edge and spill into the houses of the rich. Restlessly, and matter-of-factly, sand dunes bury anything that lies before them as they advance downwind. Yet, sand dunes provide a barrier against flooding and house important ecosystems. Dunes are important as earth science laboratories in which to study natural processes and plant and animal communities found within their limits. Knowledge of sand transport, wind dynamics, and resulting land features gives scientists valuable tools with which to interpret the face of both the earth and Mars, with its distinctive dunes.

As the earth's population burgeons, careful planning for previously unused land becomes more important. The mechanics of sand dune mobility are important to the engineer and land-use planner, who try to satisfy the needs of more and more people. Understanding that dunes, planted with grasses and shrubs, can be stabilized provides more options for the development of land contiguous to sand dunes. Yet dune stabilization exacts a price if adjacent systems are not considered. The fence-stabilized dunes of several barrier islands on the eastern United States coast disturbed the balance of the island itself as the ocean-facing beach in front of the dunes and the fertile wetlands behind the dunes disappeared. Other stabilization programs using vegetation have been more successful as long as dune users understand the importance of protecting the growth.

Sand dunes preserved in the geological record provide valuable clues to the climates of the geological past. Geologists use the positions of cross-bedded sandstone in buried dune deposits to reconstruct wind directions of ancient weather systems and to understand climatic fluctuations. Long after dunes have been buried and eroded, they continue to affect the drainage pattern of an area. Numerous, parallel streams flow southeast across thousands of square kilometers of the northern Great Plains. Geologists believe that the streams have cut between long, parallel dunes that record the direction of the prevailing winds of the time. Barrier islands are thought to be remnant sand dunes from the last ice age, in which rising seas and later winds worked through beached piles of mud and sand.

One of the realizations of the late twentieth century is that no land, even if completely surrounded by water or desert, is an island. Earth houses interconnected and interdependent systems. What is desert dune today may have been lush tropical forest yesterday; what is fertile agricultural ground in one generation may become windblown dunes in the next. Indeed, in the span of a lifetime, arid giants such as the Sahara are expanding their boundaries as, patch by patch, sand dunes encroach on rangeland and forest. Arresting this process of desertification is one of the challenges of modern society. Increasing knowledge about the man-made causes of desertification and accompanying dunes provides insight into the repercussions of disturbing the careful balance of earth processes.

Bibliography

Anderson, Robert S. "The Attraction of Sand Dunes." *Nature* 379, no. 6560 (January 4, 1996): 24-26.

Bascom, Willard. *Waves and Beaches: The Dynamics of the Ocean Surface.* Rev. ed. Garden City, N.Y.: Doubleday, 1980. A classic text by a well-recognized oceanographer, this book gives many specific examples of ocean and onshore processes. Written in easily understood style, with diagrams.

Dunbar, Carl O., and John Rodgers. *Principles of Stratigraphy.* New York: John Wiley & Sons, 1966. A basic text in stratigraphy, or the study of layered rocks, written for the serious student of geology. Excellent sections on recognition of ancient dunes and processes of sedimentation.

Holmes, Arthur. *Principles of Physical Geology.* 3d ed. New York: Van Nostrand Reinhold, 1978. A revised version of an often-used general geology text that details processes of land formation and topography as well as tectonic and crustal phenomena. Many examples are provided.

Page, Jake. *Arid Lands.* Alexandria, Va.: Time-Life Books, 1984. A well-written and photographed volume of a series of earth science topics (called the Planet Earth series) for the general reader. This issue describes and locates deserts of the world and adjacent lands and relates the dynamic processes of their formation.

Press, Frank, and Raymond Siever. *Earth.* 4th ed. San Francisco: W. H. Freeman, 1986. If a student newly introduced to the field of geology were to buy a single text, this book should be the one. A geology text for the beginning college student, it covers with thoroughness major topics and includes methods of study as well as current information.

Sackett, Russell. *The Edge of the Sea.* Alexandria, Va.: Time-Life Books, 1983. Like other volumes in this series for the general earth science reader (the Planet Earth series), this issue presents easily read information accompanied by excellent graphics. Particularly pertinent is the section on barrier islands and dune stabilization.

Thornbury, William D. *Principles of Geomorphology.* 2d ed. New York: John Wiley & Sons, 1969. This frequently used text for college-level geomorphology courses gives thorough coverage to landscape features of all kinds. Good section on applied geomorphology and geological engineering.

Cory Samia

Cross-References

Coastal Processes and Beaches, 71.

Sedimentary Rocks

Because sedimentary rocks are formed by several different processes—for example, precipitation, crystallization, and compaction—no single classification scheme is applicable to all of them. Used in various combinations, the main elements of classification are mode of origin, mineralogy, the size of the individual mineral grains that make up the rock, and the origin of these grains.

Field of study: Petrology

Principal terms

ARKOSE: a sandstone in which more than 10 percent of the grains are feldspar or feldspathic rock fragments; also called feldspathic arenite

CARBONATE ROCK: a sedimentary rock composed of grains of calcite (calcium carbonate) or dolomite (calcium magnesium carbonate)

CLASTIC ROCK: a sedimentary rock composed of broken fragments of minerals and rocks; typically a sandstone

CLAYSTONE: a clastic sedimentary rock composed of clay-sized mineral fragments

GRAYWACKE: a sandstone in which more than 10 percent of the grains are mica or micaceous rock fragments; also called lithic arenite

LIMESTONE: a carbonate sedimentary rock composed of calcite, commonly in the form of shell fragments or other aggregates of small calcite grains

ORTHOQUARTZITE: a sandstone in which more than 90 percent of the grains are quartz

SANDSTONE: a clastic sedimentary rock composed of sand-sized mineral or rock fragments

Summary

Because there are several very different processes that lead to the formation of sedimentary rocks, no single classification scheme is suitable to all sedimentary rocks. The main elements of classification, however, are mode of origin, mineralogy, the size of the individual mineral grains making up the rock, and the origin of the individual grains. These elements are used in various combinations to categorize several major groups of sedimentary rocks.

One of the major groups is the evaporites. All natural waters contain some dissolved solids that will precipitate when the water evaporates. The crust that forms in a teakettle that has been used for a long time is an example. Seawater contains about 33 parts per thousand dissolved solids and is the major source of the sedimentary rocks classified as evaporites. When a body of seawater in an area with low rainfall is cut off from the sea, as when a sandbar builds up across the mouth of a bay, the trapped seawater tends to evaporate. During this evaporation, several minerals are precipitated in a predictable order. The first to precipitate is the mineral gypsum, hydrated calcium sulfate. If the water is very hot, the precipitating calcium sulfate will not be hydrated, and the mineral anhydrite will form. Further evaporation will cause the precipitation of halite (sodium chloride, or ordinary table salt), and still further evaporation will lead to the precipitation of a complex series of potassium and magnesium salts.

In some environments, most commonly closed depressions in desert areas (the Dead Sea for example), fresh water evaporates to produce evaporite minerals that are quite different from the minerals produced by the evaporation of seawater. The natron (hydrated sodium carbonate) that was used by the Egyptians in embalming mummies and the borax that originally made Death Valley, California, famous are freshwater evaporite minerals.

From the standpoint of classification, the rocks of evaporative origin—that is, masses of individual crystals of minerals produced by evaporation of seawater or fresh water—are not usually given distinctive names. A fist-sized piece of evaporative sedimentary rock composed of gypsum is normally called gypsum. When it is necessary to indicate clearly that a rock, rather than a mineral, is being mentioned, the term "rock gypsum" is used. An exception is a rock composed entirely of crystals of halite, which is almost invariably referred to as rock salt.

A second major group of sedimentary rocks is the clastic rocks. "Clastic" comes from the Latin for "broken"; the individual grains of clastic rocks are the product of the mechanical and chemical breakdown, or weathering, of older rocks. Beach sands are composed of such grains. They are composed of residue from weathering of granites and many other kinds of igneous, sedimentary, and metamorphic rocks. Common soil, or mud, also is a residue of the weathering of rocks. Mud differs from beach sand primarily in its content of fine-grained clay minerals. Clay minerals are similar to the mineral mica, but the individual grains are very small, by definition less than 0.004 millimeter in size. When subject to prolonged attack by water and the atmosphere, many minerals that are common in igneous and metamorphic rocks—principally feldspars—are changed into micalike clay minerals. Between beach sands and the clay-sized component of muds are an intermediate size of mineral grains called silts. Silts range in size, again by definition, from 0.0625 millimeter to 0.004 millimeter. Muds are mixtures of silt-sized and clay-sized mineral grains. Grains larger than 2 millimeters in diameter are classified as granules, pebbles, and boulders with increasing size.

The primary basis for classification of clastic sedimentary rocks is grain size. Coarse-grained rocks composed of pebbles and granules are called conglomerates, rocks composed of sand-sized grains are called sandstones, rocks composed of mud-sized materials are mudstones, and rocks composed only of clay-sized grains are called claystones. Mudstones and claystones that split readily along flat planes, the bedding planes, are referred to as shales.

Sandstones are further classified by mineral composition. In most sandstones, more than 90 percent of the sand grains are quartz, but some sandstones contain appreciable amounts of feldspar grains, volcanic rock fragments, mica, and micaceous rock fragments. These grains are the basis for further classification. Sandstones that are nearly pure quartz sand (more than 90 percent quartz grains, by one common definition) are called orthoquartzites, or quartz arenites ("arenite" is from the Latin for "sand"). Sandstones containing more than 10 percent feldspar grains and volcanic rock fragments are called arkoses or feldspathic arenites, and sandstones with more than 10 percent mica flakes and micaceous rock fragments are called graywackes or lithic arenites.

Limestones are another major group of sedimentary rocks. Limestones are composed predominantly of the mineral calcite (calcium carbonate), although some limestones may include some clastic material, typically quartz sand or clay. A similar group of rocks is the dolomites. Dolomites consist predominantly of the mineral dolomite (calcium-magnesium carbonate) and appear invariably, or nearly invariably, to form by chemical alteration of preexisting limestone. Therefore, many workers prefer to lump the limestones and dolomites together in a rock group named carbonates.

The greater part of the calcite in limestones is secreted by marine organisms that make their shells from the mineral; clams and oysters are good examples of such organisms. Once these organisms die, their shells are washed about by waves and currents and are broken and abraded into fragments. The fragments may range in size from pebbles to mud, and most limestones are composed of this biogenic detritus.

Two common components of limestones appear to be of inorganic origin. Oölites are round grains, composed of very small crystals of calcite, that have a superficial resemblance to fish eggs. A shell fragment or quartz grain, the nucleus, at the center of the oölite, is surrounded by a coating of fine-grained calcite crystals layered like tree rings. It appears that oölites grow by inorganic deposition of calcite, directly from seawater, on the surface of the nucleus. Field observation of modern oölites suggests that they form only on sea bottoms that are shallow and are periodically agitated by strong waves or currents. It also appears that mud-sized calcite crystals precipitate directly from seawater in some circumstances.

Finally, small organisms, principally marine worms, ingest calcite mud, to extract whatever useful organic matter it may contain, and excrete it as fecal pellets. The fecal pellets, held together by mucus from the gut of the organism, survive if the bottom currents are not too strong.

Sedimentary Rock Types

Type	Defintion	Subcategories	Explanation	Examples
Clastic	Rocks that consist of fragments of other rocks	Conglomerates	Grains are boulder-, cobble-, or gravel-sized	Breccia
		Sandstone	Grains are sand-sized	Quartz sandstone, arkose, graywacke
		Mudstone	Grains are silt- or clay-sized	Mudstone, shale
Precipitates	Chemically precipitated or replaced; inorganic in orgin	Evaporites	Solids have precipitated after evaporation of water in which they were dissolved	Salt, gypsum, anhydrite, borax, potash
		Carbonates	Compounds of calcium or magnesium	Calcite (inorganic limestone), dolomite
		Siliceous	Chemically precipitated silicas	Chert: flint, jasper
Organic (biochemical)	Remains of plant or animal organisms	—	—	Coal, organic limestone

Most limestones consist of aggregates of the materials described above. One classification, originally introduced by Robert L. Folk in 1959 and the most widely used, is based on the nature of the aggregates and the material that occurs between the grains and cements them together. Mud-sized calcium carbonate tends to accumulate in quiet waters, and a limestone consisting of only mud-sized material is called micrite. Sand-sized calcite grains are deposited in areas with stronger currents, generated by waves, winds, or tides. After final deposition, open spaces between sand-sized grains are often filled by inorganic calcite cement, called spar. In Folk's classification, the abbreviated name of the sand-sized material followed by the name of the material between the sand-sized grains is the rock name, with the traditional rock-name ending "-ite." Typical examples are oösparite, pelmicrite, and biosparite (where "bio-" refers to shell fragments).

A fourth major group of sedimentary rocks is the chemical rocks. These rocks are divided into two subgroups: chemical precipitates and chemical replacements. Chemical precipitates are sediments that accumulate directly on the sea bottom as a result of chemical reactions that do not involve evaporation. Deposits of iron minerals (most typically the iron oxide mineral hematite) and phosphate minerals (most typically calcium phosphate, or the mineral apatite) are the most common, and economically important, examples. The process of formation of these rocks is not well understood, but research suggests that in most cases bacteria are involved in producing the proper chemical environment for formation of these important, although rather rare, deposits.

The second subgroup is the chemical replacements. In some cases, the original sediment is dissolved and a new mineral takes its place. Typical examples are the solution of calcite and its replacement by dolomite and the solution of calcite and its replacement by fine-grained quartz. The replacement of calcite by fine-grained quartz in limestones is especially common. The replacement product, the very fine-grained quartz replacement, is called chert, but most people are more familiar with the popular term "flint."

Methods of Study

More than twenty thousand minerals have been identified, but, perhaps fortunately, only twenty-two of them are common in rocks at the earth's surface; these are easily identified in the field by observation and simple tests of physical properties. Even most fine-grained rocks can be identified in the field with the aid of a twelve-power hand lens. Gypsum, for example, is easily identified by its satiny sheen and the fact that it can be scratched by a fingernail. Calcite and dolomite have very similar appearances and physical properties, but a drop of dilute hydrochloric acid will cause calcite, but not dolomite, to effervesce, or fizz. A pocket-sized dropper bottle of dilute hydrochloric acid is standard equipment for the sedimentary geologist.

More detailed studies are done on samples brought to the laboratory. A very common method is the study of thin sections of rocks. To prepare a thin section, a thin slice, about a centimeter thick, is cut from the sample with a diamond saw. The slice is glued to a glass slide and further thinned with a diamond saw. The slice is thinned still further by grinding with progressively finer abrasives. When the thickness has been reduced to about 0.03 millimeter, a thin cover glass is glued to the top of the slice, and the thin section is complete.

Thin sections are studied with a microscope, usually a petrographic microscope that has a polarized light source. The effect of passage of the polarized light through individual mineral crystals can be analyzed and much information about the arrangement of the atoms in the crystals obtained. For example, passage of the light through halite does not affect the planar vibration of the polarized light, but when it passes through quartz, it is forced to vibrate in two planes that are perpendicular to each other and parallel to the two crystallographic axis directions of quartz. Therefore, even microscopic-sized grains of halite and quartz are easily distinguished.

Clay minerals, which are too small to be studied effectively by optical methods, are commonly studied by X-ray diffraction. X rays have very short wavelengths and can be diffracted by the regularly arranged planes of atoms in a crystal in the same way that light is diffracted by a diffraction grating. The sample to be analyzed is irradiated with X rays of a single wavelength at steadily varying angles of incidence. The angles at which diffraction occurs represent planes of atoms at different spacings. If the clay mineral kaolin, for example, is present in the sample, a strong diffraction will occur at an angle that corresponds to an atomic plane spacing of 7.2 par @215 10^{-10} meters (usually expressed as 7.2 angstroms), and weaker diffractions will occur at angles corresponding to several other spacings that are characteristic of the mineral.

Scanning electron microscopy is a powerful tool for the study of all types of sedimentary rocks but is especially useful for the very fine-grained varieties. Photographic images at 50,000 magnifications are routinely obtained, revealing remarkable details of individual grains and the openings between grains. In addition, semiquantitative chemical analyses of individual grains for sodium and elements heavier than sodium can be made.

Very early in the nineteenth century, James Hutton laid down the basis for the modern science of geology when he pointed out that the processes responsible for all features of the earth are still in operation and can be observed today. This concept is commonly expressed by the saying "The present is the key to the past." Much of what is known of sedimentary rocks has been learned by studying modern sediments. Some examples are the Bahama Banks and the Great Barrier Reef of Australia, where limestones are now being formed, and the Mississippi Delta and seacoasts the world over, where sandstones and mudstones are being deposited.

Context

Sedimentary rocks cover 75 percent of the land surface of the earth and are of great economic importance. People everywhere are surrounded by the products of sedimentary rocks. The coating that makes magazine and some book pages slick is the sedimentary clay mineral kaolin. The pigment in white paint is titanium oxide, which comes from some types of sandstone. The brick used in construction is made from mudstone, and the cement that binds concrete together is made from limestone mixed with mudstone. Glass is made from sandstones mixed with the sodium-rich variety of the mineral feldspar. The list can be made very long.

Less familiar, but probably more dramatic, is the fact that petroleum and natural gas are generated in organic-rich mudstones and accumulate, in economically recoverable concentrations, in the pores between the grains of sandstones and limestones. In addition, coal, very important in electric power generation and steelmaking, is a sedimentary rock.

Classification of sedimentary rocks has led to the recognition of predictable associations of sedimentary rock types with particular geologic conditions. Sandstones that are derived from, and deposited near to, mountains with granitic cores—the Front Ranges of the Rocky Mountains, for example—most commonly contain more than 10 percent feldspar, mostly of the potassium-rich variety, and are classified as arkoses. Sandstones derived from mountains with metamorphic rock cores, such as the Appalachian Mountains, contain more than 10 percent mica and are classified as graywackes, or lithic arenites. Sandstones deposited far from any mountain chain normally are nearly pure quartz, in some cases more than 99.9 percent quartz. For quality-control purposes, glass manufacturers prefer sand that is very nearly pure quartz, and the sedimentary geologist who specializes in industrial minerals will begin the search for glass sands far from any large mountain chain.

The organic-matter content of muds that lie on the sea bottom for long periods of time tends to be destroyed by scavenging organisms. Sedimentary geologists who specialize in petroleum exploration know that one of the requirements for the generation of petroleum is a mudstone with a high organic-matter content. They look for mudstones that were buried by new mud shortly after deposition—that is, mudstones that had a high rate of sedimentation. As a general rule, such mudstones will be associated with arkoses or graywackes rather than with nearly pure quartz sandstones.

The classification of sedimentary rocks serves first as a tool for rapid communication. When one geologist tells another that arkose occurs in a given area, he has communicated information about the grain size of the rock (sand-sized) and its mineral composition (more than 10 percent of the grains are feldspar, mostly of the potassium-rich variety). Classification is based on factors that have meaning in terms of the geologic background of the rocks and is, in part, a method of organizing accumulated knowledge of the rocks. Finally, classification is a factor in developing strategies for exploration for vital mineral resources.

Bibliography

Blatt, Harvey. *Sedimentary Petrology.* San Francisco: W. H. Freeman, 1982. A very well illustrated work, offering complete coverage of the subject of sedimentary rocks. Intended to be a college-level textbook but perfectly accessible to the interested high school student or layperson. The final two chapters cover the design and conduct of research projects.

Blatt, Harvey, Gerard Middleton, and Raymond Murray. *Origin of Sedimentary Rocks.* 2d ed. Englewood Cliffs, N.J.: Prentice-Hall, 1980. A graduate-level university textbook, 782 pages long, useful to one who wants detailed information on a particular aspect of the subject. Very well illustrated and full of references to research papers. Undoubtedly one of the best books in the field.

Carver, Robert E., ed. *Procedures in Sedimentary Petrology.* New York: John Wiley & Sons, 1971. Covers the methods commonly used to study sedimentary rocks, including the methods of grain-size analysis and mineralogical analysis that are the basis of their classification. Where appropriate, a chapter on obtaining the data is followed by a chapter on mathematical or statistical analysis of the data. The mathematics involved is not difficult.

Ehlers, E. G., and Harvey Blatt. *Petrology: Igneous, Sedimentary, and Metamorphic.* San Francisco: W. H. Freeman, 1982. Stated to be intended for the college sophomore or junior but perfectly suited to the interested high school student or general reader. The section on sedimentary rocks is clear and very well illustrated. An excellent introduction to the subject of sedimentary rocks in general and classification in particular.

Fairbridge, Rhodes W., and Joanne Burgeois, eds. *The Encyclopedia of Sedimentology.* New York: Van Nostrand Reinhold, 1982. An excellent reference for the reader who needs more information on any specific aspect of the classification of sedimentary rocks. For example, the information on clay-pebble conglomerates would not be easily found in any other reference. Extensively and usefully cross-referenced.

Ham, W. E., ed. *Classification of Carbonate Rocks: A Symposium.* Tulsa, Okla.: American Association of Petroleum Geologists, 1962. A classic work on the title subject by the originators of various schemes of classification of limestones and dolomites, written at a time when there was still much discussion of what was the most appropriate classification. Required reading for all students of carbonate rock classification.

Hatch, F. H., R. H. Rastall, and J. T. Greensmith. *Petrology of the Sedimentary Rocks.* Rev. 4th ed. London: Thomas Murby, 1965. An older reference from the time of widespread fascination with classification. Therefore, it contains details of composition and classification of some rock groups—for example, the carbonaceous group (coal and related rocks)—that cannot be found elsewhere. The illustrations are mostly line drawings of thin sections (a lost art), but these are very informative.

Pettijohn, F. J., P. E. Potter, and R. Siever. *Sand and Sandstone.* New York: Springer-Verlag, 1973. A thorough, well-illustrated, and clearly written treatment of the subject. Some preliminary study of the general subject is advised, because a general familiarity with the geology and mineralogy of sandstones is required.

Tucker, M. E. *Sedimentary Petrology: An Introduction.* New York: John Wiley & Sons, 1981. A clearly written and well-illustrated introductory text aimed at British college undergraduates, who learn principally through independent study. The greater part of the text is devoted to classification in terms understandable to the general reader.

Williams, Howel, F. J. Turner, and C. M. Gilbert. *Petrography: An Introduction to the Study of Rocks in Thin Sections.* 2d ed. San Francisco: W. H. Freeman, 1982. An introductory college text that is yet accessible to younger students and to general readers who have some familiarity with the literature. As in the work by Hatch, Rastall, and Greensmith (above), the sedimentary rock illustrations are primarily line drawings of thin sections, but they make the point very well.

Robert E. Carver

Cross-References

Stratigraphic Correlation, 593; Weathering and Erosion, 701.

Shield Volcanoes

Shield volcanoes are the products of eruptions of low-viscosity materials. They have been identified on five solar system bodies. The most active volcanoes on earth are shields.

Field of study: Volcanology

Principal terms

CORONAE: ring structures on Venus consisting of alternating concentric ridges and valleys, higher than the external terrain; possibly volcanic in origin

HOT SPOTS: areas of anomalously hot mantle

IGNIMBRITE: rock formed by widespread deposition and consolidation of block, pumice, and ash flows

MONOGENETIC: pertaining to an eruption in which a single vent is used but once

PATERAE: inverted, saucer-shaped features considered to be of volcanic origin

POLYGENETIC: pertaining to volcanism from several physically distinct vents or repeated eruptions from a single vent punctuated by long periods of quiescence

PYROCLASTIC MATERIALS: broken rock formed by volcanic explosion or aerial expulsion from a volcanic vent

THOLUS: an inverted, bowl-shaped feature considered to be of volcanic origin

Summary

The shield volcano is so called because of its similarity to the ornate shields of the ancient Nordic warriors. More specifically, the name is derived from the Icelandic word *dyngja* (shield). The typical shield volcano is circular to ellipsoidal when viewed from above; its "ornamentation" can take the form of superposed cinder cones, radiating lava flows, a summit or flanking crater or craters, and rift valleys. Viewed from the side, the shield can take forms ranging from inverted saucers to inverted mixing bowls. In this respect, shields differ from the familiar volcano shape represented by the symmetrical, conical Mount Fuji in Japan.

Most shields are basaltic in composition and formed primarily of lavas (more than 98 percent). Exceptions are found in shields which have lavas that are richer in silica and of andesitic composition, such as Hayli Gub in Ethiopia, and those which are poorer in silica, such as the carbonatitic volcanoes of the East African Rift zone. There are also shields composed primarily of pyroclastic materials, although frequently they are separately classified.

One of the earlier attempts to produce a systematic classification of volcanoes was made by Alfred Rittmann in 1936. This scheme did not achieve wide recognition until the second German edition of *Vulkane und ihre Tätigkeit* (1960; *Volcanoes and Their Activity*, 1962) was translated into English. In this text, Rittmann recognized a Hawaiian type and an Icelandic type of shield volcano. A similar subdivision was recognized by Sir Charles Cotton in his text *Volcanoes as Landscape Forms* (1944). In the most influential of the English geomorphology textbooks of the twentieth century, *Principles of Physical Geology* (1944), Arthur Holmes classified the shield volcanoes as "domes of external growth" to distinguish them from the domes of viscous lava built by the addition of material to the interior.

With the acquisition of images of volcanoes on other bodies in the solar system, there began a new phase in volcano classification. These new images provided views of volcanoes produced on bodies with different gravities, atmospheres, compositions, and other important parameters. The volcanologist now needed a database with which to make objective comparisons of the morphologies of the volcanoes on the different bodies. One such attempt was made by James Whitford-Stark in the book *Volcanoes of the Earth, Moon, and Mars* (1975). This classification was made on the basis of the basal diameter versus the height of the shield volcanoes and resulted in the identification of four classes of terrestrial shield volcanoes: Hawaiian, Icelandic, Galápagos, and scutulum. (The last is from the Latin *scutum*, or "shield.") Lunar domes were shown to be similar to Icelandic shields, whereas Martian volcanoes had dimensions exceeding those of most terrestrial volcanoes. A more detailed analysis was undertaken by Richard Pike and Gary Clow, who employed a statistical technique called principal component analysis. They first distinguished between polygenetic and monogenetic shields and then erected subgroups on the basis of size, shape, composition, and location. The monogenetic shields, shields produced by only one eruption phase, included the large Icelandic shield, a smaller steep shield, and a low shield. The polygenetic shields included a Hawaiian type, a Galápagos type, and a locally varied type, all of which are characterized by tholeiitic basalts; an oceanic and continental type dominated by alkali-rich basalts; and a group represented by seamounts, or submarine volcanoes. The more silicic of the shields were grouped into alkalic and calcalkalic ash-flow plains categories. Pike and Clow also recognized three distinct groups of Martian volcanoes and a group of lunar domes. Volcanic landforms have since been recognized on the Jovian satellite Io and on Venus. Unfortunately, there is a paucity of topographic data for these features, so objective comparisons cannot yet be made.

In reality, shield volcanoes do not fall into isolated morphological groups; rather, they are transitional. Furthermore, the processes of erosion and deposition serve to modify rapidly the morphologies of terrestrial volcanoes. The figure illustrates a classification of shields based on volume and slope.

Small low and small intermediate-slope shields are monogenetic and variously described as scutulum-type shields, lava cones, or low shields. These shields are abundant in the Snake River plain volcanic field of the western United States and are represented by the volcano Mauna Iki, which erupted between 1919 and 1920 on the flanks of Kilauea. Typically these shields have other slopes of about 0.5 degree and steepen to about 5 degrees near the summit. This slope change has been attributed to a change from extensive, thin, fluid, pahoehoe flows early in the eruption to more viscous, aa flows during the waning eruption. Many have summit craters formed by collapse, and a few have spatter ramparts around the vent. In the Snake River plain, the shields are aligned along fractures and probably represent initial fissure-type eruptions that subsequently contracted to point source eruptions.

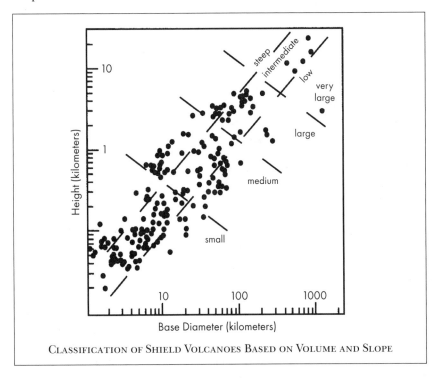

CLASSIFICATION OF SHIELD VOLCANOES BASED ON VOLUME AND SLOPE

Small steep shields are typified by Mauna Ulu, which was produced in an eruption lasting from 1969 to 1974 and punctuated by a three-and-a-half-month period of quiescence in late 1971 and early 1972. The early phase was marked by lava fountains reaching heights of up to 540 meters, whereas in the second phase, the fountains rarely exceeded 80 meters. Nearly 350×10^6 cubic meters of lava contributed to the final structure. The eruption was characterized by a summit lava lake. Sustained overflows of this lake led to the production of tube-fed lava flows which traveled up to 12 kilometers from the vent. Short-duration overflows resulted in growth of the shield structure. The final structure rose 121 meters above the pre-1969 base and had a basal diameter of slightly more than 1 kilometer. The smaller Icelandic and seamount shields also fall within this category.

Medium low shields are polygenetic constructs represented by calcalkalic ash-flow plains and the smaller of the alkalic ash-flow plains. The calcalkalic ash-flow plains are composed primarily of rhyolitic or dacitic ignimbrites, rather than lavas, surrounding a caldera ranging from 7 to about 60 kilometers in diameter. Toward the caldera, lavas and pyroclastic fall deposits tend to dominate, and the flank slopes increase from about 1 degree to about 8 degrees. These structures are the products of extremely powerful eruptions which, fortunately, have not occurred within historic times. The Venusian coronae, if of volcanic origin, would probably fall within this group. Medium intermediate-slope shields consist of a wide variety of the groupings established by Pike and Clow. Included are the steeper of the ash-flow sheet shields, the majority of the continental alkalic basaltic shields, and the smaller of the oceanic and Martian shields. Medium steep shields are represented primarily by the Icelandic shields, the larger of the seamounts, the steeper continental shields, and the small steep oceanic shields. The small-volume members can be monogenetic, whereas the large-volume members are invariably polygenetic.

Large low shields are represented by the terrestrial Toba volcano, the Martian paterae, and the larger of the Venusian volcanoes, such as Colette and Sacajewea. Large intermediate-slope shields include the majority of the Hawaiian volcanoes, other terrestrial oceanic shields, and the Martian tholii. The mature Hawaiian shields typically have radiating rifts, a summit collapse caldera, and superposed small shields and cinder cones of a more evolved composition than the mass of the volcano. The large steep shield group includes the Galápagos volcanoes and other oceanic shields. The Galápagos shields differ from the Hawaiian shields in being much steeper in the summit region, having concentric rifts, and having a greater portion of more alkaline basalts.

The very large low shield category is currently occupied by one Martian volcano, Alba Patera. It is probable that a number of the Io volcanoes will be found to fall in this category. Very large intermediate shields are represented by the giant Mons volcanoes on Mars. Data suggest that these very large shields are dominated by lava flows.

Methods of Study

The morphologies of terrestrial shield volcanoes are readily obtainable from topographic maps; their compositions are obtained via standard geochemical techniques; their ages can be ascertained by radiometric dating; and their interior structures can be determined by observation of eroded structures. On other bodies in the solar system, these quantities are less easily obtained. Topography can be determined by measuring shadow lengths or determining the time delay between the sending and receiving of electromagnetic radiation by such instruments as laser altimeters. Since there are rock samples only from the moon (excluding an exotic group of igneous meteorites found on the earth), the compositions of other extraterrestrial materials have to be determined by analysis of their spectral characteristics. Relative ages of the outermost layers of extraterrestrial volcanoes can sometimes be determined from the density of superposed impact craters—the more craters, the older the surface. The internal structure of these volcanoes cannot be determined, since none of their parent bodies has as powerful agents of erosion as those found on the earth.

One of the reasons for assembling morphometric data for terrestrial volcanoes is to ascertain any relationships between the size and shape of the volcano and the forces that led to its construction. Then, by making corrections for the different gravities, atmospheres, and other parameters on the nonterrestrial bodies, one can infer the conditions that led to the construction of their shield volcanoes.

In spite of the wide range in the volumes of shield volcanoes, all appear to be constructed of low-viscosity material, whether it be basaltic lava or silicic ignimbrite. A major difference exists in the volume eruption rates between the lava and ignimbrite shields. The lava shields are formed primarily by an eruption style called Hawaiian, which is characterized by volumetric eruption rates of about 100 cubic meters per second for short durations but less than 1 cubic meter over time spans of a few years for the smaller structures, to about 1 million years for the larger Hawaiian shields. The ignimbrite shields are products of a plinian or ultraplinian eruption style and are thought to have had volumetric eruption rates of the order of 1 million cubic meters per second for eruptions lasting less than one day.

A feature which serves to distinguish the majority of terrestrial shield volcanoes from cone volcanoes is that the shields are commonly not found at plate margins; that is, they are predominantly intraplate volcanoes. It has been suggested that many shields are associated with hot spots, or areas of anomalously hot mantle. The paucity of terrestrial continental shields with heights in excess of 2 kilometers implies that the earth's oceans are acting as buttresses to prevent the collapse of the oceanic shields. The great heights of the Martian Mons volcanoes is probably, in part, a function of the lower gravity on that planet.

Context

Shield volcanoes include the most active volcano on earth (Kilauea), the tallest volcano in the solar system (Olympus Mons on Mars), and perhaps some of the highest eruption columns (more than 200 kilometers high, on

Io). They have undergone some of the most violent eruptions ever to take place on earth, such as the eruption at Toba Caldera, Sumatra. On earth, shields are present on the continents and on the ocean floor, and they occur as oceanic islands. From the sea floor to its summit, Mauna Loa is the largest mountain on earth, rising some 10 kilometers above its base. Small shields, resembling those of Iceland, are found on the moon. Gigantic shields have been identified on Mars. Active shield volcanoes are present on Io, and numerous shieldlike structures occur on Venus.

Chains of terrestrial shield volcanoes extending across a plate and formed by a single, fixed hot spot have been employed to determine the direction of motion of the plate, and the ages of the various volcanoes along the chain have been used to infer the plate's rate of motion. At the large and very large end of the shield volcano scale, inferences can be drawn as to the nature of the planet's interior, such as the internal heat transfer as a function of time and the strength of the outer layers of the planet required to support such huge masses. The heights of shield volcanoes have been employed to determine lithosphere thicknesses, and the horizontal separations between volcanoes have been used to infer the depths in the interior at which magma originated.

The high eruption frequency of terrestrial shield volcanoes, such as those of Hawaii, is of importance to those living in the paths of lava flows, which may destroy their homes. The enormous scale of eruptions associated with the formation of the ash-flow shields has fortunately not been witnessed within historic times, but such eruptions have been relatively frequent throughout geologic history and may recur in the future.

On a more positive note, shield volcanoes can provide a source of tourist revenue; furnish new, rich arable land; and endow geothermal energy.

Bibliography

Cotton, C. A. *Volcanoes as Landscape Forms*. Christchurch, New Zealand: Whitcombe and Tombs, 1944. One of the earliest modern attempts at the systematic analysis of volcanic landforms. Suitable for high school readers.

Decker, Robert W., Thomas L. Wright, and Peter H. Stauffer, eds. *Volcanism in Hawaii*. U.S. Geological Survey Professional Paper 1350. Washington, D.C.: Government Printing Office, 1987. Superbly illustrated and researched volume prepared for the seventy-fifth anniversary of the Hawaiian Volcano Observatory.

Fielder, Gilbert, and Lionel Wilson. *Volcanoes of the Earth, Moon, and Mars*. New York: St. Martin's Press, 1975. An early attempt at a systematic analysis of volcanic features on several planetary bodies. For general audiences.

Holcomb, Robin T. "Evidence for Two Shield Volcanoes Exposed on the Island of Kauai, Hawaii." *Geology* 25, no. 9 (September, 1997): 811-815.

Holmes, Arthur. *Principles of Physical Geology*. Rev. ed. London: Thomas

Nelson and Sons, 1965. A classic, pre-plate-tectonic account of geologic processes. Chapter 12 is on volcanoes. Out of date, but still worth reading. For college-level readers.

Lunar and Planetary Institute, Houston, Texas. *Basaltic Volcanism on Terrestrial Planets.* Elmsford, N.Y.: Pergamon Press, 1982. A compendium, by several dozen expert authors, of the majority of information published in English relating to basaltic volcanism. Chapter 5 is particularly pertinent to shield volcanoes. Aimed at a college-level audience.

Rittmann, Alfred. *Volcanoes and Their Activity.* Translated by E. A. Vincent. New York: Interscience, 1962. An early, thought-provoking synthesis of volcanology. For the reader with some knowledge of geology.

Simkin, Tom, et al. *Volcanoes of the World.* Stroudsburg, Pa.: Hutchinson Ross, 1981. A summary of locations, dimensions, and eruption histories of all terrestrial historic eruptions. For the reader desiring a detailed source.

James L. Whitford-Stark

Cross-References

Calderas, 43; Lava Flow, 367; Plate Tectonics, 505; Pyroclastic Rocks, 512; Volcanoes: Recent Eruptions, 675; Volcanoes: Types of Eruption, 685.

Siliciclastic Rocks

Siliciclastic rocks, which include siltstone, sandstone, and conglomerate, are second only to shales in abundance among the world's sedimentary rock types. They form major reservoirs for water, oil, and natural gas and are the repositories of much of the world's diamonds, gold, and other precious minerals. In their composition and sedimentary structures, siliciclastic rocks reveal much about paleogeography and, consequently, earth history.

Field of study: Petrology

Principal terms

CLASTIC ROCK: a sedimentary rock composed of particles, without regard to their composition; this term is sometimes used, incorrectly, as a synonym for "detrital"

DETRITAL ROCK: a sedimentary rock composed mainly of grains of silicate minerals as opposed to grains of calcite or clays

DIAGENESIS: chemical and mineralogical changes that occur in a sediment after deposition and before metamorphism

LITHIC FRAGMENT: a grain composed of a particle of another rock; in other words, a rock fragment

SHALE: a rock composed of abundant clay minerals and extremely fine siliciclastic material; diameter less than 0.004 millimeter

SILICICLASTIC ROCK: a sediment made of grains whose main ingredients are particles or fragments of silicate minerals

Summary

Weathering and erosion constantly strip the earth's surface of its rocky exterior. Rocks experience fluctuating temperatures, humidity, and freeze-thaw cycles that physically disaggregate (break up) mineral grains. Simultaneously, these mineral grains are exposed to water, oxygen, carbon dioxide, and dissolved acids or bases that chemically attack them. As a result, a hard, rocky surface is eventually transformed to a collection of mineral grains, clay minerals, and ions dissolved in water.

Weathering and erosion are slow, but continuous, agents of destruction. Because mountains and other high regions are constantly being uplifted, weathering and erosion provide a more or less steady supply of sediment to streams and rivers and, from there, into the ocean for deposition. Deposited

584

sediment is eventually compacted and cemented into rocks. The rocks that are formed by this process are known collectively as siliciclastic rocks because they are made mostly of particles of silicate minerals. The other main types of sedimentary rocks are carbonates (such as limestone) and mudrocks (such as shale).

Siliciclastic rocks differ in the sizes of their grains and their composition. Three main size classes are recognized: silt, sand, and gravel. Silt includes particles between 0.004 and 0.0625 millimeter in diameter; sand is composed of particles between 0.0625 and 2 millimeters in diameter; and gravel generally includes particles larger than 2 millimeters. Most sedimentologists refer to sand and sandstone when discussing siliciclastic rocks.

Siliciclastic rocks are composed of three broad classes of material: framework grains, matrix, and cement. Framework grains are particles of minerals or small fragments of other rocks that usually make up the bulk of a siliciclastic rock. Matrix is extremely fine-grained material such as clay that is deposited at the same time as framework grains. Cement is any material precipitated within the spaces, or pores, between grains in a sediment or rock. Framework grains and matrix are primary deposits, whereas cement is a secondary deposit because it is precipitated in pores after the primary material.

The main minerals that compose siliciclastic rocks are quartz, various types of feldspar, and micas. The ferromagnesian minerals, such as olivine, pyroxene, and amphibole, are not as common in sedimentary rocks as they are in igneous and metamorphic rocks. Ferromagnesian minerals are more easily dissolved during weathering, and they fracture more readily during erosion, than do the more durable quartz and feldspar. Consequently, quartz and feldspar (and, to a lesser extent, the micas) are concentrated in sediments because the other minerals are selectively removed beforehand. Carbonate rocks dissolve very easily and seldom form much residue to contribute to sand.

Quartz is the most chemically and physically stable mineral that forms siliciclastic rocks. Although quartz is uncommon in some igneous and metamorphic rocks, it is sufficiently abundant in granite and gneiss to contribute a large amount of bulk to almost all sands. Subtle differences may provide useful clues as to the origin of some quartz grains. For example, quartz from plastically deformed metamorphic rocks (schist, gneiss) usually occurs as polycrystalline aggregates or displays a distorted crystal structure, revealed by the petrographic microscope. Quartz from volcanic rocks often possesses planar crystal faces or embayments (deep rounded indentations) that were formed as the crystal grew in magma. Quartz grains eroded from older sedimentary rocks sometimes retain the previous rock's quartz cement as a rind, which is called an inherited overgrowth.

Feldspar minerals are also common in most siliciclastic rocks. Two main groups of feldspar are recognized: plagioclase feldspar and the potassium feldspars (microcline, orthoclase, and sanidine). The plagioclase group is actually a collection of many similar minerals with different chemical compositions. Anorthite is a plagioclase feldspar composed of calcium, silicon, and oxygen, whereas albite is a plagioclase composed of sodium, silicon, and oxygen. Calcium and sodium substitute rather easily for each other, so most plagioclase feldspars have some calcium and some sodium. In general, plagioclase with more calcium is more susceptible to chemical weathering than is sodium-rich plagioclase. Consequently, sodium-rich albite is more common than anorthite in most siliciclastic rocks.

The potassium feldspars all have the same chemical composition of potassium, silicon, and oxygen. They differ from one another in their chemical structure. Microcline is the most highly organized, crystallographically, followed by orthoclase and then by sanidine. Microcline is formed in igneous and metamorphic rocks that crystallize very slowly, permitting the greatest amount of crystallographic ordering. At the other extreme, sanidine is formed in volcanic rocks where little time is available for ions to get into their "proper places." During chemical weathering of detrital potassium feldspars, sanidine dissolves much more readily than does microcline or orthoclase. Microcline is usually about equal in abundance to orthoclase in siliciclastic rocks, whereas sanidine is usually rare.

The mica minerals, biotite and muscovite, are common in some siliciclastic rocks and rare in others. In general, muscovite (white mica) is more durable and, consequently, more abundant than is biotite (brown mica). Chlorite is a mineral with a sheetlike crystal structure similar to the micas. It is common in a few sandstones but generally less abundant than the micas. A large variety of grains is often present in siliciclastic rocks as "accessory minerals." Altogether, they seldom constitute more than 1 percent of any siliciclastic rock. Sometimes these accessory minerals are called heavy minerals because most of them have a much greater density than does quartz or feldspar. This group includes zircon, tourmaline, rutile, garnet, the ferromagnesian minerals pyroxene, amphibole, and olivine, and iron oxides such as hematite, magnetite, and limonite. If one examines a handful of sand, the accessory minerals are usually the dark grains.

Accessory minerals are important in several ways. First, they may reveal information about the source rock from which the sand was eroded. For example, the accessory mineral chromite is generally formed in basalt, so its presence in a sand indicates that basalt (usually from the sea floor) had previously been uplifted and eroded nearby. Similarly, garnets of particular composition may be representative of particular metamorphic source rocks. Second, accessory minerals may tell geologists how "mature" a sand is: in other words, how great has been its exposure to chemical and physical weathering. Zircon, tourmaline, and rutile are far more durable than are the other accessory minerals. If they are the only accessory minerals in a sand, then the sand probably experienced severe weathering in the period before its final deposition. Third, accessory minerals may be economically valuable. Diamonds and gold are accessory minerals in a few cases. Much of the

United States' titanium comes from unusual concentrations of rutile and ilmenite in beach sands along the coast of South Carolina. Zircon in sands provides an important source of zircomium, used in high-temperature ceramics.

Pieces of rocks (lithic fragments) may form an important part of coarse-grained sandstone and conglomerate. Lithic fragments may be composed of pieces of almost any igneous, metamorphic, or sedimentary rock. Their presence is a powerful indicator of the source rock from which the sand or gravel was weathered and eroded. As one might expect, not all rock fragments are equally durable in the sedimentary environment. Limestone and shale fragments disintegrate very rapidly, whereas granite and gneiss rock fragments are quite robust. Consequently, both the presence and the absence of certain rock fragments may reveal information about the origin of a siliciclastic sediment.

Matrix in siliciclastic rocks is usually clay or fine silt. This substance is the "mud" that often accompanies sand or gravel during deposition. It is commonly deformed between the more rigid framework grains as the sediment is compacted. Some types of matrix were not originally deposited as fine-grained matrix. This "pseudomatrix" is actually squashed fragments of soft grains of, for example, shale or schist. During compaction of the sediment, these fragments are more easily deformed than are durable quartz and feldspar, and the fragments superficially resemble original matrix. Considerable skill is required to distinguish between true matrix and pseudomatrix.

Sand-sized siliciclastic rocks are often classified based on their mineral composition. Sandstones may be named with the term "arenite" (after the Latin *harena*, sand). Relative abundances of quartz, feldspar, and rock fragments determine the name. One of the major types of sandstone is quartz arenite, which is composed almost entirely of quartz, with less than 5 percent of other framework minerals and less than 15 percent clay matrix (usually much less). This rock is usually white when fresh. Another is feldspathic arenite, which is sometimes called arkose. This rock has abundant feldspar and up to 50 percent lithic fragments. Typical composition is perhaps 30 percent feldspar, 45 percent quartz, and 25 percent lithic fragments. Up to 15 percent detrital clay matrix may also be present. This rock is often pinkish-gray in color. A third type is lithic arenite, a sand composed largely of lithic fragments and up to 50 percent feldspar. A typical composition might be 60 percent lithic fragments, 30 percent feldspar, and 10 percent quartz. Up to 15 percent matrix may be present as well. Almost all conglomerates are lithic arenites. Lithic arenites are often dark gray, sometimes with a salt-and-pepper appearance from the lithic fragments. A fourth type is wacke, sometimes called graywacke; this rock has between 15 percent and 75 percent clay matrix at the time of deposition. The term "wacke" may be modified by prefixing with the name of the dominant nonmatrix component, such as "quartz wacke," "feldspathic wacke," or "lithic wacke." Some wackes may be "formed" diagenetically from lithic arenites by alteration of shale or schist fragments so that they appear similar to originally detrital clay matrix. This rock is usually gray in color (thus, the name "graywacke").

The composition of sandstones holds important clues as to the kinds of rocks from which they were derived. These clues may help petrologists to reconstruct the plate tectonic setting of the sand's source region even though the source region has long since disappeared because of erosion or tectonic destruction. For example, quartz arenite often reflects sedimentation on or near a stable craton (piece of the earth's crust), where physical and chemical weathering are permitted to eliminate unstable minerals for an extended period of time. At the other extreme, a lithic arenite with abundant basaltic rock fragments, calcic plagioclase, and rare quartz may represent detritus shed from an island arc during plate collision. Careful observation of the mineral composition of sandstones has become an important part of reconstructing the earth's history.

Diagenesis of siliciclastic rocks is a complex collection of processes that include cementation, replacement, recrystallization, and dissolution. Diagenesis includes chemical reactions that begin at the time of deposition and may continue until metamorphism takes place. For many sedimentary petrologists, low-temperature cementation is part of diagenesis, whereas for others, diagenesis includes only those reactions that occur at temperatures exceeding 100 degrees Celsius. The upper range of diagenesis grades into metamorphism; sedimentary petrologists and metamorphic petrologists share some common territory in diagenesis.

Siliciclastic sediments are transformed to rocks by compaction and cementation. Common cements in siliciclastic rocks are quartz, calcite, clays, and iron oxides such as limonite or hematite. It is common for a rock to have several cements, each deposited at different times or under different conditions of pore-water chemistry. Cement is introduced into sediment in the form of elements or compounds dissolved in water. For example, silica may be dissolved from quartz grains or given off during low-grade metamorphism and may saturate water in the pores of a rock. One means of dissolving silica from quartz grains is called pressure solution. Quartz grains exert very high pressures upon one another at their points of contact. Quartz is more soluble under high pressure, so it is more easily dissolved where grains touch each other. Well-compacted sands often display flattened grain-to-grain contacts as a consequence of pressure solution. The quartz lost from these grains is carried away, dissolved in pore water, to be precipitated elsewhere.

As the water moves slowly through the pores of the rock, it may encounter different temperatures, pressures, acidity, or gas concentrations. Changes in these conditions may cause the water to lose some of its ability to dissolve silica, and the silica may therefore precipitate as crystals. Similar factors influence the dissolution and precipitation of other cements in siliciclastic rocks. Cement not only holds sediment grains together, forming a rock, but it also fills the pores between grains, which may inhibit further flow of fluids.

As a consequence, the porosity and permeability of most rocks are reduced by cementation. Rocks with abundant cement generally make poor reservoirs for petroleum or water.

In some cases, the pore-water chemistry conducive to precipitation of one cement is capable of dissolving another cement. A common example is the generally inverse relationship between quartz and calcite cement. Pore water that precipitates quartz often dissolves calcite at the same time, or the other way around. Sometimes, dissolution occurs without immediate precipitation of another mineral type in the void that is formed. Pores formed by dissolution are called secondary porosity; such porosity may be the main cause of porosity in many sandstones. Minerals commonly dissolved include carbonates such as calcite and aragonite, silicates such as feldspars and ferromagnesian minerals, and, though rarely, accessory minerals such as garnet. In some cases, earlier cements may be dissolved to form secondary porosity.

Recrystallization is a common feature in diagenesis. In recrystallization, one mineral is transformed into another mineral with the same (or nearly the same) chemical composition or into different-sized (usually larger) crystals of the original mineral. The most common example is the transformation of aragonite to calcite in limestones. In siliciclastic rocks, microcrystalline quartz may recrystallize into more coarsely crystalline quartz, or rutile may recrystallize into anatase.

A more prevalent form of diagenesis among ancient sandstones is replacement. In this process, one mineral is replaced volume-for-volume by another mineral. In some cases, it appears that the replacing mineral has combined material from its host grain with dissolved ions from pore water. The most common replacement examples involve clay minerals. For example, kaolinite may replace potassium feldspar, forming a grain of kaolinite that looks superficially like a potassium feldspar grain.

Methods of Study

The principal means of studying siliciclastic rocks is with petrographic microscopes. This microscope is similar to other microscopes except that light passes through the specimen rather than illuminates its surface. Rocks are normally not transparent, however, and they must be cut into very thin slices in order for light to pass through them. Using a diamond-edged rotary saw, rocks are sliced into 1 × 5 × 2.5 centimeter rectangular chunks and mounted with epoxy to glass microscope slides. Almost all the rock chunk is cut and ground away, leaving a thin slice of rock remaining on the glass. This slice, approximately 30 micrometers thick, is known as a thin section. It is nearly transparent yet reveals a large amount of information when viewed in polarized light under a microscope. Magnifications of up to four hundred times are routinely used. In thin section, it is possible to view closely individual grains and their relations to other grains, to identify the mineral grains in the rock, and to view matrices and cements. Thin sections are routinely used to estimate the amount of pore space in rocks, revealing their potential as reservoirs for fluids such as water, oil, and natural gas.

When illuminated by polarized light, different minerals display characteristic interference colors that reveal the identity of a mineral to the experienced observer. Other clues are the existence and pattern of cleavages in the minerals, alteration products resulting from diagenesis, and overall grain shape. In addition, it is possible to view cements that surround grains and partially or completely fill pores. The sequence of cementation can be interpreted by the relative positions of cements filling pores. For example, if a sandstone shows small grains of ankerite cement (an iron carbonate mineral) close to detrital quartz, with quartz cement deposited over them, followed by a zone of calcite cement, then it can be said that the sequence of cementation was ankerite, quartz, and then calcite last. Knowing the sequence of cements can help geologists understand how deeply buried the rock was when it was cemented and how the chemical composition of the pore water changed through time.

Another means of examining siliciclastic rocks is with the scanning electron microscope (SEM). The SEM permits examination of clastic grains with much greater magnification than with the conventional petrographic microscope. Magnifications of up to ten thousand times are easily accomplished, with details as small as 0.001 micrometer being easily resolved. It is possible to view the interiors of individual pores and to examine the shape of clay crystals growing on the surfaces of single sand grains. Very simple sample preparation techniques are used to examine rocks with the SEM. A small fragment of rock is glued to an aluminum "stub" that holds the sample in the SEM. The glued sample and its stub are lightly coated with either carbon or an alloy of gold and palladium. This coating permits dissipation of the electric charge developed by the SEM's electron beam. The sample (on its stub) is then placed in the SEM. After the sample holding chamber is evacuated of nearly all air, an electron beam scans the sample surface. The image viewed by the geologists is formed by collecting secondary electrons released by the sample when the electron beam strikes it. Most SEMs cannot clearly identify the particular mineral being viewed. Identification is usually based on the shape of the mineral grain and its expression of cleavage. Some SEMs, however, are equipped with detectors for the X rays also given off by minerals when struck by a beam of electrons. These X rays reveal the elemental composition of the mineral being viewed. Thus, for example, a mineral with abundant potassium, aluminum, and silicon is probably potassium feldspar, whereas a similar-appearing mineral (similar in the SEM) with abundant iron, magnesium, and silicon is probably pyroxene. The SEM is very useful in examining the surface texture of minerals. The presence of pits, grooves, and cracks can reveal clues as to the environment under which the sand grains were deposited. Were the sand grains blown by wind, carried by water, or transported by glaciers? In some cases, this question may be answered by examination under the extreme magnification of the SEM.

When accurate data concerning chemical composition of minerals are needed, the electron microprobe is used. The microprobe is actually an elaborate version of a SEM. It differs in that its beam does not normally scan but instead remains fixed in one spot, and its detectors are carefully calibrated to give accurate readings of the elemental composition of the material being examined. Microprobes collect X rays and can determine from them the amounts of different elements in almost any mineral. It is often essential to know the amount of sodium or calcium in detrital plagioclase feldspars or the magnesium content of calcite cement. In some instances, the abundance of trace elements in minerals such as zircon, tourmaline, garnet, or rutile may provide clues as to the source region from which the grains were derived, giving an indication of the source for a sedimentary formation. The electron microprobe is a very useful tool in these studies.

There are limitations to the microprobe. It cannot analyze for elements lighter than sodium, so the amount of oxygen or carbon cannot be determined. Furthermore, rock material must be very carefully prepared for readings to be accurate. Also, electron microprobes are rather expensive and require considerable maintenance; however, X rays are employed in the study of rocks in other ways. X-ray diffraction of powdered or whole specimens can identify minerals and their degree of crystal ordering. X-ray fluorescence identifies the chemical composition of minerals. Both techniques are commonly used in the study of clays in the matrix of siliciclastic rocks as well as in shales composed entirely of clay-sized material. Sample preparation is relatively easy, and beginners can operate most modern X-ray machines. X-ray analysis usually requires destruction of the sample by grinding, however, so the sizes, shapes, and relationships between grains are lost.

Context

Sandstone and its cousins siltstone and conglomerate are common sedimentary deposits. They form in a variety of environments, including alluvial fans, rivers, deserts, beaches, and submarine fans thousands of meters below sea level. These rocks, collectively known as siliciclastic rocks, are important for a number of reasons. For example, they are porous—that is, they contain small spaces between grains that may be filled with valuable fluids. Siliciclastic rocks are the major type of aquifer in most of the world. Also, they form reservoirs for oil and natural gas. Sand and gravel are essential parts of modern construction because they form the "bulk" in concrete. In this role, they represent the most economically significant mineral resource in the United States, ahead of petroleum, coal, and precious metals.

Bibliography

Blatt, Harvey. *Origin of Sedimentary Rocks.* 2d ed. Englewood Cliffs, N.J.: Prentice-Hall, 1980. A standard textbook on sedimentary rocks, cover-

ing more than siliciclastics. Includes a complete discussion of the sources of grains, their transportation, deposition, and deformation.
_____. *Sedimentary Petrology*. San Francisco: W. H. Freeman, 1982. A useful introduction to sedimentary petrology. Suitable for general readers.
McDonald, D. A., and R. C. Surdam, eds. *Clastic Diagenesis*. Memoir 37. Tulsa, Okla.: American Association of Petroleum Geologists, 1984. A thorough discussion of sandstone diagenesis, complemented by well-documented case studies. Superbly illustrated.
Pettijohn, F. J., P. E. Potter, and R. Siever. *Sand and Sandstone*. 2d ed. New York: Springer-Verlag, 1987. Perhaps the most comprehensive, authoritative, and readable book on sandstone in English. Very well illustrated, with copious references. The basic reference for all sandstone studies.
Scholle, Peter A. *Color Illustrated Guide to Constituents, Textures, Cements, and Porosities of Sandstones and Associated Rocks*. Memoir 28. Tulsa, Okla.: American Association of Petroleum Geologists, 1979. An excellent guide to the major types of grains, matrices, and cements common in sandstones. Includes microphotographs, complete with short description of the features they show.
Scholle, Peter A., and D. R. Spearing, eds. *Sandstone Depositional Environments*. Memoir 31. Tulsa, Okla.: American Association of Petroleum Geologists, 1982. Well-illustrated compendium of the environments in which siliciclastic rocks may be found.
Tucker, M. E., ed. *Techniques in Sedimentology*. Oxford, England: Blackwell Scientific Publications, 1988. A very useful collection covering the major techniques of sedimentary petrology. Well referenced.

Michael R. Owen

Cross-References

Alluvial Systems, 1; Coastal Processes and Beaches, 71; Minerals: Physical Properties, 415; Sand Dunes, 559; Sedimentary Rocks, 568.

Stratigraphic Correlation

Stratigraphic correlation is the process of determining the equivalence of age or stratigraphic position of layered rocks in different areas. It is critical to understanding earth history because stratigraphic correlation is one of the principal methods by which the succession and synchrony of geological events are established.

Field of study: Stratigraphy

Principal terms

BIOSTRATIGRAPHY: the identification and organization of strata based on their fossil content and the use of fossils in stratigraphic correlation

CORRELATION: the determination of the equivalence of age or stratigraphic position of two strata in separate areas; more broadly, the determination of the geological contemporaneity of events in the geological histories of two areas

INDEX FOSSIL: a fossil that can be used to identify and determine the age of the stratum in which it is found

LITHOLOGY: loosely used by many geologists to refer to the composition and texture of a rock

SEDIMENTARY ROCKS: rocks formed by the accumulation of particles of other rocks, organic skeletons, chemical precipitates, or some combination of these

STRATIGRAPHY: the study of layered rocks, especially of their sequence and correlation

STRATUM (pl. STRATA): a single bed or layer of sedimentary rock

Summary

Stratigraphic correlation is the process of determining the equivalence of age or stratigraphic position (position in a vertical sequence of rock layers) of strata in different areas. Strata thus determined to be of the same age or in the same stratigraphic position are called "correlative" or "correlated."

The most obvious method of stratigraphic correlation is to demonstrate that strata are continuous over a given area. This is achieved by tracing the layers (often by walking along them) throughout the area to demonstrate

593

their physical continuity. In areas where strata are continuously exposed (the Grand Canyon in Arizona is a good example), this direct method is simple and best. Over much of the earth's surface, however, the strata that scientists wish to correlate are covered by younger rocks or vegetation, or erosion has removed them over some portion of their geographic extent, making this method impractical.

A commonly used method of stratigraphic correlation that is not subject to this drawback is correlation by lithologic similarity, or similarity in rock composition and texture. Thus, strata of the same lithology (rock type) in different areas may be considered correlative. In effect, this method attempts to identify the same layer in different regions without demonstrating the continuity of (that is, without tracing) the layer. The major problem with this method, however, is that there are many layers in the earth's crust that are of similar lithology. In eastern Colorado, for example, it is not easy to tell a 90-million-year-old sandstone layer from a 70-million-year-old sandstone from lithology alone. Thus, mistakes in determining age equivalence are easy to make when correlating only by lithologic similarity.

A third method of stratigraphic correlation relies on the idea that two layers in different areas that occupy the same stratigraphic position are correlative. This method can be referred to as correlating by position in the stratigraphic sequence. For example, if a limestone layer overlies sandstone A in one region and a shale layer overlies sandstone A elsewhere, scientists may conclude that the limestone and shale are correlative. This method, however, is not without its pitfalls. In the example, it might be that the limestone layer at the first location was deposited right after sandstone A was deposited, whereas at the second location, the shale layer was deposited many millions of years after sandstone A. To correlate the limestone and shale in this case is to recognize their equivalence of stratigraphic position but not to demonstrate their age equivalence.

Age equivalence in stratigraphic correlation is almost always based on fossils. Stratigraphic correlation by fossils depends on biostratigraphy, the recognition of strata by their fossil content. If a fossil represents an organism characteristic of a particular interval of geologic time, then it is referred to as an index fossil. Thus, if an index fossil is collected from a stratum, then that stratum is the same age as strata elsewhere that contain the index fossil.

Besides fossils, there are some other, less often used, methods of determining the age equivalence of strata. The most common of these methods is to obtain a numerical estimate (usually in millions of years) of the age of the strata. This can rarely be undertaken unless the strata contain volcanic material, which contains chemical elements needed to obtain a numerical age. These elements are not normally present in sedimentary rocks; volcanic ash beds are not nearly as common in sedimentary rocks as are fossils. That is why fossils have been and continue to be the primary means for determining the age equivalence of strata. To undertake stratigraphic correlation, it thus is necessary to define the lateral extent of strata in a given region, characterize their lithology, determine their stratigraphic sequence, and collect and establish the stratigraphic ranges of fossils. Once these data have been collected in one or more regions, a geologist can attempt to correlate the strata within an area or between disparate areas.

The development of stratigraphic correlation began in 1669 when Nicolaus Steno, a Danish naturalist, recognized that the sequence of strata is directly related to their relative ages. Steno's principle of superposition thus identified the oldest layers as those at the bottom and the youngest strata as those at the top of a sequence of strata. Steno's principle and the recognition of fossils as the remains of past life allowed William Smith in England and Georges Cuvier and Alexandre Brongniart in France to undertake the first stratigraphic correlations based on fossils during the early 1800's. Smith's correlations resulted in the first geologic map of England (1815). Cuvier and Brongniart were able to correlate stratigraphically the rock layers in the vicinity of Paris and thereby reconstruct the geological and biological history of this area. Previously, during the late 1700's, the German mining engineer Abraham Werner and his students had laid the basis for stratigraphic correlation by lateral continuity and lithologic similarity by arguing for the continuity of strata of a particular lithology over a broad area. Thus, by the early 1800's, stratigraphic correlation by lateral continuity, lithologic similarity, position in the stratigraphic sequence, and fossils was already being practiced by European geologists.

Stratigraphic correlation plays a central role in understanding geological history. By allowing geologists to determine the synchrony or diachrony of strata in different areas, the geological and biological events recorded in these strata can be ordered in time. This ordering is the basis of the

Vast stretches of exposed rock strata are visible in the Grand Canyon. *(R. Kent Rasmussen)*

chronology of geological and biological events during the last 3.9 billion years of earth history. Stratigraphic correlation is thus one of the methods by which the relative geological time scale of eons, eras, periods, epochs, and ages was constructed.

Applications of the Method

Stratigraphic correlation is extremely important in deciphering geological history, for it reveals the sequence of geological events in one or more regions. Understanding geological history is of interest for its own sake to scientist and layperson alike, but it is also crucial to the discovery of mineral deposits and energy resources within the earth's crust.

A good example of stratigraphic correlation helping to decipher geological history and leading to the discovery of a giant oil field comes from the Guadalupe Mountains of West Texas and eastern New Mexico. There, in the years before World War II, geologists and paleontologists used stratigraphic correlation (especially biostratigraphy) to unravel the complex geological history of the strata that form these mountains. This history revealed that huge barrier reefs had developed as the sea encroached from the south about 260 million years ago. When it was later learned that hydrocarbons often accumulate around reefs, interest in the petroleum potential of these rocks was aroused. Because drilling the rocks in the rugged Guadalupe Mountains was impractical, an effort was made to trace the reef strata into the nearby lowlands. There, drill cores brought up rocks and fossils from deep beneath the surface. Stratigraphic correlation by lithologic similarity, position in stratigraphic sequence, and fossils was used to identify the strata adjacent to the reef in what is now called the Delaware basin, south and east of the Guadalupe Mountains. Successful drilling for petroleum soon turned the Delaware basin into one of the world's giant oil fields, thanks to the identification of petroleum source rocks through stratigraphic correlation.

By ordering events in the history of the earth, stratigraphic correlation has been critical to the development of a global geological time scale. Such a time scale provides geologists with a shared temporal framework within which to view their observations. A good example of the use of stratigraphic correlation in the development of the global geological time scale comes from the concept of the Cambrian period. The British geologist Adam Sedgwick coined the term "Cambrian" (from "Cambria," the ancient Roman name for Wales) in 1835 to refer to rocks in northern England that he believed contained the oldest fossils. Geologists now recognize the Cambrian period (505-570 million years ago) worldwide because stratigraphic correlation proved that an interval of geologic time that corresponds to Sedgwick's original Cambrian rocks can be identified across the globe. One of the most distinctive aspects of Cambrian strata is the abundance and types of trilobites they contain. Correlation of the Cambrian strata of England with strata elsewhere has been mostly undertaken on the basis of these trilobite fossils. Indeed, Sedgwick himself first conducted such stratigraphic correlation in continental Western Europe. By the 1880's, when American geologist Charles Doolittle Walcott identified Cambrian trilobites in North America, the Cambrian was well on its way to becoming a geologic time period recognized worldwide. Stratigraphic correlation, especially using fossils (biostratigraphy), thus played a significant role in the recognition of the Cambrian. All the geologic time periods now recognized worldwide are similarly rooted in stratigraphic correlation.

Context

Stratigraphic correlation, the recognition of the equivalence of stratigraphic position or age of strata (or both) in different areas, is critical to understanding geological history. Stratigraphic correlation allows geologists to establish the sequence of events in an area and match them with synchronous events elsewhere. Without an understanding of geological history, the search for mineral deposits and energy resources is almost impossible. Moreover, understanding the history of geological disasters—earthquakes, volcanic eruptions, meteorite impacts—and thereby the ability to use that history to predict the timing of future disasters requires knowledge of the sequence and timing of geological events, knowledge rooted largely in stratigraphic correlation. Deciphering the history of life on this planet, the myriad appearances, changes, and extinctions of the earth's biota during the last 3.9 billion years, largely relies on the timing and sequence established by stratigraphic correlation.

The global geological time scale is the temporal framework within which all geological and biological events during earth's history are ordered. The current time scale is the outgrowth of more than 150 years of stratigraphic correlation. Intervals of geologic time first conceived of in local sequences of strata have gained worldwide acceptance when stratigraphic correlations identified strata of the same age across the globe. The resultant time scale provides not only a chronological framework within which to view the last 3.9 billion years but also a language of geological time that facilitates communication among paleontologists, geologists, and other scientists from all countries.

Bibliography

Ager, Derek V. *The Nature of the Stratigraphical Record.* 2d ed. New York: Halsted Press, 1981. A witty and unabashed look at stratigraphy. Much of the discussion focuses on problems of stratigraphic correlation. The bibliography is extensive, there is an index, and a few, well chosen illustrations illuminate the text.

Berry, William B. N. *Growth of a Prehistoric Time Scale.* Rev. ed. Palo Alto: Blackwell Scientific Publications, 1986. Provides an excellent review of the principles of stratigraphic correlation but is mostly devoted to the history of how the global geological time scale was formulated. As such, it well relates many of the stratigraphic correlations upon which the time scale is based. Well illustrated, with a good bibliography and an index.

Brenner, Robert L., and Timothy R. McHargue. *Integrative Stratigraphy Concepts and Applications.* Englewood Cliffs, N.J.: Prentice-Hall, 1988. Chapters 11-13 provide a comprehensive look at all facets of stratigraphic correlation. Well illustrated, with extensive reference lists and an index.

Eicher, Don L. *Geologic Time.* 2d ed. Englewood Cliffs, N.J.: Prentice-Hall, 1976. Chapters 4 and 5 provide a somewhat less technical look at stratigraphic correlation than do Brenner and McHargue. Very readable, well illustrated, with some references and an index.

Harbaugh, John W. *Stratigraphy and Geologic Time.* Dubuque, Iowa: Wm. C. Brown, 1968. Although this book is somewhat dated, chapter 2 provides one of the most succinct reviews of stratigraphic correlation available. Well illustrated, indexed, but few references cited.

Hedberg, H. D., ed. *International Stratigraphic Guide.* New York: John Wiley & Sons, 1976. The international "rule book" for stratigraphy. It sets procedures and standards to be met when naming stratigraphic units. It also defines many terms used in stratigraphy and has an extensive bibliography. Chapter 6 is devoted to biostratigraphy.

Rainbird, R. H., and C. W. Jefferson. "The Early Neoproterozoic Sedimentary Succession." *Geological Society of America Bulletin* 108, no. 4 (April, 1996): 454-471.

Spencer G. Lucas

Cross-References

Biostratigraphy, 37; The Fossil Record, 257; The Geologic Time Scale, 272.

Stratovolcanoes

A volcano that erupts both cinder and lava is a composite volcano, or stratovolcano. These volcanoes are found at subducting plate margins and are the most abundant of the large volcanoes. The tallest and most famous stratovolcanoes of the world, such as Vesuvius, Fuji, and Mount St. Helens, have produced the most devastating and violent eruptions.

Field of study: Volcanology

Principal terms

ANDESITE: a gray volcanic rock with a silica content of about 60 percent

ASH: fine volcanic ejecta (less than 2 millimeters in diameter)

CALDERA: a large circular basin with steep walls that resulted from the collapse of a summit or of an earlier volcanic cone

CINDER CONE: a small volcano composed of cinder or lumps of lava containing many gas bubbles, or vesicles; often the early stage of a stratovolcano

PUMICE: a solidified volcanic froth of a glassy texture, light enough to float on water

PYROCLASTICS: fragmented ejecta released from a volcanic vent

RHYOLITE: a light-colored volcanic rock, composed of a viscous lava containing about 70 percent silica

TEPHRA: all pyroclastic materials blown out of a volcanic vent, from dust to large chunks

VISCOSITY: the resistance of lava to flow; it depends upon the chemical composition, temperature, and crystalline nature of the magma

Summary

The most abundant type of large volcano on the earth is the stratovolcano, also referred to as a composite volcano. These classic "poster volcanoes" constitute 80 percent of all active volcanoes, including Mount Fuji, Mount Vesuvius, Mount St. Helens, and Krakatoa. Stratovolcanoes vary in elevation from a few hundred to 4,000 meters or more above base levels, with diameters approaching 40 kilometers. Their slopes vary from 10 to 35 degrees, with an increase toward the summit because of larger amounts of volcanic material deposition there. In humid regions, much of this volcanic material, in the form of tephra, is washed downslope in mudflows, but when transported aerially, very little tephra actually reaches the base.

The shapes of most of these volcanoes are symmetrical, with eruptions from a single pipe vent at the center, although some build up around long fissure vents. Mount Helka in Iceland is an example of the latter, with a 5-by-10-kilometer oval shape. In the former case, a central crater, usually a funnel-shaped hole, is found on the summit. The central vent tends to remain in the same position despite the explosive events. Often, the crater may be enlarged by wall collapse because of magma withdrawal. A flat central floor in the crater may be more prevalent in humid areas, where heavy rains wash fine sediments downward.

Stratovolcanoes may reach great heights, but these large, symmetrical cones require a riblike structure of lava for support. Pure cinder cones, in contrast, rarely exceed 500 meters in elevation, for their symmetry tends to be destroyed by slumping. One eruptive cycle may be an effusive flow of lava, followed in some later period by an explosive event that produces cinder and ash layers. The relative proportions of tephra and lava in stratovolcanoes' structures vary considerably, from pure ash or cinder to pure lava flows. Rarely do volcanoes exhibit the classic "layer-cake" geology—that is, alternating layers of lava and ash. Some mountains (for example, Taal in the Philippines) consist of 70-80 percent pyroclastics, but others are dominated by lava eruptions.

Eruption patterns vary from one stratovolcano to another; three basic types are described here. The Vulcanian type of eruption, characteristic of many active volcanoes, produces a solid crust that lasts until the next eruption. Gas pressure builds up in the magma column, eventually blowing out the solidified obstruction. A great explosion may follow, accompanied by a large, dark, cauliflower-shaped eruption cloud. With the reduction of pressure, gas-charged magma is replaced by pumice and ash. After the vent has cleared, lava flows may issue forth from the crater. Vesuvius, the stratovolcano east of Naples, falls within this group of eruptions.

The Strombolian eruption is distinguished by glowing fragments of lava, accompanied by a white eruption cloud. In contrast to the Vulcanian eruption, which contains much ash, the Strombolian cloud contains very little ash. The crust that forms over the lava column is very thin, allowing for frequent, mild eruptions. Stromboli, a stratovolcano located on an island west of Italy, is the classic example, but some volcanoes may exhibit Strombolian activity during some portion of their history.

The Peléan eruption gives rise to the expulsion of nuées ardentes. Pumice and ash are characteristic; no liquid lava develops. The magma is under such intense pressure that it is shattered into a fine dust and mixed with superheated steam. This mixture rolls over the top of the crater as an avalanche of red-hot dust. In the final stages of eruption, the gas content of the magma is greatly reduced, no longer breaking the surface but instead pushing upward to form a dome with spinelike projections, as in the case of Mount Pelée on the island of Martinique.

Stratovolcanoes tend to be positioned at the margins of descending or subducting masses of ocean floor or plates beneath the continents. The western United States has thirty-five volcanoes that have erupted and may erupt again in the future. One region of volcanic activity is found in the Cascade Range, extending from Washington and Oregon to northern California with a continuation into Nevada and through Idaho to Yellowstone National Park. Another volcanic zone winds through southeastern Utah into Arizona and Mexico.

In terms of frequency of eruption, volcanoes are divided into two groups. The first comprises those that have experienced eruptions on average every two hundred years and have last erupted within that period of time. Examples in this group are Mount St. Helens, Lassen Peak, and Mount Shasta, Rainier, Baker, and Hood, all in the Cascade Range. The other group includes those whose last eruption was more than a thousand years ago; their frequency of eruption is greater than one thousand years. Examples are Crater Lake Volcano and Mounts Adams and Jefferson, also located in the northern Cascade Range.

The stratovolcano gains its heat from subducting-plate friction. The magma tends to have a lower temperature and a higher viscosity than does the basaltic magma of shield volcanoes (found at rift zones and located over hot spots). The eruptions are explosive because of the high silica and gas content and high viscosity. Rhyolite and andesite are often present in the flows, along with volcanic products called pyroclastics. Not all stratovolcanoes, however, erupt lava containing such high amounts of silica; Mount Etna in Sicily and Mount Fuji in Japan tend to have a silica composition of 50-60 percent in a basaltic lava.

Several stratovolcanic eruptions have had significant consequences for people living nearby. Located in the southeastern state of Chiapas, Mexico, El Chichón experienced a series of explosive eruptions on March 28, 1982, killing 187 people and leaving another 60,000 homeless. A large dust cloud, which was monitored by satellite, rose to an altitude of 25 kilometers and lasted for one month after the eruption. Sixteen kilometers away, 50 centimeters of ash fall was recorded; at 80 kilometers' distance, it had decreased to 20 centimeters. In Palenque, 120 kilometers east of the volcano, ash fall was measured at 40 centimeters.

On the eastern shore of Sicily, Mount Etna, the largest and highest of the European volcanoes, rises 3,320 meters above sea level. Small subsidiary cones around its flanks, up to 1,000 meters high, erupt rather frequently, but only rarely does sufficient energy accumulate to cause an eruption from the summit crater. Early eruptions, several centuries B.C., were recorded by the Greeks. Thousands of lives have been lost and several towns destroyed during Etna's history. In 1669, twenty thousand died in an eruption. The town of Mascati was destroyed in 1853, and in 1928 the village of Nunziata was nearly leveled. Explosive eruptions near the summit in 1979 claimed nine lives. Mount Etna is classified as a transitional volcano, transforming

Japan's Mount Fuji, one of the world's tallest stratovolcanoes, last erupted in 1707. (Japan Air Lines)

from shield to stratovolcano, and has had more than 150 eruptions since the first recorded one, in 1500 B.C.

On June 6, 1912, one of the greatest eruptions of the twentieth century occurred on the Alaska Peninsula in the Valley of Ten Thousand Smokes. Mount Katmai is a complex stratovolcano, with both a caldera and a lake on its summit. The explosion was heard in Juneau, 1,200 kilometers away. Twenty-five cubic kilometers of pumice, ash, and gas flooded the valley, filling it to a depth of nearly 200 meters. In Kodiak, some 160 kilometers away, pumice and ash nearly blocked out the sun, reducing visibility to about 2 meters. The upper 1,900 meters of the mountain collapsed, forming a giant caldera more than 2 kilometers wide, which filled with water and became a crater lake. A new vent, Mount Trident, on the west flank of Katmai became active in 1949 and has since experienced several thick lava flows, the last erupting in 1974.

Stromboli, one of the most active volcanoes, lies on a Liparian island 60 kilometers north of Sicily. It has been in a continuous state of eruption its entire history, although the intensity varies considerably. It was known for centuries to sailors as "the Lighthouse of the Mediterranean," for eruptive flashes from its summit were visible far out to sea. Even during World War II, it was used by Allied bombers as a navigational aid. The more explosive eruptions occur at about fifteen-minute intervals, with moderate activity in between. The principal crater, located 600 meters up the 900-meter mountain, is always full of semifluid lava and thus does not greatly resist magma pressure from beneath.

Mount Tambora, on the island of Sumbawa, Indonesia, began a series of eruptions on April 5, 1815, that did not end until July of that year. The early explosions sounded like cannon fire and could be heard 720 kilometers away. The greatest eruptive activity happened on April 11 and 12, when explosions were heard in Sumatra, 1,600 kilometers to the west. There were only a few survivors of a population of twelve thousand on the island, and forty-five thousand other lives were lost on surrounding islands. Volcanic ash was so heavy that there was darkness at noon in Java, 480 kilometers distant. The dust sent into the atmosphere obscured sunlight more than had any other volcanic dust within the previous four hundred years. The year 1816 was known as the "year without summer."

The famous eruption of August 24, A.D. 79, was described by Pliny the Younger, from a distance of 30 kilometers. His account describes a strange cloud that rose up and away from the mountain as well as roars and explosions coming from deep within the mountain. The residents of Pompeii were overcome with searing ash and toxic gases. Some sixteen thousand lives were lost, mainly by suffocation; the residents of nearby Herculaneum, for example, were overcome by boiling mudflows. Vesuvius is the only active volcano on mainland Europe. Another violent eruption took place in 1631, causing the loss of eighteen thousand lives. During World War II, on March 18, 1944, an eruption forced the evacuation of the city of Naples, and lava flows caused extensive damage at an air base close to the mountain. Mount Vesuvius is a complex stratovolcano, with lava flows generally following explosive eruptions.

Methods of Study

A major goal in volcanology is the prediction of time, place, and nature of an eruption, so as to minimize loss of lives and property. A variety of methods are employed, for no single factor is sufficient to warn of impending eruption.

Geophysicists monitor the earthquakes and tremors that accompany volcanic eruptions with seismographs that measure the intensity of earthquake activity. Almost every eruption or increase in volcanic activity is preceded by earthquakes—often very small earthquakes, or micro-earthquakes, that occur in swarms of hundreds or thousands. A volcanic tremor is almost always present during an eruption and often begins before the surface outbreak. This motion has a natural frequency from 0.5 to 1.0 hertz and produces a very low hum. The source of this noise is not clearly understood, but it may be related to the formation of gas bubbles in the lava.

Small changes in the slopes, shown by distances between markers on the flanks and summits of active volcanoes, are another precursor of volcanic activity. Techniques for detecting such changes include conventional leveling, reflected light beams, and instruments called tiltmeters capable of measuring changes in slope of less than one part in a million. One type of tiltmeter uses brass pots filled with water and fastened to support posts connected by hoses. The posts are placed a few meters apart. Micrometers that measure fractional water-level change are placed inside the pots.

With the rise of magma toward the surface, the earth's magnetic field is disturbed by molten lava, which loses its magnetic properties at high temperatures. Such changes may be detected by a magnetometer carried aerially over the volcano. Moving magma within the volcanic chambers may cause changes in the electrical currents within the ground. Sensitive resistivity meters placed under the surface are used to measure electrical conductivity. An increase in temperature of hot springs, fumaroles, and groundwater may signal an eruption; several months before the 1965 eruption of Taal in the Philippines, the temperature of the water inside the crater increased by 12 degrees Celsius. The chemical composition of waters and volcanic emissions may also change during eruptions. An increase in sulfur dioxide or hydrochloric acid in groundwater or surface emissions can precede an eruption.

Direct visual observations are helpful in monitoring volcanoes. Ice and snow on a volcano may melt from the intense heat. A volcano may bulge from movement of magma along one of its flanks, as was observed prior to the 1980 eruption of Mount St. Helens. Active volcanoes are observed from nearby observatories or sentry outposts. Valuable contributions are made by volcanologists in these facilities, but close observation can be very dangerous.

Context

The majority of the world's six hundred active volcanoes lie within the Pacific Ocean basin, with about half in the western Pacific region. They form an almost continuous pattern, known as the Ring of Fire, along the edges of the Pacific. Stratovolcanoes that form along the ocean margins are termed marine stratovolcanoes. Their initial structure resembles basaltic seamounts, with pillow basalts building up on the sea floor, followed by pyroclastics and less effusive eruptions.

Deep marine volcanoes are less explosive than volcanoes on land, because the pressure of the seawater retards the expansion of steam within the magma. Subduction-zone and island-arc volcanoes, on the other hand, contain a high concentration of gases in the upper parts of their magma chambers, making them more explosive than are volcanoes in other locations. The lava tends to have a greater amount of silica, so it is more viscous and resistant to flow. The lava that solidifies within the vent, or throat, forms a hardened plug that is blasted into pieces, or pyroclastics, by the pressure of the gas trapped below. Lava flows that are rough and blocky are termed aa, while those with a smooth, ropy texture are called pahoehoe.

Island-arc stratovolcanoes have had an explosive and dangerous history. Volcanoes such as Krakatoa and Tambora in Indonesia as well as Rabaul in Papua New Guinea are examples. It is believed that these volcanoes are economically important for the deposition of metallic sulfides. The magma chambers beneath may be regions where copper, molybdenum, and gold are concentrated during the final consolidation.

It is important to study volcanoes to minimize volcano damage, including loss of lives. Various methods have been devised to reduce volcano damage; these include the digging of channels and construction of levees to divert lava flows. Lava streams have been doused with water in attempts to slow down and solidify the lava. In Java, dams have been constructed to direct volcanic mudflows away from cities and agricultural lands. In Hawaii, the U.S. Air Force has bombed lava flows emerging from Mauna Loa, with limited success.

Some volcanoes eject considerable amounts of ash and dust into the atmosphere. Certain massive volcanic eruptions had a significant effect on global temperatures for months afterward. The 1963 eruption of Mount Agung, a stratovolcano in Bali, Indonesia, ejected clouds of gas and dust 10 kilometers into the stratosphere. The average worldwide temperature dropped 0.5 degree Celsius for three years after the eruption. Whether volcanic eruptions do effect significant changes in the weather and climate cannot be satisfactorily answered until scientists can obtain accurate measurements of the volcanic dust and debris entering the stratosphere and the heating and cooling of the earth's atmosphere that result from such dust.

Bibliography

Bullard, F. M. *Volcanoes of Earth.* Austin: University of Texas Press, 1984. A definitive study of volcanoes that includes sections on mythology and early speculation, how volcanoes are classified, the types of volcanic eruptions, and the distribution, current activity, theory, and environmental effects of volcanoes. Many diagrams, photographs, and tables make this a valuable resource for the geology student.

Cas, R. A. F., and J. V. Wright. *Volcanic Successions, Modern and Ancient.* Winchester, Mass.: Allen & Unwin, 1987. A very detailed work on the classification of modern and ancient volcanic rocks and volcano source types. Detailed geologic columns and cross sections show types and thickness of volcanic rock successions. A chapter on stratovolcanoes includes tables of dimensions and average lifetime rates. Suitable for the student of geology.

Chester, D. K., A. M. Duncan, J. E. Guest, and C. R. J. Kilburn. *Mount Etna: The Anatomy of a Volcano.* Stanford, Calif.: Stanford University Press, 1985. A comprehensive study of the major volcano on the island of Sicily. An interesting historical chronology of both flank and summit eruptions is included. Another table presents the effects of eruptions and earthquakes on settlements in the region. Includes many black-and-white photographs. Some chapters are technical and are best suited to the geology student, while other sections, especially the historical activity, would be of interest to the general reader.

Decker, Robert, and Barbara Decker. *Volcanoes.* San Francisco: W. H. Freeman, 1981. A short, easy-to-read book on volcanoes for the layperson. Chapters deal with where and how volcanoes form; spectacular eruptions, including the Krakatoa explosion and the 1980 eruption of Mount St. Helens, are described. Volcanoes' effect on world climate and the forecasting of eruptions are also discussed.

Erickson, Jon. *Volcanoes and Earthquakes.* Blue Ridge Summit, Pa.: TAB Books, 1987. An excellent resource for the general reader. A chronology is given for the major eruptions and their toll in human lives; more detailed descriptions of fourteen different volcanic eruptions follow this table. Chapters have been added on the formation of the solar system and the evidence for plate tectonics and continental drift. Volcanoes that have been detected on the other planets and moons are also described.

Lambert, M. B. *Volcanoes.* Seattle: University of Washington Press, 1978. Features a large format with many illustrations of volcanoes and associated structures and landforms. Various types of lava and other volcanic products are described. Full page black-and-white photographs of volcanoes and spectacular eruptions are included. A very good source of information for the general public as well as the geology student.

Macdonald, Gordon A. *Volcanoes.* Englewood Cliffs, N.J.: Prentice-Hall, 1972. A comprehensive, definitive study of volcanic activity, suitable for all levels of reader. The growth of the field of volcanology is traced. The composition of volcanic rocks and lava is analyzed, and the structures of various stratovolcanoes are illustrated, with explanations of how such mountains attain their symmetrical shape and large size.

Time-Life Books. *Volcano.* Alexandria, Va.: Author, 1982. A most highly recommended general resource on volcanoes. Many full-page color plates and diagrams clarify the text. Spectacular eruption photographs are provided for Vesuvius, Pelée, Mount St. Helens, and others, with major chapters devoted to famous eruptions.

Van Wyk de Vries, B., and P. W. Francis. "Catastrophic Collapse at Stratovolcanoes Induced by Gradual Volcano Spreading." *Nature* 387, no. 6631 (May 22, 1997): 387-391.

Michael Broyles

Cross-References

Calderas, 43; Pyroclastic Rocks, 512; Shield Volcanoes, 577; Subduction and Orogeny, 615; Volcanoes: Recent Eruptions, 675; Volcanoes: Types of Eruption, 685.

Stress and Strain

Stress and strain have to do with why and how a solid body deforms. Each point within a body under a load will have a set of stresses associated with it, varying in direction, magnitude, and the planes on which they act, according to the intensity of the forces acting within the body at that point. Each point within a deformed body will have a set of strains associated with it that indicate the translation, rotation, dilatation, and distortion experienced by the material at that point during the deformation.

Field of study: Structural geology

Principal terms

DILATATION: the change in the area or volume of a body; also known as dilation

DISTORTION: the change in shape of a body

NORMAL STRESS: that component of the stress on a plane that is acting in a direction perpendicular to the plane

ROTATION: the change in orientation of a body

SHEAR STRESS: that component of the stress on a plane that is acting in a direction parallel to the plane

STRAIN: a measure of deformation, including translation, rotation, dilatation, and distortion

STRESS: the intensity of forces (force per unit area) acting within a body; may refer to a particular stress or to the collection of all stresses acting on all planes at a given point

TRANSLATION: the movement of a body from one point to another

Summary

Stress and strain are concepts that help to explain how and why rocks deform. Strain describes the deformation, and stress pertains to the system of forces that produce it. In considering strain, first it will be helpful to review some aspects of the physics of rigid bodies in order to appreciate the significance of these concepts.

When dealing with many problems in mechanics, it is common to assume that the bodies involved are perfectly rigid; that is, they do not deform. Such problems usually involve the balancing of forces, or, if forces do not balance, determining the resulting accelerations. Movement of a rigid body can involve translation, rotation, or both, but the individual points within the body do not move relative to one another. In a translation, all points within the body move the same linear distance and in the same direction. In a rotation, all points within a body rotate through the same angle around the same center of the rotation. Any rigid body motion can be described in terms of a translation plus a rotation. Deformation introduces further complications. A volume-conserving, shape-changing deformation is called distortion. A change in volume, without a change in shape, is called dilatation (or dilation). Strain combines all four of these possibilities: translation, rotation, distortion, and dilatation.

If the beginning and ending locations and orientations of a rigid object are known, it is easy and straightforward to determine the net translation and the net rotation of the object and of every point within the object. For example, if an airplane begins in New York facing north as it is loaded and ends up in Madrid facing northwest as it is unloaded, then it can be said to have moved 5,781 kilometers to the east and rotated 45 degrees counterclockwise. Similarly, every piece of luggage on that airplane had a net translation of 5,781 kilometers to the east and a net rotation of 45 degrees counterclockwise. Very little information is needed to determine such net displacements and rotations, but the path that the airplane took is not well represented by them. The plane probably flew along a great circular route, changing its bearings constantly, and it very likely circled a bit after taking off and again before landing. To describe the path of the plane, one would need much more data. These data might consist of a series of translations and rotations taken at one-minute intervals. Each item of luggage, rigidly fixed within the hold of the aircraft, would move through an exactly identical series of translations and rotations. Furthermore, by applying the basic laws of mechanics, one could attribute each acceleration (linear or angular) to the forces resulting from the interplay of the thrust of the engines, the force of gravity, air resistance, prevailing winds, and other relevant factors.

In much the same way that net translations and rotations can be determined by knowing original and final locations and orientations, net distortions and dilatations can often be determined relatively easily when initial and resultant shapes and volumes are known. Analysis is simplified if the area of study can be divided into subareas such that straight, parallel lines within each area remain straight and parallel after deformation. Such deformation, called homogeneous strain, is often assumed in the study of strain. Under these conditions, initially circular objects deform into ellipses.

Determining the strain path requires a series of known translations, rotations, distortions, and dilatations, while, in order to tie the strain to the series of forces and stresses that produced it, one needs to know the strain path. Just as there are an infinite number of ways to fly from New York to Madrid, there are an infinite number of strain paths that could result in identical net strains. As an indication of the problem, consider a circle 1 centimeter in radius that deforms into an ellipse with a semimajor axis of

2 centimeters and a semiminor axis of 0.5 centimeter. Although there is no net dilatation, the deformation may have consisted of stretching in one direction and shrinking in the direction perpendicular to it. Alternatively, this deformation could have been produced entirely by distortion, as can be seen by drawing a circle on the edge of a deck of cards and then moving each card slightly to the right of the one below it. Each card will have two spots on it, one from each side of the circle. Since the distance between the spots on an individual card does not change, and the number of cards does not change, the area inside the resulting ellipse will remain constant. Continuing to deform the deck in this manner (a process called shearing) will result in the ellipse getting longer and thinner.

Strains are produced by stresses somewhat as movements of rigid bodies are produced by forces. More specifically, unbalanced forces acting on a rigid object cause it to accelerate. The amount of acceleration can be calculated if the net force and the mass of the object are known. The intensity of the forces acting within a body cause it to deform, and this force intensity is called stress. The units used to measure stress are the same as those used to measure pressure and are given in terms of force per unit area. Data may be presented in terms of atmospheres, pounds per square inch, bars, or similar units. The appropriate unit (based on the International

The enormous stresses within the earth's crust produce earthquakes such as the 1994 tremor that damaged structures in Southern California. *(Reuters/Fred Prouser/Archive Photos)*

System of Units) is the pascal, defined as 1 newton per square meter, or 1 kilogram per meter-second-squared. It is important to note that stress measurements contain an area term, and therefore they cannot be added, subtracted, or resolved as if they were forces. By multiplying a stress by the area over which it is applied it can be converted to a force, which can then be treated like any other force. It is customary to resolve it into forces parallel and perpendicular to a plane of interest. Finally, by dividing by the area of this plane, stresses can be obtained once again, yielding the shear stress and normal stress, respectively.

These factors can be demonstrated with a simple case in which a cube, 1 square meter on a side, has two forces acting on it in the vertical direction: One force of 10 newtons is pushing down on the top, another of 10 newtons is pushing up on the bottom. The forces balance, so there will be no acceleration. Any horizontal plane within this cube will have an area of 1 square meter. It is subject to stresses of 10 pascals, perpendicular to the plane, acting on each side of it. A diagonal plane through this cube, cutting the cube in half from one edge to the other, has an area of 1.414 square meters. The component of the vertical downward force acting perpendicular to this plane (the normal force) will be 7.071 newtons, and there will also be a component of the vertical downward force acting parallel to this plane (the shear force) of 7.071 newtons. When these forces are divided by the area over which they act, it is apparent that there will be a normal stress of 5 pascals and a shear stress of 5 pascals acting on the upper surface of this plane. Similar stresses can be shown to exist on the lower surface of the plane. Planes with different orientations will have other combinations of normal stresses and shear stresses, even though the forces responsible for those stresses remain the same. There are equations to manipulate the general situation, which give the normal and shear stresses acting on any plane as functions of the size and directions of the boundary forces. These result in what is called the stress ellipse, in two dimensions, or the stress ellipsoid in three. A graphical way of representing these equations (and the equations for strain also) was developed by Otto Mohr in 1882 and is now called Mohr's Circle.

The results obtained above may be compared with those for a hydrostatic condition, where stresses are the same in every direction. If stresses of 10 pascals are acting on all six sides of the cube, no matter what plane one considers inside the cube, there will be normal stresses of 10 pascals acting on each side of it. The stress ellipses and ellipsoids one might construct would be circles and spheres, and there would be no shear stresses anywhere.

Different modes of deformation are favored by different combinations of stresses. Movement on a fault plane, for example, is favored by low normal stresses and high shear stresses on that plane. Through the simple analysis described above, it becomes clear that faulting is much more likely to occur along diagonal planes than along horizontal ones.

Methods of Study

The study of stresses often involves determining the stress field in a particular area, either at present or at some time in the past. After some simplifying assumptions are made concerning the geometry, mechanical properties, and boundary stresses of the area, a model is constructed that will indicate certain aspects of the stress field. Sometimes the model can be a physical one, produced from photoelastic plastic, for example. Such models can display the magnitudes of shear stresses when viewed appropriately with polarized light. More often, though, the model is constructed on a computer, and the stress ellipses are calculated for points of interest throughout the area. If geological stress indicators exist, such as the igneous sheet intrusions called dikes, the results of the model can be compared with the observed indicators, and the model can be adjusted until it fits the observations as closely as possible.

Determining at least parts of the strain field is in some ways more direct. Objects are sought in those rocks of the area that have net strains that can be determined. Frequently the distortion experienced by such an object can be easily observed and measured. A fossil that is elliptical but is known from its appearance in other areas to have been circular when it was alive provides a simple example. Strain ellipses showing the distortion of such objects can be constructed by measuring the shapes of these objects in the field.

These ellipses can then be plotted on a map. (A map with ellipses on it is a way of representing a tensor field, for example, a stress field or a strain field.) Most of the time, however, the initial size, location, and orientation of the objects are not known. One cannot tell whether a particular fossil is small because it never grew very large or whether it was once large and became smaller by deformation. If all the deformation occurred within a limited period of time, this map will represent the distortion part of the net strain field for that deformation. When similar data obtained from rocks deformed at different times are combined, a partial strain path can be obtained. The effects of more recent deformations are removed from the effects of earlier ones to isolate the earlier distortions.

The next step is to determine the stresses responsible for each increment of strain observed. To do so, it is necessary to know how each of the rocks responded to stress. Such data on mechanical behavior come from studies of experimental rock deformation. With these data, estimates can be made of the stress field present at different times in the history of the area. Finally, all this work can be applied to the known geological history of the area, permitting quantitative assessments of the various forces thought to have been active in the past.

Context

The stresses in a body are a function of the geometry of the body and the distribution of loads acting on it and within it. Determining the distribution of stresses is usually considered to be an exercise in statics, a branch of mechanical engineering, but it also plays a significant role in the earth sciences. Earthquakes occur when rock fails suddenly. Mine collapses, landslides, and dam failures are other catastrophes that occur when stress exceeds the strength of the material involved. In discussing the strength of a material, similar geometries are compared. When one says that steel is stronger than hemp, one means that a steel cable would support more weight than would a hemp rope of the same diameter. Consequently, strength is described in terms of force per unit area. Stress is measured this way as well.

When a load is placed on a solid, the distribution of stresses within that solid is usually uneven. If the solid deforms, the deformation will usually also be uneven. Striking a piece of pottery with a hammer will cause it to shatter, but the number of cracks will usually be greatest near the point of impact. Hitting a lump of clay with a hammer will leave an imprint of the hammer head where the clay deformed in a ductile fashion by moving out of the way. To measure such deformation, one examines strain, which includes movements and changes in size and shape. As with stress, strain usually varies throughout the region being deformed.

A structural geologist is often concerned with determining how a region of the crust of the earth became deformed, and then why it deformed that way. Rocks often contain objects that are presently deformed but whose original shapes are known; such strain indicators include fossils, raindrop impressions, and bubbles. Using these indicators, a geologist seeks to reconstruct the strain field that existed at some time in the past. With enough indicators, along with dates for each, it may be possible to construct a strain history for the area in question. The next step is to guess, using the known mechanical behavior of the rocks involved, what the stresses were that produced the reconstructed strains. A final goal might be to seek causes for those stresses in terms of a larger picture of earth history, perhaps involving plate tectonics.

Bibliography

Davis, George H. *Structural Geology of Rocks and Regions.* New York: John Wiley & Sons, 1984. Chapter 4, "Kinematic Analysis" (44 pages), deals with strain, and chapter 5, "Dynamic Analysis" (30 pages), deals with stress. The treatment is the most descriptive and least technical of the references listed here, although some knowledge of stereonets is assumed. The double subscripts often used in the field are not employed here, and calculus is carefully avoided. Suitable for the general reader.

Hobbs, Bruce E., Winthrop D. Means, and Paul F. Williams. *An Outline of Structural Geology.* New York: John Wiley & Sons, 1976. The first section, "Mechanical Aspects" (71 pages), covers stress, strain, and the response of rocks to stress. The approach is largely descriptive, with no calculus, but double subscripts and some linear algebra are used. Although presumably written by Means, this section is sufficiently different from his book to complement it. Suitable for college students.

Johnson, Arvid M. *Physical Processes in Geology*. San Francisco: Freeman, Cooper, 1984. Chapter 5, "Theoretical Interlude: Stress, Strain, and Elastic Constants" (43 pages), develops a fairly rigorous treatment of stress and strain. Includes partial differential equations and other elements of calculus. Suitable for technically oriented college students.

Marshak, Stephen, and Gautam Mitra, eds. *Basic Methods of Structural Geology*. Englewood Cliffs, N.J.: Prentice-Hall, 1988. Chapter 15, entitled "Analysis of Two-Dimensional Finite Strain," by Carol Simpson (26 pages), presents an overview of strain, with easily understood analogies, followed by an excellent survey of the techniques that have been used to measure strain in the field. The approach is definitely "how-to," with each step of each method clearly spelled out. Suitable for college students.

Means, W. D. *Stress and Strain*. New York: Springer-Verlag, 1976. This 339-page text is intended to be used for self-study by college undergraduates. Each of the twenty-seven chapters is followed by a number of problems, and solutions to the problems are provided. Although several of the chapters go into partial differential equations and other technical subjects, most of the book is easily read by nonspecialists. Suitable for college students.

Ragan, Donal M. *Structural Geology: An Introduction to Geometrical Techniques*. 3d ed. New York: John Wiley & Sons, 1985. Chapter 7, "Stress" (18 pages), chapter 9, "Concepts of Deformation" (24 pages), and chapter 10, "Strain in Rocks," introduce the concepts in a mathematically straightforward way. The use of a card deck as a means of demonstrating strain was pioneered by Ragan and is included in this text. Suitable for college students.

Ramsay, John G. *Folding and Fracturing of Rocks*. New York: McGraw-Hill, 1967. The classic book on strain and strain analysis. Chapter 3, "Strain in Two Dimensions" (71 pages), chapter 4, "Strain in Three Dimensions" (64 pages), and chapter 5, "Determination of Finite Strain in Rocks" (70 pages), have provided the background in these subjects to generations of geologists. Although much of what is presented is fairly technical, the abundant photographs and line drawings and some of the descriptions in the text are useful to the general reader. A quick browse through this book furnishes valuable insight into what is involved in strain analysis. Suitable for college students.

Suppe, John. *Principles of Structural Geology*. Englewood Cliffs, N.J.: Prentice-Hall, 1985. Chapter 3, "Strain and Stress" (33 pages), presents an elegant development of the subject. It is the only reference listed here that uses the Einstein summation convention and relies heavily on matrix manipulations. Thus, mathematically, this work may be above the level of most nontechnical college students. The eight photographs and the discussion of strain versus displacement are excellent, however, and useful to the general reader.

Otto H. Muller

Cross-References

Earthquakes, 148; Folds, 248; Igneous Rock Bodies, 307; Joints, 340; Normal Faults, 448; Rocks: Physical Properties, 550; Thrust Faults, 631.

Subduction and Orogeny

Subduction and orogeny are fundamental consequences of plate tectonics and are the two processes that build mountains on the edges of continents. Through the recognition of subduction, scientists have been better able to determine regions where risks of earthquakes and volcanic explosions are significant.

Field of study: Tectonics

Principal terms

CONTINENTAL MARGIN: the edge of a continent that is both exposed on land and submerged below the water that marks the transition to the ocean basin

CRUST: the outermost layer of the earth, which consists of materials that are relatively light; the continental crust is lighter than oceanic crust, which allows it to float while oceanic crust sinks

FAULTING: the process of fracturing the earth such that rocks on opposite sides of the fracture move relative to one another; faults are the structures produced during the process

FOLDING: the process of bending initially horizontal layers of rock so that they dip; folds are the features produced by folding and can be as small as millimeters and as big as kilometers long

GEOSYNCLINES: a major depression in the surface of the earth where sediments accumulate; geosynclines lie parallel to the edges of continents and are long and narrow

INTRUSION: the process of forcing a body of molten rock generally derived from depths of tens of kilometers in the earth into solidified rock at the surface

MAGMA: molten rock that is the source for volcanic eruptions

Summary

Subduction and orogeny are two processes that are fundamental to the evolution of continents. All continents contain long, narrow mountain chains near their edges that are composed of folded and faulted rocks that are younger than the rocks in the continental interiors. The event that

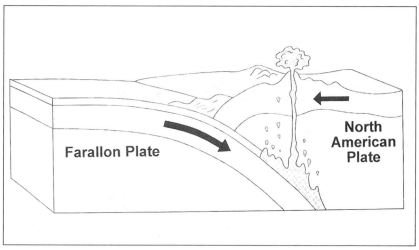

North
American
Plate

Farallon Plate

Subduction zones are m rked by linear belts of volcanoes on the overriding plate. *(Pinnacles National Monument)*

formed the mountains is termed an orogeny, and the mountain chain itself is called an orogenic belt. Because of the proximity of mountain chains to the edges of continents, scientists have believed for centuries that orogenies reflected movements localized along continental margins. It is only recently, however, that orogeny was coupled with subduction, the process in which sea floor descends below a continent or another piece of sea floor. Earlier views of orogeny were part of the theory of geosynclines. Geosynclines are linear basins that form on subsiding regions of the earth's surface adjacent to continental margins, fill with sediment, and evolve into mountains composed of folded and faulted sedimentary strata. The origin of the compressive forces responsible for the creation of the mountains was not known. Erosion of the newly created mountains provides sediment for new geosynclines that develop seaward of the mountain belt, thereby completing one geosynclinal cycle, which typically lasts on the order of a few hundred million years. The advent of the theory of plate tectonics in the 1960's led the majority of the scientific community to abandon the geosynclinal cycle in favor of the subduction process as an explanation for orogenies. Subduction was attractive because it readily provided a mechanism by which the large compressive forces needed to form mountains could be produced.

Orogenic belts are characterized by the folding and faulting of layers of rock, by the intrusion of magma, and by volcanism. Folds and faults form parallel to the continental margin and extend hundreds of kilometers toward the continental interior. Folding bends layers of rocks, whereas faulting takes rocks that were side-by-side and stacks them on top of each other in sheets up to 20 kilometers thick. Both processes significantly shorten the horizontal and thicken the vertical dimensions of the continents. At the same time as they are folded and faulted, the rocks are intruded by magmas derived from tens of kilometers below the surface. Some of the magmas eventually erupt, building volcanoes on the deformed rocks. An additional feature of orogenic belts is the juxtaposition of sequences of rock that have nothing in common with each other. The rocks in the two sequences may be different in age, composition, or style of folding. The origin of this juxtaposition was unreconciled by the theory of geosynclines, which holds that all rocks in a mountain belt were originally deposited near one another and were derived from the same source. The theory of plate tectonics, however, easily explains the juxtaposition.

In order to understand subduction and orogeny, one must have a clear grasp of the theory of plate tectonics. The theory states that the surface of the earth is composed of about twelve rigid plates which are less than 100 kilometers thick. Plates are either oceanic or continental. Below the plates is a partially molten layer that allows the rigid plates to float and move relative to each other at speeds between 2 and 10 centimeters per year. The motions are defined primarily by the oceanic plates; the continental plates drift passively. The relative motions of the plates define three types of boundaries: convergent, where plates move toward one another; divergent, where plates spread apart; and transcurrent, where plates slide smoothly past each other. Convergent boundaries are frequently along the margins of continents, and divergent boundaries are commonly in the ocean basins. For example, the west coast of South America is a convergent boundary, and the Mid-Atlantic Ridge, the mountain range that runs down the middle of the Atlantic Ocean, is a divergent boundary. Divergent boundaries are zones along which two plates separate. This type of plate boundary is typically demarcated by a linear ridge system in an ocean basin where magma rises from deep in the earth to fill the gap created by the diverging plates. When the hot magma contacts the cold seawater, it solidifies into new oceanic crust. As the plates continue to separate, additional magma wells up from the earth's interior, allowing the continuous creation of oceanic crust at the ridge. This process is known as sea-floor spreading and is responsible for the drifting of the continents on the surface of the earth. Convergent boundaries are where two plates move toward each other and one plate subducts, or descends below, the other. The subducting plate is always oceanic, but the overriding plate may be either oceanic or continental. This reflects the greater density of oceanic crust relative to continental crust, which allows the oceanic plates to sink readily into the earth's interior whereas the continental plates remain afloat. When two continental plates collide, neither plate subducts—they are too light—but the plates push against each other with tremendous force such that their edges buckle and huge mountain ranges grow. This process built the world's tallest mountains, the Himalaya, which are the result of the collision between the subcontinent of India and the continent of Asia.

Subduction zones are characterized by a progression from the subducting to the overriding plate of deep trenches, high mountains, and many volcanoes that occupy an area hundreds of kilometers wide and thousands of kilometers long. The deep trench, frequently filled with sediments eroded from the adjacent mountains, marks the point in the ocean floor where the subducting plate bends to descend below the overriding plate. As the oceanic plate descends, these sediments are scraped onto the overriding plate. Slivers of oceanic crust may also scrape off and mix with the sediments. The offscraped rocks form an intricately folded and faulted region tens of kilometers wide and several kilometers high at the edge of the overriding plate. These complexly deformed mixtures of sediments and slivers of oceanic crust are called melanges and are characteristic of most ancient subduction zones now exposed on land.

Another important feature of subduction zones is the linear belt of volcanoes on the overriding plate that parallels the plate boundary. The volcanoes grow from the eruption of magma that is generated at the interface between the subducting and overriding plates at depths between 100 and 200 kilometers. At these depths, the temperature of the earth is hot enough to melt small areas of either the subducting or the overriding plate. The magma rises, intruding the rocks at the surface and eventually erupting to build the volcanic belt. Some of the magma, however, may solidify between the top of the oceanic plate and the surface.

The similarity of features in orogenic belts and subduction zones is striking and forces the obvious conclusion that subduction leads directly to orogeny. An orogeny can occur either during subduction of an oceanic plate below a continental plate, such as on the west coast of South America, or during the collision of two continental plates, such as in the Himalaya. Because continents do not subduct, the compressive forces are much greater in a continent-continent collision than in seafloor subduction. The mountains produced during collision (Himalaya), therefore, are much taller than those generated during subduction (Andes).

The theory of plate tectonics elucidates important differences between the oceans and the continents and provides a mechanism by which different rock sequences can be juxtaposed in orogenic belts. The ocean basins are transient features that are constantly modified by the growth and destruction of new sea floor at divergent and convergent boundaries, respectively. In contrast, the continents are too light to be subducted and are permanent features of the earth's surface. This consequence of plate tectonics is supported by the 200 million year age of the oldest sea floor and the 4 billion year age of the most ancient rocks on the continents. Continents, therefore, drift, fragment, and collide as relative plate motions change through geologic time. The collision of continents that were once widely separated allows the bringing together of rocks that have had very different histories. As the collision leads to orogeny, these different sequences of rock may be juxtaposed in the same mountain belt.

The difference between the age of orogenic belts and the interiors of continents implies that the continents have evolved through time by the addition of material at their edges during orogenies. Orogenic belts are also of different ages, ranging from a billion years to zero (actively forming). Two or three belts whose ages decrease away from the continental interior may define one edge of a continent. This suggests that orogenies have occurred repeatedly through geologic time and that continents have added material continuously to their margins since the formation of their interiors. Because the ocean floor is so young, orogenic belts are the only record of subduction and collision events prior to 200 million years ago. If subduction is the only mechanism responsible for orogeny, plate tectonics must have been active since early in the history of the earth.

Methods of Study

Subduction and orogeny are studied by hundreds of scientists, each of whom looks at only a small part of the picture. One may determine the composition of volcanic rocks that are characteristic of subduction zones; another may examine the styles of folds and faults in orogenic belts. Three techniques, however, are dominant in the study of subduction and orogeny: the analysis of the locations and sizes of earthquakes; the discrimination of relationships between different types of rocks in the field; and the investigation of features in deep-sea trenches and in the submerged region of folded and faulted rocks. The first defines where subduction and orogeny occur today, whereas the second determines what the physiographic expressions of these processes are, how they are preserved in the rocks, and where they were active in the past. The third technique provides a direct link between subduction and orogeny and illustrates the early stages of development of a mountain belt.

One of the most important discoveries of plate tectonics was that earthquake zones define plate boundaries. Earthquakes occur when a fracture, or fault, forms in the earth's crust, and the two pieces on either side of the fault move, or slip, past each other. For large earthquakes, the slip is on the order of 10-20 meters. The forces responsible for faulting are simply the result of the relative motions of the plates at the plate boundaries. The motion can accumulate in the rocks for hundreds of years prior to causing a rupture. When the crust finally breaks, the energy stored by the rocks is released suddenly as waves that travel through the earth and generate the intense vibrations associated with an earthquake. The rupture continues for as much as 1,000 kilometers and moves at speeds in excess of 10,000 kilometers per hour.

The energy carried by the waves is recorded on seismographs, which are instruments that monitor ground motion. Seismographs are composed of a mass attached to a pendulum. The mass remains still during an earthquake, measuring the amount the earth moves around it. The motion is recorded on a chart as a series of sharp peaks and valleys that deviate from the background value measured during times of no earthquake activity. The arrival of the waves at different times at different places allows the geophysicist to calculate the location, or epicenter, of the earthquake. The amount of the deviation of the peaks and valleys from the background noise is an estimate of the magnitude of the earthquake.

Earthquakes near mountain belts define zones that extend at an angle from the surface of the earth at the deep-sea trench to depths of hundreds of kilometers below the continents. This zone corresponds to the subducting plate at a convergent boundary. As a result, the locations of subduction zones that are currently active are very well known. The descent of a subducting plate below an overriding continent has triggered some of the deepest and largest earthquakes ever recorded. Continued motion of the plate and rupturing of the earth's crust in response translates into mountain ranges on the earth's surface.

Analysis of earthquakes is essential to evaluate the modern plate tectonic setting of the earth but reveals nothing about the geologic past. Information about plate tectonics of the past must be obtained from looking at ancient mountain belts. Recognition of relationships among rocks in the field involves determining the ages, compositions, and histories of the rocks. This process led to the discovery that mountain belts on different continents contained rock sequences that were very similar. For example, rocks in the Appalachian Mountains of the east coast of North America were found to match closely those in the Atlas Mountains of the west coast of Africa. Conversely, recognition of relationships determined that dissimilar rock sequences frequently are adjacent to each other in the same orogenic belt. Both phenomena are most readily explained by continental drift, sea-floor spreading, and subduction.

The critical link between ancient orogenic belts and modern subduction zones identified by earthquake activity was provided by deep-sea trenches. Using highly sophisticated techniques to "see" the ocean floor, scientists discovered the region of offscraped rocks that lies on the overriding plate in a subduction zone. These regions sometimes continue to the continental margin where they are exposed on land as mountains. Thus, subduction was observed to cause folding and faulting in rocks and to build mountains, both important processes in orogenic belts.

Context

The theory of plate tectonics provides scientists with a process that can be observed—subduction—to explain the origin of mountains. Because young mountain chains are the locus of most of the large earthquakes that occur today, understanding subduction yields insight into the potential for destructive earthquakes in any given area. This is extremely important; most of the global population lives along convergent plate boundaries. The identification of subduction zones at the margins of the Pacific Ocean has explained the so-called "ring of fire": a region of abundant earthquakes and volcanoes that had long puzzled the scientific community. Restriction of most earthquakes to plate boundaries allows the assessment of earthquake hazards anywhere in the world if the location of plate boundaries is known. For example, the city of Santiago in Chile, which is above a subduction zone, has a high risk, whereas the city of Chicago in the United States, which is in the continental interior, has a low risk.

Additional information can also be gathered about the type of earthquakes that may occur. In subduction zones, the piece of the crust that is above the rupture typically moves upward relative to the piece below, which generates waves that shake the ground in certain directions. At divergent plate boundaries, however, the piece of crust that is above the rupture moves downward relative to the piece below. This motion produces waves that vibrate the ground in directions different from those generated by earthquakes in subduction zones. Additional differences between convergent and divergent boundaries that may affect ground motion include the depth and size of the earthquakes. Subduction zones generate the deepest and largest earthquakes; earthquakes at divergent plate boundaries are more frequent, smaller, and shallower. Knowledge of the way the ground may move helps civil engineers to design and construct buildings able to withstand large earthquakes.

Eruptions of volcanoes that lie above subduction zones can be devastating. These volcanoes typically erupt violently and explosively in contrast to volcanoes near mid-ocean ridges, which erupt quietly and smoothly. This reflects the greater viscosity (resistance to flow) of magmas at convergent boundaries relative to those at divergent boundaries. Because of their greater viscosity, the magmas above subduction zones tend to plug the volcanoes at the surface, preventing any eruptions. Finally, when the pressure below the plug is great enough, the volcano erupts with such force that cities nearby are damaged considerably. For example, in A.D. 79, the entire city of Pompeii, Italy, was destroyed, and hundreds of people were killed by the volcano, Vesuvius. Clearly, the investigation of subduction and orogeny is beneficial to understanding the forces of nature that are harmful to humankind. Perhaps someday in the future, large earthquakes and violent volcanic eruptions may be predicted far enough in advance that precautions can be taken to prevent the loss of human life.

Bibliography

Press, F., and R. Siever. *The Earth.* 4th ed. New York: W. H. Freeman, 1986. An excellently illustrated introductory text on geology, the book has five chapters that deal with folding, faulting, plate tectonics, earthquakes, and orogeny. A map of the major plates is on the inside back cover. The glossary is huge and indispensable. Recommended for senior high school and college students.

Shelton, J. S. *Geology Illustrated.* San Francisco: W. H. Freeman, 1966. This book has a superb collection of photographs and sketches drawn from the photographs that illustrate specific geologic features such as folds, faults, and volcanic landforms. Although it does not talk about subduction and orogeny directly, this source does help the reader in visualizing the features representative of convergent plate boundaries.

Short, Nicholas M., and Robert W. Blair. *Geomorphology from Space: A Global Overview of Regional Landforms.* Washington, D.C.: National Aeronautics and Space Administration, 1986. This book contains beautiful pictures taken by various satellites that orbit the earth. Text accompanies each picture and explains the tectonic setting. Many mountain belts and volcanic chains are shown. Although the text is fairly technical, the photographs are worth examining by anyone. Recommended for college-level students.

Uyeda, Seiya. *The New View of the Earth: Moving Continents and Moving Oceans.* Translated by Masako Ohnuki. San Francisco: W. H. Freeman, 1978. The book discusses the historical context of the theory of plate tectonics in addition to explaining the theory very well. Nontechnical and designed for the nonscientist. Interesting stories about the people responsible for the theory abound. Suitable for anyone interested in plate tectonics.

Wilson, J. T., ed. *Continents Adrift.* San Francisco: W. H. Freeman, 1972.

_____. *Continents Adrift and Continents Aground.* San Francisco: W. H. Freeman, 1977. These two volumes are collections of articles originally printed in *Scientific American* magazine. The amount of overlap is very small, which amply illustrates the rapid advances in plate tectonics in the late 1960's and early 1970's. The articles are very well illustrated, and the introductions in each volume are very helpful. Suitable for anyone interested in any aspect of plate tectonics.

Pamela Jansma

Cross-References

Continental Growth, 102; Earthquakes, 148; Ocean Basins, 456; Plate Tectonics, 505.

Supercontinent Cycle

Supercontinent cycles, which recur over periods of 400 million to 440 million years, are helpful in understanding the distribution of certain natural resources, fluctuations throughout geologic time of sea level and climates, the process of mountain-building, and the evolution of life.

Field of study: Tectonics

Principal terms

CONTINENTAL DRIFT: an early theory of continental fragmentation and displacement causing the creation of new ocean basins and the formation of mountain ranges

CONVERGENCE: the second-half process of a supercontinent cycle, whereby crustal plates collide and intervening oceans disappear as a result of plate subduction

DIVERGENCE: the process of fracturing and dissecting a supercontinent, thereby creating new oceanic rock; divergence represents the initial half of the supercontinent cycle

OPHIOLITE SUITE: a unique vertical sequence of peridotite (very basic) rock overlain by gabbro, basalt (lava), and oceanic sediments representative of ancient sea-floor material

PLATE TECTONICS: a modern theory describing the earth surface as composed of rigid plates continually in motion over the interior, causing earthquakes, mountain-building, and volcanism

SEA-FLOOR SPREADING: the continual creation of new sea-floor bedrock along mid-ocean ridges through the process of ascending thermal currents

SUPERCONTINENT: a single vast continent formed by the collision and amalgamation of earth crustal plates

WILSON CYCLE: the creation and destruction of an ocean basin through the process of sea-floor spreading and subduction of existing ocean basins

Summary

One of the most persistent questions relating to the historical development of Earth concerns the processes whereby extensive mountain chains, such as the Himalaya, Appalachian, and Ural ranges, have been formed.

During the nineteenth century, the planet-contraction hypothesis, supported by the doctrine of permanence of continents and ocean basins, suggested that such linear features of the earth crust were comparable to the wrinkles within the skin of a dried apple. After the discovery of radioactivity in 1896, studies suggested heat formed by radioactive decay of rock minerals approximately equaled heat lost to the atmosphere by the gradual cooling of earth from its supposed original liquid stage. This balance of heat gain and loss did not support a shrinking-earth concept. With the additional knowledge that linear mountain ranges had formed during different stages of geologic history, rather than simultaneously, the contraction hypothesis gradually lost favor.

Between the publication of his 1915 book *The Origin of the Continents and Oceans* and his death during an expedition to the Greenland icecap in 1930, Alfred L. Wegener, a German meteorologist and geologist, gained international repute as the father and chief advocate of the theory of continental drift. Wegener postulated that mountain ranges were created by the collision of large blocks of continental crust moving through oceanic crust, following fragmentation and dispersion of the vast supercontinent he called "Pangaea" (Greek for "all lands"), surrounded approximately 200 million years ago by the universal ocean Panthalassa (from "thalassa," Greek for sea). An intriguing set of evidences supports this theory, including the unusual degree of geometric fit of present-day continents (especially South America with Africa) and, through the reassembly of Pangaea, the reconstruction of truncated salt, fossil-reef, glacial-deposit, and mountain-range trends. While many scientists at the time became "pro-drift," many others questioned what possible mechanism could displace solid continental crust through equally solid oceanic crust. Various earth forces, ranging from centrifugal to tidal to rotational axis-wobble, were investigated and rejected as inadequate by physicists and mathematicians. By the 1930's, continental drift as a theory was no longer considered viable and was increasingly mentioned only within the context of the history of science.

Following World War II, a new era of earth investigation began, focused primarily on the ocean basins. Employing newly developed military technology, including methods of water-depth sounding (fathometry) and magnetic-body detection, surplus military aircraft and surface-vessel equipment began to collect a wide array of information. By the International Geophysical Year of 1957-1958, ocean-floor depth-profile analyses confirmed the existence of a previously unknown, 65,000-kilometer-long global submarine mountain range system traversing the Atlantic, Pacific, Arctic, and Indian Oceans. This feature, identified almost five hundred years following the discovery of the Americas and the Pacific Ocean, defied immediate explanation.

During the same period, ocean-evaluation programs sponsored by both private and U.S. government interests began the routine task of measuring the magnetism of the ocean floor. By the mid-1950's, sufficient data had been collected off the west coast of the United States to reveal a repetitive north-south pattern representing alternate zones of above-average and below-average magnetism. Soon similar patterns were shown to exist in the Atlantic and Indian Oceans.

The discovery of these unexplainable phenomena prompted the collection of any form of additional data that would help to explain the existence of ocean floors dissected by a universal mountain range and masked by symmetrical magnetic patterns. New oceanographic programs gathered information on bedrock temperature, radiometric age, and ocean-sediment thickness and studied the worldwide distribution of earthquakes and volcanic eruptions.

By the early 1960's, broad-based analyses of these various forms of oceanographic data were being conducted in Canada, the United Kingdom, and the United States. Gradually, consensus began to form that perhaps at least some of the ideas of Alfred Wegener were worthy of reconsideration. Paramount among these resurrected ideas was that of the assembly of the supercontinent Pangaea. Rather than postulating continental crust as floating through oceanic crust, the revised continental drift theory, termed "plate tectonics," envisioned the outer layer of the earth as divided into a series of major plates, each composed of both continental and oceanic bedrock. As examples, the North American plate is made up of the continent of North America (including Greenland), the western half of the north Atlantic Ocean, and eastern Siberia, while the Indian-Australian plate is composed of the continent of Australia, the country of India, and portions of the Pacific and Indian Oceans. Major plate displacement was considered possible because of the movement of global thermal-convection cells, which form within the mantle of the earth and rise toward the surface until they are blocked by the presence of a supercontinent. The blockage of further transmission of heat, caused by the insulating nature of continental rock, divides the convection current into lateral, horizontally directed segments, which gradually dome by thermal expansion and then dissect the supercontinent.

As the new subcontinent begins to diverge, the separating void fills with high-density gabbro and basalt-type rock, which forms the floor of a newly developed ocean. Because the earth is neither expanding nor contracting in size, continuing divergence cannot proceed indefinitely without experiencing resistance caused by the convergence of antipodal plates. The effect of plate convergence depends on whether such plate margins are oceanic (basaltic) or continental (granitic) in nature. Where margins are oceanic, the more dense margin will subduct, or plunge under, the less dense margin. Where margins are both continental, subduction is unlikely, and the result is massive folding and faulting (earthquaking). Finally, where a continental margin converges with an oceanic margin, the latter, being more dense, will subduct beneath the former. In all three possible convergence cases, earth-crust shortening is accomplished. Rock volume harmony results, as the formation of new oceanic crust through plate divergence is matched by the destruction of older crust through the process of subduction.

Processes of divergence and convergence are believed to have been ongoing throughout a large portion of geologic time and to continue today. Divergence, accompanied by new-ocean development in region A, continues simultaneously with convergence in region B, resulting in the gradual destruction of region A ocean by way of subduction. Eventually, region A ocean will cease to exist, and the cycle of ocean birth and death will be complete. This sequence of events, during which it is estimated 2.6 square kilometers of ocean floor rock is created and destroyed each year, is termed a "supercontinent cycle" (also known as a "Wilson cycle," after the Canadian geologist J. Tuzo Wilson, an early advocate of the plate-tectonics theory). Conversely, the dispersion and amalgamation of continental crustal masses, as opposed to oceanic masses, constitutes the principle phases of what has been termed the "Pangaean cycle." The operation of plate tectonics continuously creates and recycles ocean basins, while continental regions increase in age geologically even as they are agglomerated and dissected by ongoing sea-floor spreading.

A maxim of geology states that the validity of any hypothesis or theory is determined by the degree to which that concept can be examined through the analyses of extant geology. Where, then, might there be modern-day examples of a supercontinent cycle in its various stages of tectonic development? The Great Basin of the western United States has been portrayed as a model of very early continental rifting that may in the future separate the North American plate from a newly constituted Pacific plate. The approximately one hundred block-faulted mountains composing this geographic terrane are caused by the same extensional forces responsible for the early continental- rift stage of dissection of a supercontinent. Similar forces have formed the more structurally advanced rift structures of East Africa, the most illustrative of which are those broad-basin and steeply dipping escarpment topographies of Tanzania, Kenya, and Ethiopia, which continue to yield an ever-revealing record of mammalian and hominid evolution. The fresh waters that partially cover these rift valleys, such as Lake Tanganyika, are evidence of a late stage of continental rifting. To the north, the central valley or rift of the Red Sea is filled with salt water characteristic of an incipient ocean developing during an early stage of oceanic rifting separating the Arabian from the African plate. Finally, the Atlantic Ocean is the often-cited mature example of oceanic rifting representative of the midpoint of a supercontinent cycle. According to the concepts of plate tectonics, the Atlantic Ocean has been created through some 200 million years of sea-floor spreading driven by extensional rifting of Pangaea. This stage, representing the first half of a supercontinent cycle, will in theory be followed by assembly of the world's continents over the next 200 million years into a new supercontinent. This assembly may have already begun, as India, an island subcontinent up to approximately 35 million years ago, has since that time been colliding with the Eurasian plate, resulting in the formation of the Himalaya Mountains.

If the above are examples of contemporary stages of a supercontinent cycle, what of past supercontinent cycles? Following the general acceptance of the plate-tectonic theory in the 1960's, many ideas have been advanced regarding the existence and nature of pre-Pangaea supercontinents. Since the supercontinent cycle not only creates but also destroys oceans, pre-Pangaea supercontinents must be reconstructed from continental geologic data that becomes, with increasing geologic age, more difficult to interpret. Certain criteria have been developed in the attempt to discern a pattern of supercontinent cycling since the earliest periods of geologic time.

Methods of Study

The convergence of continents will largely eradicate any intervening ocean. Such loss by subduction is seldom complete, as attested by the presence of remnants of pre-existing ocean floor rock contained within the deformed (suture) zone caused by plate collision. Basalt, gabbro, and olivine-rich rocks, termed an "ophiolite suite," are scraped off, or "obducted" from, the subducting ocean floor and thus preserved in the developing mountain belt. Obducted ophiolitic rock from the former Tethys Sea, which lapped onto the eastern shores of Pangaea, is present in the Alps and the Himalayas, while ophiolites within the Ural Mountains of central Asia are evidence of an ocean that was destroyed by the collision of Baltica and Siberia, continental masses that preceded the formation of Pangaea. The presence, location, and age-dating of ophiolite suites are helpful in the identification of former supercontinents.

Paleontological and high-pressure rock evidence collected from the Appalachian Mountains, created by the convergence of proto-Africa (the earliest form of Africa) with proto-North America, suggest the existence approximately 500 million years ago of a proto-Atlantic Ocean. The existence of this body of salt water, the Iapetus Ocean, attests the existence of an associated supercontinent.

The presence of exotic terranes (also known as a "melange," from the French for "mixture") forming the collisional edge of former crustal plates in Japan, New Zealand, and the Apennines of northern Italy is further evidence of former supercontinents and their cycles. These terranes, packages of rock possessing similar mineral and fossil character, are accreted to enlarging supercontinents by the same obduction process as ophiolites.

Using the above and other lines of reasoning, the existence of various pre-Pangaea supercontinents has been postulated. Obviously, the older the proposed supercontinent, the more conjectured its existence. As the theory of plate tectonics is but several decades old, little agreement exists on the naming and geologic age of pre-Pangaean supercontinents. The following discussion, however, presents two contemporary schools of thought regarding supercontinent cycles.

The oldest Earth specimen found to date is the 3.96-billion-year-old metamorphic continental rock forming a portion of the Northwest Territories of Canada. This and slightly younger rock terranes from Greenland, Antarctica, and Australia may have combined to form the first amalgamated continental masses. John Rogers, a professor of geology at the University of North Carolina, has proposed the existence three billion years ago of an early subcontinent he calls "Ur" (from the German for "original"). Five hundred million years later, a second subcontinent, Arctica (predecessor to Canada, Greenland, and eastern Russia) formed, followed another 500 million years later by Baltica (proto-Western Europe) and Atlantica (eastern South America and western Africa). One-and-one-half billion years ago, plate-tectonic forces formed the sub-supercontinent Nena (from the Russian for "motherland") through the merging of Baltica and Arctica. This lengthy chain of events culminated one billion years ago in the formation of the first supercontinent, Rodinia (also known as "proto-Pangaea"), as a result of the joining of Ur, Atlantica, and Nena. After a period of stability lasting some 300 million years, Rodinia subdivided, forming numerous proto-continents and Iapetus, the proto-Atlantic. Finally, the reassembly of these proto-continents and subsequent subduction of Iapetus brought about the creation of the second supercontinent Pangaea about 250 million to 300 million years ago.

A second school of thought differs principally in the suggestion that plate-tectonic processes did not begin until some 2.5 billion years ago, on the occasion of the development of the first distinct oceanic and continental crust. Prior to that time, the very high temperature of the earth and the relative thinness of primordial crust forestalled the onset of supercontinent-cycle processes such as divergence, convergence, and subduction. While differing in detail, both schools of thought generally recognize Rodinia and Pangaea as supercontinents.

The length of a typical supercontinent cycle is variously estimated at from 400 million to 440 million years. Such a cycle would constitute three phases. Once formed, a supercontinent would exist for 100 million to 120 million years before the accumulation of thermal convection heat initiated crustal dissection. During the second phase, lasting from 150 million to 160 million years, maximum dispersal of subcontinents would take place, with resultant development of new oceans. Finally, subduction would gradually destroy the intervening oceans over a period of 150 million to 160 million years, creating a new supercontinent and terminating the cycle.

Context

The concept of plate tectonics and the supercontinent cycle has been termed the contemporary unifying theory of earth history. The dynamics of the supercontinent cycle have encouraged geologists to create new exploration models establishing a scientific link between solid, liquid, and gaseous natural-resource deposits and plate-tectonics processes. Knowing the location and specific geologic conditions under which mineral deposits are formed is paramount to discovery and economic evaluation. Many metallic and nonmetallic minerals are formed by magmatism (relating to magma, naturally occurring mobile rock) at divergent and convergent crustal plate boundaries. The formation of fossil fuels is dependent upon changes in earth structure, sedimentation rates, and climate attributed to the varying stages of a supercontinent cycle. Even the role of water in crustal plate movements is necessary to an understanding of hydrocarbon migration and hydrothermal (subsurface hot water) processes.

The discovery in 1977 of "black smokers," mineral-rich hot springs with temperatures up to 450 degrees Centigrade, along the East Pacific oceanic ridge is believed to be evidence of metallic ore deposits in the early stages of formation. Black smokers precipitate iron-, zinc-, and copper-rich minerals on divergent plate borders during the first half of a supercontinent cycle. These deposits later become accreted onto continental borders through ophiolitic obduction during the final half of the cycle. Conversely, convergent borders are commonly associated with Kuroko-class (named from the location of discovery in Japan) sulphide ores of copper, lead, and zinc emplaced during subduction of crustal plates accompanying supercontinent amalgamation.

Eighty percent of known world oil and gas reserves are contained within sedimentary rocks deposited since the breakup of the supercontinent Pangaea some 200 million years ago. Much of this hydrocarbon is contained within offshore (shallow marine water) provinces resulting from plate-tectonic dynamics. Examples are the Gulf of Mexico, the North Sea, and the gigantic reserves of the northern coast of Alaska.

The distribution and economic grade of coal is also supercontinent-cycle-dependent. In the United States, the Appalachian coal fields are divided into a narrow eastern anthracite (high heat rank) and a wider western bituminous (lower heat rank) province. This distribution is directly related to compressional forces accompanying the assembly of Pangaea.

Finally, the understanding of the supercontinent cycle has proven most valuable in helping to provide a plausible explanation of the mass extinctions of organisms that have periodically occurred over the past 600 million years. The most famous of these extinctions marked the end of the Paleozoic era of geologic history 245 million years ago. The fossil record indicates a disappearance or severe reduction at that time of as much as 90 percent of all marine species; the supercontinent Pangaea had then completed its assembly. With the presence of one large continent, surrounded by universal sea waters, faunal diversity decreased as the result of loss of warm, shallow seas and the onset of colder, continental climates. Similarly, when a supercontinent dissects, this hypothesis would suggest an increase in faunal diversity, as newly developed shallow water seas and warmer climates expand the geographic area favorable to reproduction.

Bibliography

Dalziel, Ian W. D. "Earth Before Pangaea." *Scientific American* 272 (January, 1995): 58-63. An easy-to-read account of the nomadic wanderings of the North American plate prior to the assembly of Pangaea.

Davidson, J. P., W. E. Reed, and P. M. Davis. *Exploring Earth: An Introduction to Physical Geology.* Upper Saddle River, N.J.: Prentice-Hall, 1997. Chapters 6 through 11 are an excellent undergraduate-level introduction to the natural consequences of plate tectonics and their role in the supercontinent cycle.

Dietz, Robert S., and John C. Holden. "The Breakup of Pangaea." *Scientific American* 223 (October, 1970): 30-41. An interesting extrapolation into the next 50 million years of the present-day supercontinent cycle. Contains illustrations of the expected geographic locations of continents and ocean basins at that time.

Nance, R. Damian, T. R. Worsley, and J. B. Moody. "The Supercontinent Cycle." *Scientific American* 259 (July, 1988): 72-79. Combining specialties of tectonics, oceanography, and geochemistry, the authors discuss in a clear manner several supercontinent cycles and their effects on climate, evolution, and geologic changes.

Nicolas, A. *The Mid-Ocean Ridges.* New York: Springer-Verlag Berlin Heidelberg, 1995. A review presenting the European view of the construction and destruction of ocean basins. Written at the knowledgeable adult level.

Sullivan, W. *Continents in Motion.* New York: American Institute of Physics, 1991. The author, a science editor for *The New York Times,* presents a recommended compilation of the history and dynamics of plate tectonics. Suitable for the educated layperson.

Wegener, A. *The Origin of the Continents and Oceans.* New York: Dover, 1929. Written by "the father of Continental Drift," this volume presents the original thoughts and data sets used in the formulation of the first unifying theory of earth science.

Albert B. Dickas

Cross-References

Continental Drift, 86; The Lithosphere, 375; Plate Tectonics, 505; Subduction and Orogeny, 615.

Thrust Faults

Thrust faults are the result of compressional forces that exceed the natural strength of rocks and cause them to break and move. They can trigger earthquakes, create mountain ranges, and serve as natural traps for gas and oil deposits.

Field of study: Structural geology

Principal terms

DIP: the angle between a fault plane and a horizontal surface

FAULT: a fracture or zone of breakage in a mass of rock that shows evidence of displacement or offset

FOOTWALL: the block of rock that lies directly below the plane of a fault

HEAD WALL: the block of rock that lies directly above the plane of a fault; it is also known as a hanging wall

REVERSE FAULT: the same thing as a thrust fault, except that its fault plane dips at more than 45 degrees below the horizontal

SCARP: a steep cliff or slope created by rapid movement along a fault

SLIP: a measure of the amount of offset or displacement across the plane of the fault, relative to either the dip or the strike

STRIKE: the orientation of a fault plane on the surface of the ground measured relative to north

Summary

A mass of rock below the surface of the earth usually cracks and fractures when it loses its resistance to an applied force. Rocks break when their ability to store energy is exceeded; when a rock shows some evidence of movement or displacement along the zone of breakage, a fault is created. Thrust faults are commonly the result of strong compressional (squeezing) forces acting on relatively brittle, older subsurface rock that has moved upward and over or on top of a mass of younger, adjacent rock. It is a particular kind of fault and one of many types that exist.

The zone of breakage between the once united masses of rock is known as the fault plane. The motion of the rocks on either side of this plane and the plane itself are usually parallel to each other. The blocks of rock on both sides of a fault plane are known as walls, a term that comes from the days of the early prospectors, who were really the first field geologists. Since the presence of a fault marks a zone of weakness in the ground, either mineral-rich groundwater or hot fluid magmas will eventually find and follow this path of least resistance toward the surface and deposit ores, minerals, or gemstones. Prospectors would seek out faults, as they knew that a fault was likely to be the home of some valuable material. Once a fault was located, a mine shaft would be dug to follow the trace of the fault below ground.

The head wall, or hanging wall, was the wall above the miner's head; the footwall was the wall below his feet. Head walls and footwalls exist only in faults that are not vertical. In terms of the overall structure of the fault, the hanging wall is the block that occurs above the fault plane and the footwall is the rock below the fault plane. In thrust faults, the head wall always moves relatively upward and the footwall moves relatively downward (see figure). The term "relatively" is used because it is usually very difficult for a geologist to determine exactly which block has moved. For example, both blocks could have moved upward but the hanging wall moved farther; both blocks could have dropped but the footwall dropped farther; the hanging wall could have remained stationary while the footwall dropped; or the footwall could have remained stationary while the hanging wall moved upward.

The orientation of the fault relative to the earth's surface is of great importance. It allows the fault to be located and mapped as a place to avoid during construction, especially if it is an active fault or one with the potential for continued movement. A fault's orientation, or strike, is measured by the trace of the fault plane as it would appear on a horizontal plane and is measured in degrees from the magnetic North Pole. The plane's angle of tilt, or dip, is measured from a horizontal position down to the fault plane. The dip direction is always perpendicular to the strike direction.

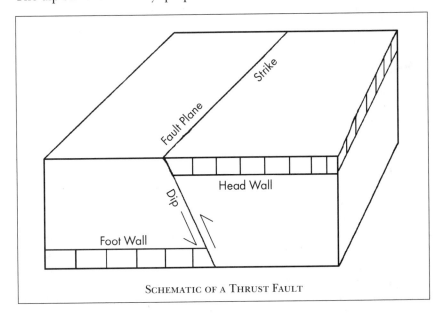

SCHEMATIC OF A THRUST FAULT

A fault can be straight in form and consist of one sharp, clean break, or it can have a highly irregular form and be composed of multiple breaks. Thrust faults of the latter type may be so closely spaced as to form a highly complex zone that may be hundreds of meters wide. Fault planes can also be curved, adding to the complexity. Geologists have located and mapped many small thrust faults at very shallow depths below the surface and have found some large thrust faults that extend down to a depth of 700 kilometers.

In general, the total displacement, or offset, in a rock along a thrust fault may be large or small, and horizontal, vertical, or oblique, depending on the strength of the compressional force and the rock type involved. An important factor in determining this displacement is the angle of the fault plane. Steeply dipping planes will show a small vertical uplift on the order of a few meters or less; shallowly dipping planes may exhibit a long horizontal displacement extending for many kilometers. Displacements are also described in terms of their relative motion. A "dip slip" occurs when the movement of the rock is parallel to the dip direction of the fault plane, a "strike slip" indicates motion parallel to the strike, and an "oblique slip" occurs somewhere between the strike and dip.

There are three main varieties of thrust fault: A "thrust fault" has a fault plane that dips at more than 10 degrees but less than 45 degrees below the horizontal; a "reverse fault" is a variety of thrust fault that dips more than 45 degrees below the horizontal; and an "overthrust" is a thrust fault that dips at less than 10 degrees below the horizontal. These faults, as well as other types, may occur in solid, massive igneous or metamorphic rocks, may be formed during the process of folding sedimentary rocks, or may occur in unconsolidated sediments. Overthrusts are very common in areas of intense folding. The offset of rock along the path of a fault is greatest near its middle and decreases at either end; therefore, a thrust fault will gradually disappear at either end.

The age of rocks may be used to determine the relative age of faults. Since a fault cuts through a block of rock, the fault must be younger than the rock. Therefore, if a rock is found to be 100 million years old and is cut by a fault, then the fault must be younger than 100 million years old.

Methods of Study

Structural geologists study faults to try to understand what was happening to the earth's crust at the time of faulting and to determine the origin of the force that caused the rocks to break and move. To study a thrust fault, the geologist must first accurately locate and measure the fault in the field. It is sometimes difficult to determine where a fault exists, especially if it is very old. The geologist, acting as a detective, must rely on direct physical evidence that he or she can gather from above and below the surface.

When a fault intersects the surface, it usually forms a fault line. The existence of this line is commonly indicated by some noticeable feature. Photographs of the land taken from high-altitude aircraft reveal these features as offsets or disruptions in rows of planted crops or trees; sharp breaks in the channels of streams; unusual linear alignments of springs flowing at the base of a mountain or along a valley; raised sections of land, such as beach and river terraces; and fault scarps. A fault scarp is a recent, sharp break in the surface of the ground that has a straight and very steep slope. The height of a scarp is directly related to the amount of upward motion along a fault. Unless the fault is still active, however, it usually is not exposed at the surface but is buried beneath the cover of more recently deposited sediments. In this case, a geologist must rely on subsurface evidence.

The subsurface location of a thrust fault is not always easy to find. The field geologist must rely on direct or indirect evidence that is not usually visible on the surface. Direct subsurface evidence of the existence of a thrust fault can be obtained from the examination of rocks within mine shafts, highway tunnels, or excavation sites dug for building foundations. Similarly, "roaducts" (highway excavations that run through mountains and valleys) and natural outcroppings of rock at the surface may indicate the presence of a thrust fault. In these situations, a geologist would look for any evidence that suggests that massive blocks of rock have moved relative to one another.

When huge blocks of rock are broken and continue to rub against one another, certain physical features are produced as a result of the friction between these moving blocks. Sure indicators of faulting are slickensides, which are recognized as highly polished and finely scratched, or striated, surfaces of rock along the fault plane. The direction of the striations is parallel to the direction of the last movement along the fault. Depending on the amount of friction generated between these blocks, their inward-facing surfaces may be crushed into a fine, soft, claylike powder known as "gouge" or may become fault breccias, rocks consisting of small, angular fragments. Microbreccias are formed when the crushed fragments along the fault plane are microscopic. Mylonites are a special type of solid microbreccia; they have a streaked appearance in the direction of motion. Pseudotachylites are a kind of microbreccia that does not appear streaked but exists as a thin glassy film because of the melting of rock from frictional heat. Other evidence of thrust fault motion would be the overlapping or repetition of the same rock units (like overlapping shingles on a roof) and an abrupt termination of a rock unit along its trend. One or more of these faulting criteria may be present, or evidence may be completely missing. If there are no accessible outcrops or underground viewpoints, the field geologist must turn to the expensive direct method of evidence-gathering: drilling.

Drilling into the ground with specialized drilling rigs allows access to the subsurface. The examination of small broken rock samples brought up by a diamond drill bit or solid "rock cores" brought up by hollow drill bits is a direct means of studying rocks that do not exist at the surface. These samples can be compared with those taken from well-known nearby regions, and any disruption or missing units of rock may indicate the presence of a fault. If a subsurface geologic map exists, the geologist can predict what rocks exist below the drill site and use this information in conjunction with the collected rock samples. If the evidence from drilling does not match the regional geology, then one or more faults may have been at work shuffling the sequence of rocks below. Sometimes large blocks of rock get caught between faults when they move and become bent, folded, or twisted, creating a highly complex pattern that is not easy to understand using drill core data alone; the uncertainty of what lies below ground increases.

Indirect evidence of thrust faulting can be had from a geologic map of an area. Geologic maps show the distribution, thickness, age, and orientation of the various types of rock that would be seen at the surface of the earth were all the soil covering removed. Any older rock formation that sits on top of a younger rock formation was most likely moved to its location by thrust faulting. Rock formations that appear out of place with the overall sequence of rocks probably suffered a similar fate.

Several notable large thrust faults exist in the United States. They can be seen in the Rocky Mountain region as a series of sharp, parallel ridges similar to the teeth on a saw—hence the term "sawtooth," as in the peaks of the Sawtooth Mountain Range in Montana. In the same state, a low-lying slab of rock known as the Lewis Overthrust shows a horizontal displacement of about 24 kilometers with a fault plane that dips at an angle of less than 3 degrees. Thrust faults occur in most mountain ranges on the earth such as the Appalachian Mountains, the European Alps, and the Himalayas.

Context

A fault is a zone of weakness in a rock; therefore, it may continue to move over a long period. Information about the rate of rock movement is very valuable, especially when the motion along a thrust fault (or any fault) is rapid enough to trigger the release of stored energy within a rock, causing an earthquake. Accurate mapping of thrust faults is also important, since the potential for future earthquakes must be carefully evaluated, especially in highly urbanized areas. A complete study of thrust faults is not easy, however, because there are many variables to consider: the fault's exact location and total horizontal extent, the orientation and strength of the rocks relative to the fault, the direction and amount of movement, and the fault's previous earthquake history. These factors are critical to decisions of where to construct nuclear power plants, dams, housing projects, and cities.

Thrust faults have great economic potential, since they have the ability to act as "traps," or rock reservoirs, for deposits of migrating oil and natural gas. In such a case, one impervious rock type is brought into fault contact with a petroleum-bearing rock. The impermeable rock now acts as a barrier to any further upward fluid migration and allows oil to accumulate beneath it. Similarly, in mineral exploration, large thrust faults have been known to harbor exploitable quantities of radioactive and other rare minerals, needed for use in industry and medicine, that were either deposited by igneous activity or precipitated by circulating, mineral-rich groundwater. Thrust faults also serve as natural underground pipelines, allowing circulating groundwater easier access to the surface of areas that otherwise might have been deserts.

Bibliography

Billings, Marland P. *Structural Geology*. 3d ed. Englewood Cliffs, N.J.: Prentice-Hall, 1972. This is an introductory college-level textbook for all aspects of structural geology. Chapters 8 through 12 discuss the variation, classification, and recognition criteria for faults. Chapter 10 is devoted to thrust faults. The book is well written and clearly illustrated with many line drawings. Also included are black-and-white photographs of faults as they appear in the field and a useful end-of-chapter bibliography. The bible of structural geologists.

Lahee, Frederic H. *Field Geology*. 6th ed. New York: McGraw-Hill, 1961. Within this thick book are described the various techniques for the measurement, mapping, relative age determination, and interpretation of rock formations in the field. A large part of chapter 8 is devoted to faulting, with a description of applied field mapping techniques that are used in the construction of surface and subsurface maps. The chapter is easy to read, and the fault types are illustrated with either line drawings or block diagrams. Although somewhat dated, this work is still a classic and the field book carried by most practicing geologists.

Meigs, Andrew J., and Jaume Verges. "Ten-Million-Year History of a Thrust Sheet." *Geological Society of America Bulletin* 108, no. 12 (December, 1996): 1608-1636.

Parker, Sybil P., ed. *McGraw-Hill Encyclopedia of Earth Sciences*. 2d ed. New York: McGraw-Hill, 1988. A source that covers every aspect of geology. Topics are arranged alphabetically. The entry on faults and fault structures discusses fault movement, procedures for locating faults, stress conditions, and examples of various types of fault. Includes references to other entries and a bibliography. Contains some equations, but they are not essential to the reader's understanding.

Spencer, Edgar W. *Introduction to the Structure of the Earth*. 2d ed. New York: McGraw-Hill, 1977. This college-level structural geology textbook was written from the "plate tectonics" point of view and covers the entire earth. Chapters 7 through 10 deal with various faults that were formed as a result of colliding continents and ocean basins and are indicators of intense rock deformation. Chapters 8 and 9 describe the many types of thrust fault. Many regional examples are provided and clear line drawings are included. There is a thorough end-of-chapter bibliography.

Tarbuck, Edward J., and Frederick K. Lutgens. *The Earth: An Introduction to Physical Geology*. 2d ed. Columbus, Ohio: Merrill, 1987. This basic geology textbook contains a section on rock deformation. It provides diagrams to illustrate the various types of fault and the concepts of strike and dip, and it includes a photograph of an overthrust formation. Review questions and a list of key terms conclude the chapter. For high school and college-level readers.

Thornbury, William D. *Principles of Geomorphology*. 2d ed. New York: John Wiley & Sons, 1968. Geomorphology is the study of landforms on the earth's surface. This basic college-level textbook describes the many types of landform that are produced by various geologic processes. Chapter 10 is an informative, highly readable, and well-illustrated chapter that discusses the landforms created by fault activity and how they are recognized.

Steven C. Okulewicz

Cross-References

Earthquakes, 148; Folds, 248; Normal Faults, 448; Plate Tectonics, 505; Stress and Strain, 607; Subduction and Orogeny, 615; Transform Faults, 638.

Transform Faults

Transform faults occur along fracture zones found at the mid-oceanic ridges. The ridges are areas of erupting ultramafic lavas, which cause sea-floor spreading, which, in turn, drives the moving lithospheric plates. Geologists have found that the offset ridges do not move in relation to each other and have been essentially fixed relative to the spreading sea floor.

Field of study: Structural geology

Principal terms

CURIE TEMPERATURE: the temperature below which minerals can retain ferromagnetism

DIVERGENT BOUNDARY: the boundary that results where two plates are moving apart from each other, as is the case along mid-oceanic ridges

FERROMAGNETIC: relating to substances with high magnetic permeability, definite saturation point, and measurable residual magnetism

FRACTURE ZONE: the entire length of the shear zone that cuts a generally perpendicular trend across a mid-oceanic ridge

HYPOCENTER: the initial point of rupture along a fault that causes an earthquake; also known as the focus

MAGNETIC ANOMALIES: patterns of reversed polarity in the ferromagnetic minerals present in the earth's crust

MID-OCEAN RIDGE: a long, broad, continuous ridge that lies approximately in the middle of most ocean basins

RIFT VALLEY: a region of extensional deformation in which the central block has dropped down with relation to the two adjacent blocks

SEA-FLOOR SPREADING: the hypothesis that oceanic crust is generated by convective upwelling of magma along mid-ocean ridges, causing the sea floor to spread laterally away from the ridge system

TRANSCURRENT FAULT: a fault in which relative motion is parallel to the strike of the fault (that is, horizontal); also known as strike-slip fault

Summary

Faults are regions of weakness or fractures in the earth's crust along which relative movement occurs. The simplest type of fault is the so-called normal fault, in which one block of crust is displaced vertically downward with respect to the other. A reverse or thrust fault is one in which the block is driven upward with respect to the other block. Yet another type of crustal displacement takes place horizontally, as one block slides laterally with respect to the other. These strike-slip or transcurrent faults are related to the most complex class of faults, known as transforms.

Actually regions of crustal transformation, transform faults are found almost exclusively along the mid-oceanic ridges that nearly encircle the globe. The ridges are the sites of newly forming crustal materials composed of very dense magmas of relatively high iron and magnesium content. As the new crust forms, lava flows act to push the oceanic crust laterally away from the spreading ridge, as the sea floor spreads out at a rate of a few centimeters a year.

If one were able to view the mid-oceanic ridge system from orbit, it would quickly become apparent that the spreading centers do not occur along a smoothly continuous line but rather are broken into scores of offset ridges. The offset is marked by a fracture zone, which serves as the border between two spreading centers. Because the ridges are displaced, it was first believed that these fracture zones were simply transcurrent faults, along which right or left lateral displacement would be observed from opposing sides. The ridges are fixed with relation to each other and appear to have been so for long periods of geologic time. Clearly, a new type of faulting was being observed.

If one were to voyage to the bottom of the mid-Atlantic Ocean to view one of these faults, its highly unusual nature would become clear only after one had traveled hundreds of miles along its entire length. Starting at the west end of the transform fault, one would find that the crust is slowly moving toward the west on either side of the fracture zone. Because the crust is essentially moving in the same direction, earthquakes are rare events in this region.

The underwater ridge mountains are marked by a distinctive rift valley, centered along the range. Volcanic eruptions emanating from the rift create pillow-shaped lavas. The ridge is a boundary between the earth's lithospheric plates; it is a divergent boundary, for the sea floor is spreading laterally in opposite directions. A view of the ridge along the transform fault would reveal that the line of mountains is broken and offset. An observer passing over the high ridge on the north side of the fault would suddenly notice that the fault had taken on the appearance of a transcurrent fault. Movement on the south side of the fault is right lateral, and although the crust is traveling west, on the north side of the fault, the movement is toward the east. An observer on either side of the fault would see right lateral displacement.

Because crustal movement is in opposing directions between the offset ridges, earthquakes occur frequently in this area. The crust's relative motion is horizontal so that the focus or hypocenter (the actual point of rupture in the rock) of transform earthquakes is shallow—typically less than 70 kilometers deep, whereas trench earthquakes can be up to 700 kilometers deep.

Magmatic eruptions and earthquakes offer convincing evidence that the earth's crustal plates are far from stationary. Past the offset ridge, crustal motion is once again in the same direction, and no lateral motion would be observed on opposite sides of the fracture. This transformation of crustal displacement along the shear zone is the derivation of the term "transform fault."

Most transform faults are oriented at right angles to the mid-oceanic ridges. Typically, they extend from ridge to ridge (the active part of the fault); if spreading has been taking place for long periods of geologic time, fracture zones extend out from the ridge systems. These transforms and fracture zones are thought to be ancient areas of weakness in the crust that formed when the ocean basin (such as the Atlantic) began forming. The discontinuous nature of the ridge system is believed to be structurally ancient. When viewed globally, the spreading axis is offset from the earth's rotational axis, with the ridges corresponding to lines of longitude and the transforms roughly approximating latitudinal lines. Spreading rates are greatest at the equatorial regions of the globe.

While the Atlantic Ocean is a basin in which new crust is forming, causing the rifting of the continents and the growth of the ocean basin at a rate of a few centimeters per year, the opposite is the case in the Pacific Ocean basin. There, the transform faults are a bit more complex. An example would be the New Zealand fault, which terminates on both ends at subduction trenches rather than at ridges. New Zealand is seismically active, because the transform fault passes directly through both of the large islands.

The most famous of all transform faults forms a distinctive type of lithospheric plate boundary. Extending from the East Pacific Rise off the coast of western Mexico to the Mendocino fracture zone and the Juan de Fuca ridge system off the coast of Washington State is the 960-kilometer-long, 32-kilometer-deep system of strike-slip faults known as the San Andreas. The San Andreas fault forms a boundary between the northward-trending Pacific plate and the North American plate. Horizontal displacement along this huge fault system is estimated at 400 kilometers since its inception nearly 30 million years ago. The rate of displacement has been measured at an average of 3.8-6.4 centimeters annually.

How did this impressive plate boundary form, and how does it fit into the transform fault model? About 60 million years ago, the east Pacific was also home to a third plate, the Farallon, which eventually was subducted under the North American plate. The Farallon plate was pushed eastward into the North American plate by a spreading ridge system and trench that were eventually overridden by the continental plate. The remnants of this ridge system are found at the ends of the San Andreas fault. Because the Pacific plate's motion was northward, the ridge system was converted into a transform fault, characterized by its right lateral strike-slip displacement.

The complexity of three plate interactions, or triple junctions, explains the complex nature of transform fault systems such as the San Andreas. The remnants of the doomed Farallon plate are presently found as the Juan de Fuca plate, which is bounded by spreading ridges and transform faults to the west and a subduction zone where the plate is being inexorably pushed into the earth's upper mantle, under the North American plate. Inland, the Cascade volcanoes, most notably including Mount St. Helens, have their volcanic fires fueled by a rising magma plume that is generated by the melting of the subducted oceanic plate.

North of the Juan de Fuca plate is another large transform system similar to the San Andreas fault, called the Queen Charlotte transform. An extension of the San Andreas system, the Queen Charlotte fault begins to the south at the Juan de Fuca ridge but becomes inactive at its northern end, whereas the San Andreas is a ridge-ridge transform that is bordered by the East Pacific Rise to the south and Gorda Ridge to the north.

The great ridge system that was overridden by the continental plate is responsible for the faulted structure of the Basin and Range province, and the triple junction of the plates may have given rise to the magma plume or "hot spot" that is responsible for the Snake River plain volcanics and the thermal activity at Yellowstone National Park. Clearly, understanding the evolution of transform faults such as the San Andreas and the Queen Charlotte is pivotal to the understanding of the complex mountain scenery of western North American.

Methods of Study

A revolution in the earth sciences had its germination in the ideas of the German meteorologist Alfred Wegener, who argued that the continents had once been joined in one supercontinent and had since drifted apart. Wegener's theories of continental drift were not taken seriously until evidence from the sea floor forced a rethinking of modern geology in the 1960's. Oceanographic research in the 1950's led to a new picture of the ocean basins. Far from the featureless abyssal plains they were once thought to be, the basins proved to be marked by dramatic mountain ranges characterized by rift valleys. In 1960, a Princeton University scientist, H. H. Hess , suggested that convection currents in the earth's upper mantle were driving volcanic eruptions along the rifts, causing the sea floor to spread and the continents to move apart.

The real breakthrough in understanding the ocean floor came as a result of numerous cross-Atlantic voyages by research vessels towing submerged magnetometers in order to measure the strength of the crustal rock's residual magnetism. As the basaltic magma erupts at the spreading rift, magnetite freezes out of the melt at 578 degrees Celsius, and the earth's magnetic field orientation is frozen into it. This temperature is known as the Curie temperature.

When the first magnetic maps of the sea floor were produced, scientists groped to explain the alternating nature of the field's polarity, which changed in a random pattern over short distances. Two research students at the University of Cambridge, F. J. Vine and Drummond Matthews, solved the riddle of the magnetic anomaly stripes by proposing that the stripes represented reversals in the earth's magnetic field over time. Because the anomaly patterns were symmetrical with respect to the axis of the spreading centers, this was proposed as evidence in favor of the Hess model of sea-floor spreading.

The huge fracture zones that appeared to offset the mid-oceanic ridges were mapped with magnetic anomaly measurements, and seismic data indicated that seismic activity was concentrated between the offset ridges along the fracture zones. Networks or seismic instruments also enabled geophysicists to study the direction of the spreading crustal movement, through geometrical solutions known as first motion studies or fault-plane solutions.

In 1965, J. Tuzo Wilson dubbed the faults "transforms" because of the changing of relative motion along the length of the fracture. He reasoned that the faults were not causing the offset of the spreading ridges, but that the offset is what caused the appearance of a transcurrent fault between the ridges and an inactive segment, in which the sea floor was spreading in the same direction beyond the zone of offset.

Geophysicists mapping the magnetic anomalies were aided by the work of terrestrial geologists, who were able to identify magnetic reversal patterns in terrestrial lavas that were identical to those found in oceanic rocks. Radiometric dating established a magnetic anomaly time scale that would allow scientists to determine the ages of rocks laterally out from the ridges and hence to deduce the rates of sea-floor spreading. Armed with the paleomagnetic data and with Wilson's notion of transform faults as directional guides, scientists were able to show that the Atlantic Ocean had indeed been opening at a rate of a few centimeters per year and that the landmasses of western Europe and eastern North America were once essentially in contact.

Wilson and others went on to study the more complex faults of the eastern Pacific. The San Andreas fault had been identified long before as a transcurrent fault. The new paradigm of plate tectonics placed it in a global tectonic scale; the fault is now accepted by most as a kind of transform fault that connects to spreading ridge centers to the north and south.

One of the most studied of all faults, the San Andreas includes a system of transform threads that trend from southeast to northwest, with a noticeable jog westward at the transverse ranges of southern California. Seismic mapping and historical earthquake data testify that California is earthquake country. Some of the most powerful evidence (other than the four great twentieth century earthquakes in California—San Francisco in 1906, Kern County in 1952, San Fernando in 1971, and San Francisco again in 1989) that the San Andreas is a moving plate boundary is geologic field evidence indicating dramatic crustal displacement. Virtually identical rock formations have been identified some one thousand kilometers apart. Seismic and laser measurements along unlocked regions of the fault indicate slippage rates of about 5 centimeters per year. The displacement rates have also been confirmed by radio astronomy techniques, using distant radio sources such as quasars as fixed points of reference to determine rates of crustal movement along faults such as the San Andreas.

While faults such as the New Zealand Alpine and San Andreas are fairly accessible to scientists, most of the planet's transform faults lie below thousands of meters of ocean. Echo sounding, radar, sonar seismographic, and magnetic anomaly maps have helped scientists to locate the earth's transform faults. The 1978 Seasat mission radar mapped the ocean floor from space, producing detailed information on the globe's mid-oceanic ridge and fault system. In addition, deep-sea submersibles have been piloted to the rifts, where eruptions of lavas and fault displacements have been observed directly.

Context

The theory of plate tectonics in the 1960's caused a revolution in the study of geology, geophysics, and even paleontology. Post-World War II oceanographic research led to the discovery that the mid-oceanic ridges are sites of newly forming crust and resulted in J. Tuzo Wilson's explanation of transform faults as ancient fracture zones offsetting the spreading ridges.

With a vast supply of paleomagnetic, seismic, geologic, and other evidence at their disposal, geologists have been able to reconstruct earth history and the relative motions of the continents, using transform faults as directional guides and magnetic reversals to determine the rates of plate movement. Like any successful theory, plate tectonics is elegantly simple; it explains nearly all the earth's diverse landforms and rock formations. Changes in continental distribution may also be linked to the extinction events that punctuate the geologic time scale.

Geologists' understanding of the interaction between ridge systems, subduction trenches, and transforms has led to a better understanding of the paleogeography of the American West and how that complex mountainous region of active faults and recent volcanic activity came into being. Aside from contributing to the fundamental understanding of earth processes, plate tectonics theory explains forces that can influence human lives. Active plate boundaries are the sites of earthquake and volcanic activity. In California, millions of people live and work astride one of the world's largest transform faults, the San Andreas. Residents of Vancouver and Seattle are similarly threatened by the offshore Queen Charlotte Islands fault. On the other side of the Pacific Ocean, New Zealand is nearly bisected by the New Zealand Alpine fault, making it a seismically active region. Whenever human beings decide to build their homes and cities near these moving regions of the earth's crust, the possibility of geologic catastrophe is very real.

Bibliography

Condie, Kent C. *Plate Tectonics and Crustal Evolution.* Elmsford, N.Y.: Pergamon Press, 1982. An excellent overview of modern plate tectonics theory that synthesizes data from geology, geochemistry, geophysics, and oceanography. Of special interest is chapter 6, on sea-floor spreading, and chapter 9's treatment of the Cordilleran system, including a discussion of the evolution of the San Andreas fault. A very helpful tectonic map of the world is enclosed. The book is nontechnical and suitable for a college-level reader. Useful "suggestions for further reading" follow each chapter.

Cox, Allan, and R. B. Hart. *Plate Tectonics: How It Works.* Palo Alto, Calif.: Blackwell Scientific, 1986. A valuable treatment of the geometrical relationships and movements of the earth's lithospheric plates. Designed for the reader who has a basic qualitative knowledge of plate tectonics but who wishes to learn more, particularly about quantitative analysis of plate movements. Filled with easy-to-follow exercises that demonstrate plate motions, particularly those associated with transform faults.

"Hotbed of Activity Lies on Ocean Floor." *New Scientist* 146, no. 1983 (June 24, 1995): 11-18.

Lambert, David, et al. *The Field Guide to Geology.* New York: Facts on File, 1988. For the beginning student of geology, this reference work is filled with marvelous diagrams that make the concepts easy to understand. Chapter 2 deals with plate tectonics and has a clear, concise treatment of sea-floor spreading and transform faults. Suitable for any level of reader, from high school to adult.

Redfren, Ron. *The Making of a Continent.* New York: Times Books, 1983. Richly illustrated with dramatic photographs, this book is a lucid discussion of plate tectonics with respect to the continent of North America. Contains excellent explanations of sea-floor spreading and transforms, along with a section on the San Andreas fault. Suitable for a general audience.

Shea, James H., ed. *Plate Tectonics.* New York: Van Nostrand Reinhold, 1985. A collection of classic and key scientific papers, mostly from the 1960's, that together constitute a sweeping overview of plate tectonic geology. Of special interest are papers by Fred Vine and J. Tuzo Wilson on the magnetic anomalies off Vancouver Island and a paper by L. R. Sykes on transform faults at the midoceanic ridges. With chapter introductions by the editor, the work is suitable for a college-level reader with an interest in the history of plate tectonics theory.

Shepard, Francis P. *Geological Oceanography.* New York: Crane, Russak, 1977. Chapter 2 addresses sea-floor spreading and faulting of the oceanic crust. Photographs, diagrams, and supplementary reading lists augment the text, which is suitable for a beginning geology or oceanography student.

Sullivan, Walter. *Continents in Motion.* New York: McGraw-Hill, 1974. Dedicated to Harry Hess and Maurice Ewing, two late pioneers of plate tectonics theory, this is the classic popular work on moving crustal plates. Well-written explanations of transform faults and their roles in sea-floor spreading and a discussion of the San Andreas fault are included in the highly readable text.

Wilson, J. Tuzo, ed. *Continents Adrift and Continents Aground.* San Francisco: W. H. Freeman, 1976. Selected readings from *Scientific American* are introduced with commentary by Wilson, a leading figure in the history of plate tectonics theory. Chapter 2 deals with sea-floor spreading and transform faults with a classic article by Don L. Anderson on the San Andreas fault. Suitable for a general audience. Contains a bibliography.

Wyllie, Peter J. *The Way the Earth Works: An Introduction to the New Global Geology and Its Revolutionary Development.* New York: John Wiley & Sons, 1976. Wyllie's book has a very informative section on transform faults and earthquake studies. An extensive list of suggested readings augments the text, which is suitable for a college-level reader.

Young, Patrick. *Drifting Continents, Shifting Seas.* New York: Franklin Watts, 1976. A good entry-level discussion of plate tectonics theory, written by a journalist with a knack for simplifying complex concepts. Contains a brief glossary, indexed with a bibliography. Suitable for high school and lay readers.

David M. Schlom

Cross-References

Normal Faults, 448; Plate Tectonics, 505; Subduction and Orogeny, 615; Thrust Faults, 631.

Transgression and Regression

Transgression and regression are two of the most common phenomena of the geologic record. The changes in sea level that they represent have a major impact on the interpretation of earth history and in the reconstruction of ancient environments. Understanding these phenomena in the geologic past is essential if present sea-level changes are to be dealt with effectively.

Field of study: Stratigraphy

Principal terms

CONTINENTAL SHELF: the margin of the continents, usually covered by shallow seas

EUSTACY: any change in global sea level resulting from a change in the absolute volume of available sea water

MEASURED SECTION: a precise field description of sedimentary rocks where the location of the units is related to an initial starting datum; the basic data for all stratigraphic studies

OCEAN BASINS: the large worldwide depressions that form the ultimate reservoir for the earth's water supply

REGRESSION: the retreat or withdrawal of the sea from land areas and the evidence of such a withdrawal

SEA-LEVEL CYCLE: one transgression and one regression; such cycles are not necessarily symmetrical in their magnitude of change or in the amount of time represented by each part of the cycle

SEQUENCE: a sedimentary rock package within which all the rocks are related to a similar genetic process

STRANDLINE: the line or level where the sea meets the land; the beach; generally marked at mean high water level

TRANSGRESSION: the extension of the sea over land areas and the evidence of such an advance

UNCONFORMITY: an erosional or sedimentary omission surface usually produced by subaerial exposure; unconformities may indicate widespread regression

Summary

Transgression and regression are two phenomena by which changes in sea level are recognized. The two are not processes as such but rather are the effects of changes in sea level. Transgression is any rise of the sea over a land area (including the continental shelf) or any change in the physical marine conditions involving a progression into a deeper water environment. Regression is a withdrawal of the sea from any land area or any change resulting in a progression from deep water to shallow water or nonmarine environments. The concepts of transgression and regression are themselves independent of a causal mechanism. They are the results of the processes that can cause sea-level change.

Transgression and regression are very diverse phenomena and are the result of many different processes. Some of these processes involve an actual change in global sea level, or eustacy. Because transgression and regression are effects, however, a transgression or regression can be produced by a local process rather than a global one. It is important to understand the various processes involved in transgression and regression, by examining first the global processes, then the local ones.

The global, or eustatic, processes resulting in transgression and regression are those that cause a change in the volume of seawater on the continents. Such global changes are accomplished by two major processes. First, an actual change in the volume of the global seawater supply and, second, a change in the volume or holding capacity of the ocean basins. A change in the amount of seawater would not seem an easy task to accomplish. The earth's environment is essentially a closed system and, although small amounts of water are added from volcanism and by human activity, that is not enough to account for wide fluctuations in global sea level. Even if enough new water could be added to the system to account for a sea-level rise, the problem of getting rid of large amounts of water to account for a sea-level fall remains. Rather than add or subtract new water, a redistribution of existing water at the earth's surface seems a more likely way to change the volume of seawater. One way that may be accomplished is by the growth and melting of the earth's continental glaciers.

Glacial ice is different from sea ice. Sea ice forms in high latitudes over open ocean. As such, there is a constant connection with the oceans and a seasonal interchange as the ice melts in the summer and freezes in the winter. Glacial ice, however, forms from accumulated snow on the continents. Once a glacier grows, the snow that forms glacial ice is no longer available to the oceans. It is locked on the continents for the duration of the glacial episode. As glaciers grow, more water is locked within them. As glaciers melt, the meltwater is available to the oceans. The affect on sea level can be dramatic. During a glacial advance (an "ice age"), sea level will fall. During a glacial melt, sea level will rise. This sea-level mechanism has characterized the geologic record of sea-level change over the past 20 million years and perhaps longer. It has also operated during various other times in earth history as long as there was continental-scale glaciation.

Another process by which transgression and regression can occur is by a change in the geometry of the ocean basins. The ocean basins are the ultimate reservoir of standing water on earth. The depths in these basins are great, and it would seem that a change in the volume would not have a great effect on sea level. Yet, if the volume of the ocean basins decreases, for example, the excess water must go somewhere. When that happens, a flooding of the continents will occur. Once the volume of the ocean basins increases, the water will withdraw from the continents. The actual mechanism in the ocean basins is related to heat flow from the earth's interior at the mid-ocean ridges. As the ridges become more volcanically active, an increase in heat flow occurs; the high heat flow causes an increase in the elevation of the mid-ocean ridge. This increase in elevation causes a decrease in the volume of the ocean basins, resulting in a transgression. As the

The study of marine fossils helps scientists to understand the processes of transgression and regression. *(Reuters\David Gray\Archive Photos)*

volcanic activity decreases, heat flow decreases, and the elevation of the mid-ocean ridge drops so that volume of the ocean basins is increased. This results in a regression. Sea-level changes brought about by this process produce large-scale fluctuations resulting in the flooding of not only the continental shelves and the coastal plains but also the interiors of the continents. Much of the geologic record contains rocks deposited in seas in continental interiors. This process may result in transgressions where ancient sea level could be as much as 300 meters above present sea level. This process is not the dominant process at the present. The last sea-level transgression apparently related to this process was during the early Paleocene epoch (about 66-58 million years ago). The sea reached from the Gulf of Mexico up the Mississippi Valley into present-day southern Illinois and reached as far west as the Dakotas.

While the global changes in sea level would seemingly result in uniform transgression or regression throughout the world, that is not apparent when the sea level record at any given place is studied. The reason for that is the effect of local processes on sea level. In fact, many transgressions and regressions are local phenomena and may not be recognizable in other regions.

Two local processes that control sea level are sedimentation and tectonics. By operating in coastal regions, these two processes can mask the global trend of sea level. Sedimentation is the process whereby sediment is deposited by water or wind. Sedimentation can affect local sea level if the accumulation rate is greater than is the global sea-level rise. As the sediment builds up in the coastal regions, the sediments displace the ocean, thus "filling in" the former sea floor. As an area that was once a part of the sea is now land, a regression has occurred. This process is presently operating in the modern environment. A very good example of this phenomenon is the modern Mississippi Delta in southern Louisiana. In this area, sedimentation rates are very high. Although global sea level is rising as the remaining continental glaciers melt, new land is being created around the Mississippi Delta. At this location, sediment is accumulating faster than sea level is rising. Thus, a regression at the delta is taking place. Away from the delta, sea level is still rising, and more land each year in southern Louisiana is being flooded by marine waters. Therefore, even though the Mississippi Delta is a site of high sedimentation rates and locally affects sea-level change, that is not enough to alter the global process.

Tectonism is another local process by which the pattern of sea-level change can be controlled. Although tectonics is a global process, the effects of this process can be very localized, and some include phenomena not related to global tectonics at all. For sea-level change, the processes of interest are uplift and subsidence. Uplift and subsidence in a coastal region can be related to many causes, but it is the result rather than the process that is important to transgression and regression. If the process of uplift is more rapid than is global sea-level rise, a regression will be apparent. Similarly, if subsidence is greater than is global sea-level fall, a transgression will be the interpretation. The two above examples illustrate the extremes of the tectonic effects, but many variations exist. In the modern environment, global sea level is rising, but the apparent change in any specific area is controlled by the tectonic setting. An example is from the Baltic sea coast in Sweden, where uplift resulting from glacial rebound is not reversing the trend of global sea level but is slowing the apparent rate of sea-level transgression. Tectonic factors are one of the most difficult effects to take into consideration when examining transgression and regression because of the geographic variability and the nature of the tectonic processes themselves.

Transgression and regression are seemingly simple phenomena. Sea level is either rising or falling. Yet, the contribution to the observed pattern by local processes is significant and, in some cases, is the dominant effect observed. In order to gain meaningful insight into the global pattern of sea-level change from local data, a thorough understanding of the geologic history of the geographic region being studied is essential.

Transgression and regression appear to be very dramatic and perhaps even catastrophic phenomena when their effects are considered. Nevertheless, both must be kept in the proper time context. Transgression and regression are not unique phenomena in the geologic record. Sea level has been in constant fluctuation throughout most of geologic time, and there is very little evidence that there were times of global sea-level stasis. Transgressions and regressions generally occur over long periods of time and, while some sea-level changes such as glacially induced ones occur over a relatively short span of geologic time (several hundred thousand years), many occur over millions of years. Thus, the observed effect of sea-level change may appear minor unless viewed over long intervals of geologic time.

Methods of Study

Transgression and regression can be studied in a variety of ways and by application of a number of methods. Because they are phenomena that have multiple effects, geologists studying transgression and regression tend to approach them from their own research specialties. Despite diverse backgrounds, all geologists study these phenomena by examining the rock record.

In order to study sea-level changes, some reference point is needed. The reference point commonly used by geologists is the strandline, which is that point where marine conditions stop and the land environment begins. On a normal coast, the strandline would be maximum high water during the tidal cycle, excluding storm surges. At this reference point, water depth is taken as zero. When the geologist is reconstructing the record of transgression and regression, it is essential to relate the data as to distance either seaward or landward from the strandline. It is also possible to relate the geologic information to depth, because as one moves seaward from the strandline, water depth increases. That is a generalization, but it is a useful one.

An example of a study of transgression and regression can be illustrated. A geologist working in a region is interested in the transgression and regression record of a sequence of sedimentary rocks. Many measured sections and detailed descriptions have been made. Particular attention is paid to the rock types and the fossils contained within these rocks. The oldest rocks are sandstones with crossbeds and some broken marine fossils. Crossbeds indicate wave and current activity, and the broken fossils are consistent with that. Additionally, the marine nature of the fossils indicates that normal marine conditions were present during the deposition of this unit. The geologist interprets this sandstone as being deposited in very shallow water under the influence of waves and currents and close to but seaward of the strandline.

The next unit higher in the section is a sandstone, but it contains no crossbeds, many burrows, and marine shells that are unbroken. The geologist believes that this unit represents a deposit formed farther from the strandline than was the underlying one and perhaps in water somewhat deeper. The interpretation is one of a transgression. The unit overlying the burrowed sandstone is a silt with a diversity of marine fossils and many burrows. The finer-grain-sized sediment indicates an increasing distance from the strandline, because the energy or strength of waves and currents diminishes with increasing distance from the strandline (and thus only progressively finer-grained sediment can be transported). The many different types of marine fossils present indicate conditions associated with an increasing distance from the strandline.

The next unit higher in the section is a shale with abundant marine fossils. To the geologist, the finer grain size and abundant marine fossils indicate a further transgression of the sea. The shale is succeeded by a limestone with marine fossils. Limestones accumulate under warm, clear water conditions far removed from a source of sand, silt, or clay. Thus, this unit represents a continued transgression. The strandline is far enough away that the influence of the continent is not felt. Water depth may or may not be significantly greater than in some of the preceding units, but because the strandline is more distant than before, the transgression is interpreted as continuing. Above the limestone, the geologist identifies another shale unit, similar to the one below. Because of the similarity, the unit is interpreted as representing a return to the conditions that were present when the lower shale was deposited. The only way to accomplish that is to move the strandline closer to the site where the rocks were deposited. Such a change represents a regression. The unit directly overlying the shale is another limestone unit, indicating an increase in the distance from the strandline, thus another transgression. Overlying the limestone unit is a coarse-grained sandstone with plant fossils. The contact between this sandstone and the underlying limestone is very sharp and irregular. The upper surface of the limestone was weathered, indicating exposure to the atmosphere. The geologist interprets this weathered surface as an unconformity representing a gap in the sedimentary record. Unconformities can be produced by nondeposition or by exposure and weathering. The interpretation of weathering in this case indicates an exposure to the atmosphere, and the only way to accomplish that is by a complete withdrawal of the sea—a regression. The overlying sandstone is interpreted as being the product of river sedimentation, reinforcing the interpretation of non-marine conditions produced by regression. This last sandstone completes the sequence studied by the geologist.

The above example illustrates several points. First, an entire sequence of transgression and regression has been described, and there were no absolute values of either distance from the strandline or water depth mentioned. Transgression and regression can be relative rather than absolute phenomena. To be sure, some insight into the absolute values of water depth and strandline may be obtained, especially if some of the fossils have known depth and distance from shore limits. Yet, these absolute values do not change the observed record in the rocks, and the concepts of transgression and regression remain. Second, a cycle of transgression and regression need not be symmetric. Too often, researchers expect a complete "wave form" pattern. Although the first transgressive cycle started in rocks deposited near the strandline, the companion regressive cycle did not return to the same environment. In fact, only a slight regression was seen. Third, in the case of transgression and regression phenomena, the missing rock record implied by an unconformity surface is significant. Finally, although the sedimentary record at the point studied by the geologist indicates an easily discernible pattern of transgression and regression, that is in fact a local record and may not bear any resemblance to the global eustatic record of sea level. In this area, no mention was made of the amount of time involved in the sequence, in the age of the rocks, or in the mechanism of sea-level change involved. For the simple interpretation of transgression and regression phenomena, these data are not important. If the phenomena are to be related to a global scale, however, time control, usually involving detailed biostratigraphy, is essential. Once the precise age and duration of the sequence is determined, the geologic history of the area can be taken into account and the relationship of the local record to the global eustatic pattern can be determined.

Global sea-level patterns themselves can be studied through the use of stratigraphic and geophysical studies on the edge of the continental shelves. These regions are so far removed from the strandline that, generally, the local effects of sedimentation and tectonics are minimal, and the pattern obtained accurately reflects the pattern of global eustasy. That does not minimize the importance of local studies, as many times the questions raised about transgression and regression are of a local rather than a global nature.

Context

The importance of transgression and regression cannot be underestimated. There is a strong economic incentive to understanding the pattern of these phenomena. For example, many stratigraphic petroleum traps occur in deposits related to transgression, and an understanding of the transgression and regression sequences in a given area may provide clues to the occurrence of undiscovered oil and gas fields. The benefits of obtaining new reserves of petroleum by application of different ideas is very important, as an entire dimension may be added to a region nearly depleted in reserves discovered by other means and with other ideas.

Another aspect of transgression and regression phenomena of critical importance to society is the relationship of sea-level change to global warming. The pollution of the earth's atmosphere with the so-called greenhouse gases such as water, carbon dioxide, and methane has resulted in a global warming trend of potentially catastrophic proportions. To be sure, the earth has been in a warming trend for at least the last ten thousand years, but that was a gradual change typical of the interglacial episodes. Industrial input to the atmosphere has speeded up the normal interglacial process and made the warming trend much faster. With this trend comes a melting of the remaining continental glaciers in the Northern Hemisphere and a decrease in the Antarctic glacier as well. The result is a global rise in sea level. With this rise in sea level, many coastal areas will become submerged. Presently, the state of Louisiana is losing land to the rising Gulf of Mexico at an alarming rate. Such global change in sea level will affect all coastal areas. The rate at which transgression is occurring is very rapid by geologic standards and even by human standards: An estimated rise of approximately 1 meter in thirty years is very sudden.

What can be done to prevent the flooding of coastal areas? It is not clear if the process of transgression can be arrested. It is, after all, a normal process of the earth that has been operating for billions of years. It may be possible to change the rate at which transgression is occurring, but the phenomenon itself probably cannot be stopped. What is necessary is to manage the coastal regions in the context of the global transgression. Rather than trying to stop the transgression, a feat which must ultimately fail, it would be wiser to plan, manage, and, finally, live with the transgression. On an optimistic note, the estimate of thirty years for a sea-level rise of 1 meter assumes a constant linear rise in sea level. Recent work on climate change over the last 2 million years indicates that changes do not occur in a linear fashion. As the climatic pattern of the last 2 million has been dominated by glacial and interglacial alternations, the nonlinear pattern of change should be reflected in sea-level change as well. Thus, the estimate of thirty years may be too short, and humankind may have more time to prepare to deal with the global transgression.

Bibliography

Cloud, Preston. *Oasis in Space.* New York: W. W. Norton, 1988. A well-written, entertaining book on historical geology. Contains excellent descriptions of transgressions and regressions through time, although not much on the mechanisms of sea-level change. Very readable at the high school level and above.

Cooper, John D., Richard H. Miller, and Jacqueline Patterson. *A Trip Through Time: Principles of Historical Geology.* Columbus, Ohio: Merrill, 1986. This book is designed for first-year college students but is written in such a way that senior high school students should be able to read it with little difficulty. Contains a specific section on sea-level change with good illustrations of concepts. Summaries of sea-level change through time are also throughout the book. A glossary is included.

Dott, R. H., Jr., and R. L. Batten. *Evolution of the Earth.* 4th ed. New York: McGraw-Hill, 1988. A book written for first-year college students in historical geology. Clearly written and well illustrated. Of special interest is a section comparing local and global transgressions and regressions. Although written at a higher level than are many other books, it will be useful to anyone interested in sea-level change.

Levin, H. L. *The Earth Through Time.* 3d ed. Philadelphia: Saunders College Publishing, 1988. This book is an excellent reference for the general reader. Although designed as a first-year college text, it has a writing style such that readers from high school on will find it a valuable reference. A well-illustrated section on sea-level change as well as summaries of transgression and regression through time are present.

Mathews, R. K. *Dynamic Stratigraphy.* 2d ed. Englewood Cliffs, N.J.: Prentice-Hall, 1984. An advanced text designed primarily for upper-level university students. Provides the best and most thorough discussion of the methods of study of transgression and regression. Clear diagrams accompany the text so that it may prove a valuable reference to anyone interested in transgression and regression.

Stanley, Steven M. *Earth and Life Through Time.* 2d ed. New York: W. H. Freeman, 1988. A beautifully illustrated book on general historical geology. Designed primarily for first-year college students, it has a clear writing style and illustrations to appeal to all readers. Provides a thorough treatment of sea-level change through time and includes data on transgression and regression from around the world. All persons interested in transgression and regression should find this book useful.

Richard H. Fluegeman, Jr.

Cross-References

Biostratigraphy, 37; The Geologic Time Scale, 272; Stratigraphic Correlation, 593.

Ultramafic Rocks

Ultramafic rocks are dense, dark-colored, iron- and magnesium-rich silicate rocks composed primarily of the minerals olivine and pyroxene. They are the dominant rocks in the earth's mantle but also occur in some areas of the crust. From a scientific standpoint, ultramafic rocks are important for what they contribute to the understanding of crust and mantle evolution. They also serve as an important source of economic commodities such as chromium, platinum, nickel, and diamonds, as well as talc, asbestos, and various decorative building stones.

Field of study: Petrology

Principal terms

CRUST: the upper layer of the earth, composed primarily of silicate rocks of relatively low density compared to those of the mantle

MANTLE: the middle layer of the earth, composed of dense, ultramafic rocks, located from about 5-50 kilometers below the crust and extending to the metallic core

OLVINE: a silicate mineral abundant in many ultramafic rocks; gem-quality, clear green olivine is called peridot

PERIDOTITE: a principal type of ultramafic rock, composed mostly of the minerals olivine and pyroxene

PYROXENE: a group of calcium-iron-magnesium silicate minerals that are important constituents of ultramafic and mafic rocks

Summary

Ultramafic rocks are dense, dark-colored, iron- and magnesium-rich rocks that constitute a volumetrically and scientifically important class of earth material. To qualify as ultramafic, a rock must contain 90 percent or more of the so-called mafic minerals, olivine and pyroxene. The term "mafic" is derived from "magnesium" and "ferric" (the latter meaning "derived from iron"); "ultramafic" is an appropriate term for these minerals as they are very rich in both magnesium and iron. Another class of common rocks in the earth's crust, the mafic rocks (including the black lava called basalt), are also rich in magnesium and iron but not to the extent shown by ultramafic rocks. A typical basalt may contain an average of about 20 percent iron plus magnesium (as the oxides ferrous oxide and magnesium oxide), but typical ultramafic rocks may contain more than twice that amount. Thus, the prefix "ultra" is well deserved. In addition, ultramafic rocks are generally—but not always—lower in silica than are mafic rocks, which is particularly true for olivine-rich ultramafic rocks, which average less than 45 percent silica; basaltic rocks average about 50 percent silica.

The Sierra Nevada of California, one of North America's largest concentrations of mantle-type ultramafic rocks. *(R. Kent Rasmussen)*

Although petrologists (geologists who study rocks) recognize several types of ultramafic rocks, the type called peridotite is the most common and, thus, considered the most important of all ultramafic rocks. Peridotites consist of more than 40 percent of the light green mineral olivine, an iron-magnesium silicate. The gem-quality variety of olivine is called peridot, from which the term "peridotite" is derived. The remaining minerals consist of various pyroxene minerals or hornblende. Pyroxene is a group of silicate minerals that contain calcium, iron, and magnesium as principal constituents; hornblende is similar to high-calcium varieties of pyroxene, but it contains water as part of its chemical structure. Variable amounts of spinel-group minerals (oxides of chromium, magnesium, iron, and aluminum), garnet, and other minerals commonly occur in minor amounts in peridotites. Peridotite is the principal constituent of the earth's upper mantle, which makes it one of the most abundant types of rock in the earth. In the crust, however, peridotite is relatively rare, occurring only in specific geological environments. It is this property of peridotites, among others, that makes these rocks so important and interesting as subjects of scientific investigation.

Peridotite is not the only type of ultramafic rock. Another important category of ultramafic rocks is the pyroxenites, which, as the name implies, are rocks composed mostly of pyroxene (greater than 90 percent). Pyroxene is not one mineral but is rather a family of related minerals, all having single chains of silicon-oxygen tetrahedra as their central structural aspect. This mineral group consists of two principal structural-chemical types: the low-calcium orthopyroxenes (with orthorhombic crystal symmetry) and the high-calcium clinopyroxenes (with monoclinic crystal symmetry). Pyroxenites can have variable amounts of either of these two pyroxene types or may have one more or less exclusive of the other. Orthopyroxenites contain mostly orthopyroxene and clinopyroxenites contain mostly clinopyroxene; the term "websterite" is used to describe pyroxenites that contain an appreciable mixture of both minerals.

Peridotites can also be subdivided into different types based on their pyroxene components. Those that contain mostly orthopyroxene as their pyroxene type are called hartzburgites. If the major pyroxene is clinopyroxene, the rock is called wehrlite; if both pyroxenes occur along with olivine, the rock is called lherzolite. One special and fairly rare type of peridotite is dunite: a nearly pyroxene-free, olivine-rich peridotite (90 percent or greater olivine). Most peridotites from the mantle contain minor amounts of an aluminum-rich mineral, generally either plagioclase (calcium-sodium aluminum silicate), spinel (magnesium-aluminum oxide), or pyrope garnet (magnesium-iron aluminum silicate). Many of the peridotites formed in the crust by igneous processes commonly contain the chromium-iron spinel-group mineral called chromite, the principal ore of chromium metal.

Geologists' knowledge of mantle rocks comes predominantly from either of two sources: chunks of rock torn from the walls of deep mantle conduits by molten magma that eventually carries these mantle "xenoliths" (Greek for "stranger rock") out of the earth enclosed in lava flows, or thick slices of ultramafic mantle material that were thrust up to the surface during mountain-building processes. Scientific examination of these mantle samples shows that most of them are metamorphic rocks that crystallized under high pressures and temperatures in the solid state. Some mantle samples, however, show evidence of an earlier history of crystallization from molten magma.

Ultramafic rocks that formed in the crust mostly originated by purely igneous processes and thus show evidence of crystallization from magma. Most crustal ultramafic rocks are found in layered complexes: crystallized bodies of magma trapped deep within the crust that display a "layer-cake" structure, with different kinds of igneous rocks composing the various layers. The production of many kinds of rocks from a single parent magma is called differentiation; the differentiation involved in the formation of layered complexes largely results from the process of gravity differentiation. In this process, early-formed minerals sink to the lower reaches of the magma reservoir, with later minerals forming progressively higher layers until the magma is wholly crystallized. The parent magma from which all layered complexes form is basaltic in composition—the same material that forms the islands of Hawaii, for example, or the steep cliffs of the Columbia River Gorge in Washington State. Because the iron-magnesium-rich (mafic) minerals, olivine and pyroxene, form relatively early in basalt magmas, these minerals collect in the first-formed, lower layers of layered complexes. Resulting rocks in these layers are typically peridotites and pyroxenites. The igneous rocks that form later, above the dark ultramafic rocks, consist of gabbros (plagioclase-pyroxene rocks) and other mafic rocks. Knowledge of the ultramafic and other rocks in layered complexes comes from the fact that some of these deep-seated, crystallized magma bodies have been uplifted in mountainous areas and uncovered by erosion.

Where one would travel to see examples of ultramafic rocks would depend upon which of the two major categories—mantle or crustal—is of interest. In North America, the best places to see mantle-type ultramafic rocks are the Coast Ranges and the Sierra Nevada of California and the Klamath Mountains of western Oregon. Belts of ultramafic rocks are also found in some areas of the Canadian Rocky Mountains and on Canada's east coast, particularly in Quebec and Newfoundland. Exposures of ultramafic rocks also occur at various locations within the Appalachian Mountains in the United States, particularly in Vermont and Virginia. Worldwide, the Mediterranean area lays claim to some of the most impressive areas of exposed mantle-derived ultramafic rocks, notably the Troodos complex in Cyprus and the Vourinos complex in Greece. Similar ultramafic belts extend in scattered exposures from France and Italy eastward to Turkey and Iran. Other prominent occurrences are known from India, Tibet, and China. Most if not all of these exposures are known as ophiolite complexes.

Ophiolites are slices of ocean crust (basalt and sediments) and underlying mantle (mostly hartzburgite and lherzolite peridotite and some ultramafic gravity-stratified rocks) that become involved in mountain-building activities where two lithospheric plates (consisting of upper mantle and crust) collide. This collision area, called a subduction zone, is commonly located near continental margins, which is why ophiolites generally occur in mountain belts (like the Coast Ranges or Appalachians) that parallel a coast. The Mediterranean ophiolites are thrust up in an area that is being deformed by the collision of Africa (on a lithospheric plate moving slowly northward) with Eurasia. Unfortunately for the scientists wishing to study them, most ultramafic rocks in ophiolites, especially the peridotites, are altered to varying degrees to serpentine (a hydrated magnesium silicate similar structurally to clay), producing a metamorphic rock called serpentinite. It is produced before or during mountain building by the interaction of olivine with water, much of which probably consists of seawater trapped in peridotites and associated rocks as they lay below the ocean floor.

Another source of mantle-derived ultramafic rocks are the xenoliths brought to the surface by lava flows. These mantle samples are particularly valuable as objects for scientific study and as potential sources of gem-quality peridot because they commonly show few if any effects of alteration to hydrous minerals such as serpentine. Well known to scientists who study ultramafic rocks are the excellent xenolith localities at San Carlos, Arizona; Salt Lake crater, Hawaii; Sunset crater near Flagstaff, Arizona; and the garnet-bearing xenoliths incorporated in diamond-bearing kimberlite deposits in South Africa. Kimberlite deposits are chaotic masses of broken fragments of ultramafic rocks and their constituent minerals mixed with serpentine, other hydrated silicate minerals, and carbonate minerals (mostly calcite and calcium carbonate). They originate in the upper mantle, between about 150 and 200 kilometers depth. Propelled rapidly upward as carbon-dioxide- and water-charged mixtures of mantle and crustal rocks encountered during their ascent, they explode violently at the surface and form large craters. The eroded remnants of these kimberlite pipes (also called diatremes) are mined for diamonds, which form at very high pressures and temperatures in the mantle. Diamond-bearing kimberlites occur in the United States in Arkansas; diamond-free kimberlites occur in Missouri, Oklahoma, and Kansas.

Large layered complexes containing ultramafic rocks are exposed in only a few places on the world's continents. Probably the best example in the United States is the Stillwater complex, located in the Beartooth Mountains of western Montana. Estimated to have originally measured 8 kilometers thick, it is now exposed as a 5-kilometer-thick strip approximately 30 kilometers long. The Stillwater has been the object of intense scientific research over several years and is a source of economic chromium deposits. Platinum deposits also occur within the ultramafic rocks of the Stillwater. An even more intensely studied layered intrusive body is the Skaergaard intrusion of eastern Greenland. Discovered in 1930, it is exposed over an area of glaciated outcrops 3.2 kilometers thick. It is believed to represent basaltic magma trapped beneath a rift zone created as the North Atlantic ocean basin opened up about 50 million years ago. This magma differentiated into a complicated layered zone consisting of various kinds of peridotites, pyroxenites, and gabbros.

No discussion of layered ultramafic complexes would be complete without mentioning the famous Bushveld complex of South Africa. Measuring 270 by 450 kilometers in area, it is estimated to have been originally about 8 kilometers thick, like the Stillwater. Much of the world's supply of chromium comes from the chromite deposits associated with peridotites in this monstrous intrusive body. The Bushveld is also the world's largest source of platinum, contained in a 1- to 5-meter-thick assemblage of olivine, chromite, orthopyroxene, and sulfides known as the Merensky Reef. This single, thin unit extends for a total of 300 kilometers. Other notable layered complexes with well-studied ultramafic rocks are the funnel-shaped Muskox intrusion in northern Canada and the Great Dyke of central Zimbabwe. With an average width of 5.8 kilometers, the Great Dyke extends for 480 kilometers. Like the Bushveld, it is a source of chromium ore.

Another type of crustal ultramafic rock is the rare but very important komatiite ultramafic lava flow. Mostly restricted to very old Precambrian terrains (most komatiite ages lie between about 2.0 and 2.5 billion years), these rocks represent nearly completely melted mantle material, a feat that requires extremely high temperatures. Present temperatures in the upper mantle are not sufficient to produce komatiites; basalt magma is produced instead by much lower degrees of melting of mantle peridotite. The restriction of komatiites to Precambrian terrains suggests that the mantle was much hotter billions of years ago compared to more recent times. Excellent exposures of these ultramafic lava flows occur at Yilgarn Block, Australia, and the Barberton Mountains, South Africa.

Methods of Study

Ultramafic rocks are studied using analytical techniques commonly applied to any kind of igneous or metamorphic rock type. These techniques include detailed microscopic analysis; analyses for major (1 percent or more by weight), minor (less than 1 percent), and trace (measured in terms of parts per million or parts per billion) elements; and analysis for various critical isotopes, such as those of rubidium and strontium, and the rare earth elements neodymium and samarium. The latter isotope systems are the most commonly used to date ultramafic rocks by radiometric methods. What kinds of analyses are performed on any given ultramafic rock depends upon the type of ultramafic rock in question and the objectives of the specific research project. For example, ultramafic rocks from the mantle would probably not be approached from the same scientific aspect as would ultramafic rocks formed in the crust. Analysis of mantle samples might reveal clues as to how the early earth formed originally and then differentiated into a core, mantle, and crust. Because many igneous magmas now residing as igneous rocks in the crust (especially those of basaltic composition) were originally produced by partial melting of ultramafic mantle rocks, the study of mantle samples provides a glimpse into how the earth's crust evolved. The detailed analysis of layered complexes, komatiites, kimberlites, and associated crustal rocks gives information about the complexities of igneous differentiation (making many kinds of rocks from a common parent) and also bears on the processes involved in crustal formation. These studies can also be applied to some ultramafic meteorites to show how other planets evolved in comparison to the earth.

Because many crustal rocks ultimately originated in the mantle, mantle ultramafic rocks have been subjected to laboratory experiments that seek to simulate the production of various magmas in the mantle. In these experiments, mantle samples (or artificially concocted facsimiles) are heated to the point of melting at pressures calculated to occur at different depths in the mantle. These experiments have shown that different kinds of basaltic magma can be produced by simply changing the pressure (thus, the depth) at which mantle materials melt. Varying the proportions of constituents such as water and carbon dioxide can also produce different magma compositions for a given pressure. These experiments have been correlated with rocks collected in the field and their particular environments of emplacement. For example, experimental evidence shows that rocks like kimberlites and alkaline lavas (high in potassium and sodium, very low in silica) originate at great depth in areas of the mantle enriched in carbon dioxide, which helps to explain the explosive, gas-rich nature of kimberlite deposits and the high carbon dioxide contents in gases released upon extrusion by alkaline lavas. More silica-rich lavas, like the basalt that floors the earth's oceans, originate at much shallower depth, and thus at lower pressures, than do alkaline-type lavas.

Mantle samples also give scientists clues as to the very early history of the planet, because their geochemistry records the earliest differentiation events, including the formation of the nickel-iron core. Trace element analyses of mantle samples show that certain elements generally expected to sink into the core along with the nickel and iron are depleted in the mantle. Some of these elements are gold, osmium, iridium, and platinum. Obviously, not all these elements sank to the core during the time about 4.5 billion years ago when the earth was much hotter than it is now; otherwise, the abundances of these elements (including iron) in the crust would not be seen. The process of core formation was not perfect; some elements that could have gone to the core stayed in the mantle, later to be erupted to the crust by volcanoes and other igneous processes. At the same time, geochemists note that laboratory analyses of some other elements show abundances similar to certain very old, "primitive" meteorites called chondrites. These meteorites are believed to have crystallized from gas that made up the solar nebula, the gaseous cloud from which the solar system eventually evolved. Chondrites are considered primitive because they contain abundances and ratios of most elements that are similar to those measured for the sun. Thus, chondrites are the rocky building blocks of most planets and asteroids. Analysis of mantle ultramafic rocks shows that their calcium-aluminum ratio, for example, is similar to that of chondrites, suggesting that mantle materials were originally composed of chondrites or chondritelike material that later differentiated to make the core and crust. Interestingly, the mineralogical makeup of chondrites is similar to that of the mantle: mostly olivine and pyroxene. Much of the mantle, therefore, probably represents highly meta-morphosed chondritic meteorites or similar precursor rocks.

Layered complexes, such as the Skaergaard, Bushveld, and Stillwater, have been extensively studied as models for how magmas can differentiate into other kinds of magmas. These processes, in turn, demonstrate why the earth contains so many different types of igneous rocks compared to other planets and satellites, such as the moon. Experimental studies have been conducted on basaltic magmas in which the liquids are allowed to cool slowly as they would at great depth in the crust. By observing the kinds of minerals that crystallize at various temperatures and analyzing the chemical compo-sitions of coexisting liquids, petrologists can attempt to explain the order and mineral compositions of the various layers in layered complexes. Experi-ments like these can also be used to show why minerals such as chromite and platinum-rich sulfides may concentrate in particularly thick layers that can be mined for these economically important minerals.

Context

Although ultramafic rocks are not commonplace constituents of the earth's crust, most people have seen examples of them, as they are used as building stone in large department stores, churches, banks, and various public buildings. The best known of these decorative stones is verde antique, which literally means "old green." Verde antique is considered to be a type of marble, but it mostly consists of dark green serpentine (a hydrous alteration of olivine) swirled together with white, gray, or pink calcite (calcium carbonate) that may also crosscut the serpentine as thin veins.

Ultramafic rocks may also serve as a source of gemstones. The semipre-cious gemstone peridot originates as large, perfectly clear, green olivine crystals in peridotites, which can be incorporated into rings, bracelets, and necklaces, some of the finest examples being the silver with peridot jewelry made by members of the San Carlos (Arizona) Apache Tribe. These Native Americans own one of the world's richest sites for peridotite nodules incor-porated in lava flows on the reservation. Some of the peridot crystals that they recover are as big as an adult thumb.

Perhaps the most important commerical use for ultramafic rocks lies in their tendency to harbor rich deposits of certain metallic ores, namely chromium, vanadium, platinum, and nickel. Chromium and vanadium (in the mineral chromite) are mined from peridotite units, usually from chromite-rich seams. The Stillwater complex in western Montana and the Bushveld complex in South Africa are some of the richest sources of this ore. Platinum is mined as native platinum (metallic) and as various platinum sulfides from the Bushveld, the Yilgarn Block area of Western Australia, and in the Ural Mountains and other sites in the Soviet Union. Nickel is commonly associated with platinum deposits in ultramafic rocks, but large quantities of nickel are also contained in so-called lateritic nickel deposits. These deposits form by intensive weathering of ultramafic rocks in tropical areas, causing the nickel to be released from olivine as it reacts with water to form serpentine and clay. Important nickel laterite deposits occur in Cuba, the Dominican Republic, Indonesia, and the southern Soviet Union.

Finally, certain nonmetallic minerals, particularly talc and asbestos, are also mined or quarried from ultramafic deposits. These include the talc deposits in Vermont and Virginia. In Virginia, the talc occurs in soapstone, a corrosion-resistant, durable stone commonly used for laboratory tables or for artistic carvings. Large talc deposits also occur in Italy and France. The principal form of asbestos fibers, chrysotile serpentine, also comes from ultramafic rocks. Large deposits occur in Quebec, as well as in the Soviet Union, Zimbabwe, and South Africa. In the United States, asbestos occurs in ultramafic deposits in California, Vermont, and Alaska.

Bibliography

"Along-Axis Magmatic Oscillations and Exposure of Ultramafic Rocks." *Geology* 25, no. 12 (December, 1997): 1059-1063.

Ballard, Robert D. *Exploring Our Living Planet.* Washington, D.C.: National Geographic Society, 1983. This book is an excellent guide to the dynamic processes that shape the earth. It does not have specific sections on ultramafic rocks, except for a brief item on the Bushveld complex. On the other hand, it is one of the best resources for the layperson on plate tectonics and volcanism, both of which involve ultramafic rocks. Extensively illustrated with color photographs and well-drafted, easily understood diagrams to show how the mantle is involved in processes of mountain building and volcanism. Includes an adequate glossary and extensive index.

Smith, David G., ed. *The Cambridge Encyclopedia of Earth Sciences.* New York: Cambridge University Press, 1981. This compendium of knowledge about the earth is a useful reference for laypersons and specialists alike. The color and black-and-white photographs are excellent, as are the color diagrams and charts. The section on peridotites and other mantle rocks is sufficiently comprehensive to include stable isotopes and trace elements. Contains an extensive glossary and index.

Symes, R. F. *Rocks and Minerals.* Toronto: Stoddart, 1988. This book, written by Dr. Symes and the staff of the Natural History Museum of London primarily with the young reader in mind, is lavishly illustrated with colored photographs and diagrams on every page. The photographs of museum rock and mineral specimens are combined with descriptive text giving characteristics, occurrences, and uses of the illustrated specimens. Most of the rocks and minerals important to the subject of ultramafic rocks are described and illustrated, including olivine and pyroxene, peridotite and serpentine, and basalt/gabbro. A visual delight, this book is highly recommended.

Wyllie, Peter J., ed. *Ultramafic and Related Rocks.* Huntington, N.Y.: Robert E. Krieger, 1979. This compendium of several articles by a number of authors is one of the most comprehensive works on ultramafic rocks for more advanced readers. Every chapter is preceded by a foreword by Dr. Wyllie, summarizing the content and significance of the article. Has an extensive list of references (all pre-1967, the original publication year) and an author and subject index. This volume is an ideal source of seminal information on ultramafic rocks for the serious student, but there are a few topics not covered concerning ultramafic and associated rocks. Most college-educated persons should gain an appreciation for the significance and complexity of ultramafic rock associations by at least a cursory scanning of this book.

John L. Berkley

Cross-References

Earth's Composition, 162; Igneous Rock Bodies, 307; Igneous Rocks, 315; Magmas, 383.

Uniformitarianism

Uniformitarianism, in its most basic form, describes a methodology used in the study of earth history. It assumes that the earth can be interpreted in terms of natural processes that are operational today. In its broader context, uniformitarianism also embraces certain conclusions about earth history: that the rates at which these processes operate are constant through time and that the earth has experienced a cyclic history.

Field of study: Stratigraphy

Principal terms

CATASTROPHISTS: adherents to the belief that the present-day slow rates of natural processes are not sufficient to explain many features of the rock record

DILUVIAL THEORY: the belief that the Mosaic flood was responsible for shaping many of the earth's surface features

DOCTRINE OF FINAL CAUSES: the belief that a purpose or design is revealed in the organization of nature

METHODOLOGICAL UNIFORMITARIANISM: that aspect of uniformitarianism that describes a procedural approach to the study of the earth, that is, that ancient phenomena can be interpreted in terms of present-day natural processes

NEPTUNISTS: adherents to Abraham Werner's belief that granite and basalt formed by chemical precipitation from seawater

PLUTONISTS: adherents to James Hutton's belief that granite and basalt formed by crystallization from molten mineral matter

ROCK CYCLE: the path of earth materials as erosional products of rock are deposited and reformed again into rock, which then can be eroded again

SUBSTANTIVE UNIFORMITARIANISM: that aspect of uniformitarianism that draws conclusions as to the history of the earth, that is, that natural processes have been taking place at constant rates through time and the earth has had a cyclic history

SURFICIAL GRAVELS: alluvium found at different places on the earth's surface, originally interpreted as the product of the Mosaic flood

UNIFORMITY OF NATURE: Charles Lyell's formulation describing a nature in which processes are both consistent and constant through space and time; a mixture of methodological and substantive uniformitarianism

Summary

Uniformitarianism owes its origin and development to three British geologists: James Hutton (1726-1797), who laid the groundwork for the concept; John Playfair (1748-1819), who elucidated and elaborated on Hutton's ideas; and Charles Lyell (1797-1875), who championed the uniformitarian cause in the mid-nineteenth century.

At the end of the eighteenth century, there was fairly close agreement between theories of the earth based on direct, though limited, observations of earth material and structure and the account of earth history revealed in the biblical book of Genesis. The earth was thought to be of no great antiquity, considered by most to be not much older than six thousand years. The surface features of the earth, its mountains with their tilted rock, its river valleys, and its ubiquitous surficial gravels, were attributed to a great debacle, assumed to be the biblical flood of Noah. These features existed in equilibrium with the elements, possibly waiting for the next great debacle. Some erosion was happening, but not enough to alter the topography greatly in the short amount of time available. It was against this intellectual backdrop that Scottish scientist James Hutton presented to the Royal Society of Edinburgh a paper entitled *System of the Earth* (1785), which contained the seeds of a controversy that continued for a half-century. The paper was published in the first volume of the Society's Transactions under the title *Theory of the Earth: Or, An Investigation of the Composition, Dissolution, and Restoration of the Land Upon the Globe* (1788).

Hutton's work essentially outlined the rock cycle. He recognized, as others had before him, that the fossiliferous strata of the continents had formed on the bottom of the sea as unconsolidated sediment. In order to be added onto the continent, in places at angles of inclination too steep to represent the original repose of water-charged sediment, a mechanism was needed to fuse the sediment into rock and uplift the sea bottom. In Hutton's scheme, the energy required to do that was derived from the internal heat of the earth. Once exposed to the elements, the rocks were eroded to form sediment again. In Hutton's view, the present land was derived from a former continent, and the disintegration and erosion of rock observed at its surface were providing material out of which a future continent would be constructed. Hutton emphasized that these processes that shaped the earth were natural and operated very slowly, hardly causing noticeable change during the history of humankind. He eschewed calling upon preternatural causes in explaining natural phenomena.

Hutton's 1788 paper contains the major elements of what was later termed uniformitarianism. The earth was to be interpreted in terms of present-day natural processes, excluding supernatural explanations. The processes operated very slowly and probably at fairly constant rates, although Hutton is not clear on this latter point. The major emphasis on constant rates was added by Charles Lyell almost a half-century later. Lastly, the earth had gone through a cyclic development, or, as Hutton phrased it, a "succession of worlds," with the destruction of one world providing the material for the next.

Hutton's view of cyclic "worlds" destroyed and reconstructed by slow processes led him to conclude that the earth is very old, in his phraseology, an earth with "no vestige of a beginning" and, in reference to the cyclicity, "no prospect of an end." Hutton's "discovery" of an earth of inscrutable age is considered his major contribution to geology. Hutton is also credited as the first to recognize the true origin of those rocks now known as igneous. His belief that these bodies were injected as molten masses into stratified rock molded his attitude toward the efficacy of heat in driving his cycles.

The prevailing view of these rocks at the time was that of the German mineralogist and natural philosopher Abraham Gottlob Werner (1750-1817), whose science of geognosy maintained that basalt and granite had been chemically precipitated from an ocean that from time to time had flooded the earth. Hutton's adherents became known as Plutonists and Werner's as Neptunists. A few years after publication of Hutton's 1788 paper, his ideas were acrimoniously attacked by the Neptunists. Although the argument ostensibly was over the origin of basalt, a major undercurrent centered on Hutton's timeless earth, a story that was not in line with that of the Scriptures. An ailing Hutton attempted to answer these attacks in 1795 with a wordy expansion of his ideas that is notorious for its unreadability. He died in 1797.

Fortunately, Hutton's good friend and sometime field companion, John Playfair, engaged the battle with the Neptunists. Playfair's prose was as facile as Hutton's was cumbersome, and it is from Playfair's *Illustrations of the Huttonian System of the Earth* (1802) that comes most of posterity's knowledge of Hutton's thoughts. In the period following Playfair's elucidation, most geologists came to accept the igneous origin of granite and basalt and to adjust their thinking in line with the unavoidable conclusion that the earth is certainly much older than the six thousand years afforded by a literal reading of Genesis.

The most important discovery in the first two decades of the nineteenth century was that the earth's strata record a progression of life forms, with primitive types in lower layers and more advanced types above. Foremost among researchers of the fossil record was the French comparative anatomist Georges Cuvier (1769-1832), who in 1812 published a monograph that was to serve as a major impetus for renewed controversy surrounding uniformitarianism. Cuvier, one of the first naturalists to reconstruct fossil vertebrates, was recognized as the premier scientist of his day. His early work had been on fossil vertebrates, the mammoth among many others, found in the surficial gravels. Because he felt that the extinction of this fauna could not be explained by processes presently acting on the surface of the earth, he concluded that their demise resulted from a sudden but prolonged and localized inundation of the land by the sea brought about by unknown

Naturalist Stephen Jay Gould has offered an influential updating of uniformitarian theory. *(Stephen Jay Gould)*

natural causes. He referred to the event as a revolution. Later research in the rocks around Paris caused him to expand his idea of earth history to include several revolutions.

With his theory of periodic revolutions, Cuvier traditionally has been considered in English-speaking countries as a proponent of an earth shaped through divinely instituted catastrophes, which in turn were followed by special creations to replace the extinguished life. (Martin J. S. Rudwick, in an excellent account entitled *The Meaning of Fossils*, points out the injustice of this characterization. Cuvier was a strict empiricist who believed that science and religion should be pursued separately.) Cuvier's research was immediately seized by William Buckland (1784-1856), England's most prestigious geologist, who took the latest of the revolutions and transformed it from its original form, a localized inundation of long duration, into a catastrophic, short-lived universal deluge—in short, the flood of Noah. Buckland's flood theory, the so-called diluvial theory, gained much popularity in England in the 1920's but was essentially abandoned toward the end of the decade as it became clear that the surficial gravels, the supposed deposits of the deluge, did not exhibit the distribution or character expected of a worldwide catastrophic event.

Still, a contingent of prominent British Diluvialists believed that the observations on which Cuvier's revolutions were based could not be explained by any of the processes in operation on the earth. In response to the threat posed to the uniformity of nature theory by the position of these individuals, who became known as Catastrophists, Charles Lyell published his landmark book *Principles of Geology* (1830-1833), a work built upon the basic tenets of Hutton's theory of the earth. The ensuing debate between the Catastrophists and Uniformitarians, as Lyell and his followers became known, was a conflict in viewpoint among scientists, not a struggle between science and the church, as is commonly assumed. In many important respects, the two sides were in agreement. The Catastrophists willingly accepted the idea of an ancient earth and were devoted in most situations to the methodology of uniformitarianism. They were firmly opposed, however, to the interpretive aspects that the doctrine imposed on earth history: namely, that the slow rates of present-day processes are representative of all time (steady rate) and that earth history is cyclic, with one cycle looking much like another (steady state).

On the surface, the Catastrophists' argument against steady-rate uniformitarianism seemed to be a simple request for accelerated rates to explain phenomena such as those on which Cuvier's revolutions were based. It is clear from their writings, however, that the Catastrophists were anxious to interpret these episodes as times during which natural law was suspended by the regulatory hand of God. Their view of God was as both creator and referee, willing to intervene in the course of His creation if it veered off track. To them, any movement to remove the punitive hand of God from the course of things would lead to moral decay and the disintegration of British society. Lyell, on the other hand, was unwilling to concede any portion of earth history, no matter how small, to a supernatural cause that fell outside the pale of scientific inquiry. As succinctly analyzed by Charles Coulston Gillispie in his revealing account of the times entitled *Genesis and Geology* (1951), the Catastrophists were apprehensive that without any cataclysms there would be no God, whereas Lyell was concerned that without the uniformity of nature there would be no science. Lyell's opinion eventually prevailed, and supernatural explanations, for at least the physical world, were expunged from the methodology of historical interpretation. As a companion to this development, the efficacy of gradual processes in explaining all historical phenomena became firmly established in geologic dogma, but not without some unfortunate side effects.

To the Catastrophists, Lyell's stubborn support of steady-state uniformitarianism not only seemed palpably false in the light of a fossil record that showed progression but also was contrary to physical laws that required the earth to evolve as its energy dissipated. In the light of today's knowledge, Lyell's arguments against the evidence of the fossil record seem foolish. He went so far as to predict that ichthyosaurs, marine reptiles seemingly extinct for more than 65 million years, would return to populate the seas when their "world" once again rolled around. Stephen Jay Gould, in his short, provocative book, *Time's Arrow, Time's Cycle* (1987), argues that differing views of history occupied a central position in the conflict between the two sides. Lyell's repetitious earth (time's cycle) was essentially ahistorical because it developed no unique configurations through time. Opposed to this was the

Catastrophists' vision of an ever-changing earth (time's arrow), unique at each stage, and thus amenable to sophisticated historical analysis. Lyell abandoned his stand on steady-state uniformitarianism shortly after Charles Darwin's convincing documentation that life on earth had evolved.

Applications of the Method

Uniformitarianism is a mixed concept, part methodology and part theory. It is almost exclusively a British construction, a product of the intellectual and social atmosphere of nineteenth century Great Britain. Although much homage has been paid to uniformitarianism in English-speaking countries, it has never been a conscious theoretical organizer in the geological studies of continental Europe. In textbooks, "uniformitarianism" is generally restricted in meaning to the methodology of interpreting the past through knowledge of the present, commonly rendered as "the present is the key to the past." More correctly, the term, which was not used by Hutton or his followers, was a title applied to Lyell's staunch support of an earth developed by gradual processes of constant rate.

Both steady-rate and steady-state uniformitarianism grew out of the peculiar way in which nineteenth century Great Britain mixed science and religion. Steady-rate uniformitarianism, which in its essence describes an earth that developed through constant, gradual processes, has had a significant influence on geologists' perspectives of earth history. Strict steady-rate uniformitarianism was developed by Lyell as he labored against what he perceived to be the Catastrophists' attempts to qualify methodological uniformitarianism. Although it was probably a battle that needed to be fought, Lyell's ultimate stand on unvarying rates served as blinders to later geologic interpretation. Too often geologists have overlooked, or from peer pressure have been forced to exclude from consideration, explanations that call on catastrophic natural events. An example is the widespread distribution of glaciers during the past ice age. Indeed, some of the early support of catastrophism came from phenomena now understood to be glacial in origin. In his book *The Nature of the Stratigraphical Record* (1981), Derek Ager uses the expression "catastrophic uniformitarianism" to emphasize that cataclysmic events also have a place in nature. It is unfortunate that many will associate Ager's terminology with "old-time" catastrophism, which had its origin in the Scriptures.

Steady-state uniformitarianism has had essentially no influence on the way in which geologists view earth history, primarily because it has little application as a historical tool. Hutton's "succession of worlds" was supposedly important in his discovery of immense time. Inasmuch as he had a large amount of rock to account for by slow processes, which is exactly the same observation that eventually convinced most later geologists of the earth's antiquity, why the emphasis on a repetitious, unchanging world? The answer would seem to lie in a practice, common in Hutton's time, of researchers to see in the results of their studies the handiwork of God. It seemed prestigious for a piece of research to add to the knowledge of final causes, which is to say it should address the question: To what purpose had God, in His infinite wisdom, caused things to be the way they were? A strictly decaying world would soon become unfit for life, especially for humankind; such a world would have to be the invention of an incompetent or indifferent creator, a conclusion that was philosophically unsatisfying. Thus, Hutton imposed on his system a cyclicity, a never-changing never-ending earth, in which he saw an "economy of nature" worthy of "Divine wisdom."

Recent debate on uniformitarianism has centered on the separation and examination of its methodological and theoretical aspects with the objective of analyzing the concept's utility in modern geological education and investigation. Gould makes a distinction between substantive and methodological uniformitarianism. He dismisses substantive uniformitarianism (steady-state and steady-rate) as untrue. In methodological uniformitarianism, he identifies two "procedural assumptions": Nature is consistent through space and time—that is, natural processes that operated in the past are the same as those that operate today—and natural phenomena should not be explained by unknown causes when known, observable causes are sufficient, which, in the context of uniformitarianism, admonishes geologists to exclude supernatural explanations from their interpretations. Gould argues that the first of these assumptions is nothing more than inductive reasoning, by which explanations for ancient natural phenomena are searched for among similar-looking present-day phenomena where causes can be inferred in terms of observable processes. Gould identifies the second assumption with the principle of simplicity, a time-honored guideline in empirical science. Because inductive inference and the notion of simplicity are common procedural aspects in all empirical sciences, Gould concludes that their inclusion in geology under the guise of procedural uniformitarianism obscures the relationship of geology to the scientific enterprise in general, falsely suggesting that geological investigation is guided by principles different from those of other sciences. He concludes that, whereas the cause of uniformitarianism served as a useful nineteenth century rallying point from which to defend inductive inference in geological inquiry against the attitudes of the Catastrophists, with the passing of those times, the term, considering its ambiguity, has essentially outlived its usefulness. Gould believes that a sounder pedagogy can be achieved by disuse of the term. He emphasizes that methodological uniformitarianism must be retained as a procedural approach to the study of the earth but recommends that it be referred to by its more universally accepted names of induction and simplicity. Opponents, while essentially agreeing with Gould's analysis, maintain either that the unique position of geology among the empirical sciences, that of investigating the causes of ancient phenomena, requires a unique label for its methodology or that abandonment of the term "uniformitarianism" might result in abandonment of the methodology itself. Because traditional uniformitarianism embodies mixed meanings, in recent years, the European term "actualism" (from the German *aktuell,* meaning "present" or "of present importance"), which refers only to methodological uniformitarianism, is gaining popularity in English-speaking countries.

Context

The role of science is to examine the natural world and to explain its phenomena in rational terms, which is to say, in terms that the human mind can understand. It is therefore required that science exclude from its sphere of consideration explanations that are miraculous or supernatural. There is no a priori reason for accepting this approach; its validity resides in the understanding that it has worked remarkably well in expanding the frontiers of knowledge. The importance of the concept of uniformitarianism to a society based in knowledge and directed by the insight that knowledge provides is obvious: It is a large part of the story of the struggle in geology to interpret the earth and its history unfettered by preconception and superstition.

In a more pragmatic sense, the methodology espoused by uniformitarianism admonishes geologists to investigate the natural world in search of better understandings of the processes that have shaped it. Geologists actively study modern-day depositional systems, such as deltas and the sediments on continental shelves, with the objective of constructing models from which the distribution of sediment types can be predicted. Rocks representing ancient systems are compared to these models, and, reasoning by analogy, interpreted as to their conditions of deposition. The predictive powers of the models are used to guide the exploration for energy and other natural resources on which society is based. These models are equally valuable in studying the effects of pollution of the environment and prescribing preventative procedures.

Bibliography

Ager, Derek V. *The Nature of the Stratigraphical Record.* New York: Halsted Press, 1981. A short, readable book emphasizing the incompleteness of the stratigraphic record (record of layered rocks) and the prominent role of natural cataclysms in earth history. Well illustrated and suitable for college-level students.

Albritton, Claude C., Jr., ed. *The Fabric of Geology.* Reading, Mass.: Addison-Wesley, 1963. A symposium sponsored by the Geological Society of America on the basic framework of the geological sciences. Of particular interest are articles by D. B. McIntyre on James Hutton and by G. G. Simpson on historical science. Suitable for college and advanced high school students.

_____. *Uniformity and Simplicity: A Symposium on the Principle of the Uniformity of Nature.* Special Paper 89. New York: Geological Society of America, 1967. A symposium devoted to uniformitarianism. Well written and accessible to advanced high school and college-level students with the aid of a good unabridged dictionary. Good coverage of the origin of uniformitarianism is provided by M. K. Hubbert, and of Lyell's contribution to the concept by L. G. Wilson.

Cope, Dana A. "Uniformitarianism and Fossil Species." *American Journal of Physical Anthropology* 95, no. 1 (September, 1994): 98-103.

Geikie, Archibald. *The Founders of Geology*. London: Macmillan, 1905. Reprint. Mineola, N.Y.: Dover, 1962. A long book, providing complete coverage on ideas about the earth and its development extending back to the Greeks. Good discussions of Hutton, Playfair, and Lyell and debates between Uniformitarians and opposing groups are provided by a geologist not far removed from the times. Well written and understandable at the high school level.

Gillispie, Charles Coulston. *Genesis and Geology*. New York: Harper and Brothers, 1951. An excellent account of the development of geology in eighteenth and nineteenth century Great Britain. Particularly valuable in illustrating the influence of social and religious attitudes on scientific explanations. Abundant quotations provide insight into the times. A recommended starting point for a study of uniformitarianism, this book is suitable for advanced high school and college levels.

Gould, Stephen J. *Time's Arrow, Time's Cycle: Myth and Metaphor in the Discovery of Geological Time*. Cambridge, Mass.: Harvard University Press, 1987. Treating what he calls the "Hutton myth," one of the country's leading geological scholars reassesses the contributions to geology of Hutton and Lyell and refocuses the debate between the Catastrophists and Uniformitarians. This short, cleverly written, provocative book, which contrasts sharply with most previously written analyses, should be read only after an introduction to the subject from other sources. It is suitable for college-level students.

Hutton, James. "Theory of the Earth: Or, An Investigation of the Laws Observable in the Composition, Dissolution, and Resolution of the Land Upon the Globe." *Royal Society of Edinburgh Transactions*. Vol. 1, 1788. Reprint, with a foreword by George W. White. Darien, Conn.: Hafner, 1970. This article began the debate over what became known as uniformitarianism. This short, readable account of Hutton on his own ideas is understandable at the high school level.

Lyell, Charles. *Principles of Geology, Being an Attempt to Explain the Former Changes of the Earth's Surface by Reference to Causes Now in Operation*. 3 vols. London: John Murray, 1830-1833. A well-written argument in support of uniformitarianism, directed toward the Catastrophists. Understandable at all levels and an excellent example of nineteenth century scientific prose, the book went through several editions, one of which should be available in most university libraries.

Playfair, John. *Illustrations of the Huttonian Theory*. Edinburgh, Scotland: William Cheech, 1802. Reprint, with an introduction by George W.

White. Urbana: University of Illinois Press, 1956. Beautiful prose describes Hutton's theory in detail. The book provides an excellent insight into the knowledge and ideas of geology of the time and is understandable at all levels.

Rudwick, Martin J. *The Meaning of Fossils: Episodes in the History of Paleontology.* 2d ed. Chicago: University of Chicago Press, 1985. A fine, well-written account by a leading scholar of science history of the development of the study of paleontology, this book is of moderate length and is easy reading. Contains a well-balanced analysis of the meaning of uniformitarianism and is suitable for advanced high school and college levels.

William W. Craig

Cross-References

Earth's Age, 155; The Fossil Record, 257; The Geologic Time Scale, 272; River Valleys, 534.

Volcanoes: Recent Eruptions

Volcanoes are dramatic expressions of the fact that the earth is a living, dynamic planet. Unfortunately, because of previously erroneous concepts regarding them as well as inadequate descriptive techniques, most written records of specific eruptions undertaken before 1800 are now of limited value. Since then, however, a number of particularly instructive examples have been witnessed.

Field of study: Volcanology

Principal terms

ASH: rocky, unconsolidated ejecta of sand-grain size

ERUPTION: volcanic activity of such force as to propel significant amounts of magmatic products over the rim of the crater

LAVA: molten rock (as opposed to ash) erupted by a volcano

NUÉE ARDENTE: a sudden basal surge of incandescent, heavier-than-air gas

VOLCANO: the natural opening from which gas-propelled magmatic products of the earth's interior are issued in such force and volume as to create sizable landforms; also, the landform itself

Summary

Although volcanic eruptions are a normal part of the earth's workings, they have always been regarded with awe because of their unpredictability and power. Only within the last two hundred years has scientific understanding of them been achieved. During that time, certain specific eruptions (and recurring activity of certain volcanoes) have been especially influential.

Located on the island of Sumbawa, about 160 kilometers east of Java, Tambora Volcano erupted in April, 1815—the exact date and preliminary activity are uncertain—creating the greatest explosion and the most powerful volcanic eruption of modern times. There were no surviving eyewitnesses, as virtually the entire population of Sumbawa died. The noise of the explosion was heard 1,500 kilometers away, and its atmospheric effects were worldwide. As clouds of volcanic ash circulated throughout the stratosphere, 1816 became in America and Europe "the year without a summer," during which normal harvests were greatly curtailed.

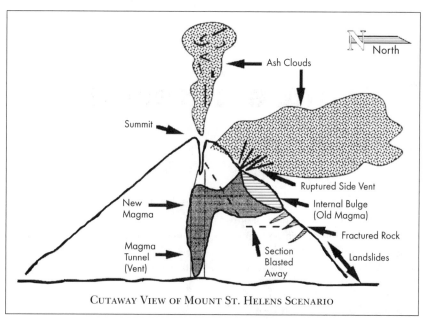

North

Ash Clouds

Summit

Ruptured Side Vent

New Magma

Internal Bulge (Old Magma)

Fractured Rock

Magma Tunnel (Vent)

Section Blasted Away

Landslides

CUTAWAY VIEW OF MOUNT ST. HELENS SCENARIO

Another otherwise routine Vesuvian eruption of October 22-26, 1822 (together with its predecessor of 1818), prompted George Julius Poulett Scrope to study the volcanoes of Europe. His *Considerations on Volcanoes* (London, 1825) inaugurated the modern science of volcanology. Graham Island was a submarine volcano that appeared off the southern coast of Sicily for a few months in 1831, only to disappear by year's end. The first volcano of any kind to be observed almost daily throughout its existence, such European geological theorists as Scrope and Sir Charles Lyell thought of it as an example of fundamental importance in their debates regarding the formation of volcanoes. The forceful but otherwise routine eruption of Vesuvius from April 24 to April 30, 1872, stimulated Luigi Palmieri, director of the Volcano Observatory on the slopes of Mount Vesuvius, to formulate his influential theory that volcanic eruptions are often characterized by predictable phases. This eruption was also the first major eruption to be photographed.

Krakatoa was a three-part volcanic island located between Java and Sumatra. After preliminary activity beginning on May 20, 1883, it erupted explosively on August 26 and 27, the reports being heard nearly 5,000 kilometers away. Tidal waves (tsunamis) generated by the eruption killed more than thirty-six thousand people. As with Tambora, subsequent atmospheric effects were worldwide, generating much scientific and popular interest. Dutch and British investigators published important studies.

Following some preliminary earthquakes, the top of domelike Mount Tarawera, on the North Island of New Zealand, split apart on June 10, 1886, exposing the internal structure of the dome and creating a series of steam-blast craters. When fissuring extended beneath Lake Rotomahana, further steam explosions followed, destroying two world-famous hot-spring formations, the White and Pink terraces. Ash and mud from the eruption buried several Maori villages and more than 140 people.

After more than a thousand years of relative inactivity, Bandai-san Volcano (on the island of Honshu) erupted suddenly on July 15, 1888, for less than two minutes initially. The north slope of the mountain was demolished, burying the villages and people of an entire valley with avalanches and mudflows. Although only a steam explosion, it was the worst volcanic disaster in the history of Japan.

La Soufrière at the northern end of the island of St. Vincent in the West Indies, became active on May 6, 1902, expelling its crater lake in a series of mudslides. Its major eruption, on May 7, was the first modern one in which glowing avalanches (nuées ardentes) were recognized. They sped down all sides of the mountain, causing little damage. The thrust of the eruption had been vertical, with the glowing avalanches on its slopes a secondary phenomenon caused only by fallout and gravity. Activity continued into March, 1903.

Nearby Mount Pelée, on the island of Martinique, was active from April 24, 1902, to October 5, 1905 (and again from September 16, 1929, to December 1, 1932). After giving ample warning, it erupted violently on May 8, 1902—a day after Soufrière on St. Vincent—but the nueés ardentes this time were ejected straight down the western slope of the volcano. Thus, searing, suffocating clouds of gas and ash rushed like an avalanche to engulf the city of St. Pierre, destroying it and all but two (some say four) of its citizens—about thirty thousand people in all. Second only to the destruction of Pompeii by Vesuvius in A.D. 79, this volcanic disaster was the most famous of modern times. It greatly stimulated interest in volcanoes, attracted two major figures (Thomas A. Jagger and Frank A. Perret) to their study, established a new type of eruption (the Peléan), and initiated a dispute regarding nuées ardentes that evidence from its own later eruptions resolved.

Accounts of a major Alaskan Peninsula eruption on June 6-9, 1912—the most powerful modern one on the North American continent—differ with age. Current understanding is that the primary effusion radiated from a newly formed volcano (named Novarupta) on the flank of Mount Katmai rather than from Katmai itself. (There were no nearby eyewitnesses, and no scientific investigations before 1916.) In any case, explosions were heard 1,000 kilometers away, ash fell thickly at a distance of 160 kilometers, and an adjacent valley floor was transformed into a steaming fumarolic basin, the so-called Valley of Ten Thousand Smokes (with rapidly decreasing activity). The old crater of Mount Katmai later collapsed, perhaps because the liquid rock (magma) underlying it had been withdrawn. Aside from its forcefulness, Katmai is significant for being the only example of an ignimbrite (fused rhyolite tuff) eruption in modern times.

Before-and-after pictures highlight the enormous destruction wrought by the 1985 eruption of Mount Pinatubo in the Philippines. *(Cascades Volcano Observatory)*

Lassen Peak, in northern California, is, like Tarawera, a volcanic dome, and one of the largest known. Long dormant, it resumed activity on May 30, 1914, becoming particularly violent a year later, from May 19 to May 22, 1915, when directional nuées ardentes coursed down its shattered slopes to devastate whole forests and create powerful mudslides. Some less important lava flows then followed, with activity continuing into 1917. Prior to the eruption of Mount St. Helens in 1980, the 1915 eruptions of Mount Lassen were the most famous recent ones to have taken place within the United States.

Parícutin emerged overnight from a cornfield in central Mexico on February 20, 1943. This fragile but rapidly growing cinder cone reached a height of 50 meters the next day; lava began flowing from its base (not its crater), destroying a nearby farm. In June, 1944, it buried an entire village. After years of spectacular but relatively harmless activity, Parícutin ended as abruptly as it had begun, on February 25, 1952, its height being 410 meters. Parícutin was the first wholly new volcano to arise in North America since nearby Jorullo (1759; poorly observed) and attracted unprecedented scientific attention.

Hekla, the best-known Icelandic volcano, erupted memorably on March 29, 1947, after more than a century of dormancy. Large amounts of ash and other debris were ejected from a fissure, and lava flows poured down the volcano's slopes from a series of fissure-related vents. After nearly thirteen months of continuous activity (ending April 21, 1948), approximately 1 billion cubic meters of lava had been extruded, in addition to immense volumes of ash. Eruptions of this type do not usually endanger human life, but Iceland's livestock and farming industries were severely affected, both from the destruction of arable land and from the widespread emissions of toxic gases.

Prior to 1951, Mount Lamington, on the northern side of Papua New Guinea, was regarded as extinct. It consisted of a deeply eroded complex of volcanic domes, with heavy forestation and no record of previous activity. On January 15, however, earthquakes and emissions began, gradually intensifying. The major eruption took place on May 21, featuring a huge eruption cloud and glowing avalanches of the St. Vincent type. These avalanches devastated an area of more than 230 square kilometers and killed about six thousand persons. Later eruptions were of the Vulcanian type and did not feature glowing avalanches but generated mudflows. Activity, lasting into 1956, then culminated in dome-building.

Bezymianny ("no name"), in Kamchatka, was also thought to be extinct. After a month of seismic activity, eruptions began on October 22, 1955. The most violent blast, on March 30, 1956, destroyed the top 200 meters of the volcano and most of one side. Nuées ardentes, followed by huge ash and pumice flows, rushed down adjacent valleys. Dome formation then concluded the volcano's activity, all of which was meticulously recorded by Soviet volcanologists at a nearby observatory.

The Azores are a group of volcanic islands in the north Atlantic Ocean. A new vent adjacent to the island of Fayal opened on September 27, 1957, creating the island volcano of Capelinhos. Called Ilha Nova ("new island") I in this manifestation, it lasted only half a month before disappearing (as Graham Island did in 1831). On November 7, Ilha Nova II appeared, soon joining the island of Fayal as a new peninsula. It then fell into dormancy after little more than a year of activity. The resulting cone resembles others (like Diamond Head in Hawaii) formed by underwater explosive eruptions.

The best-studied submarine eruption of all is that of Surtsey, a newly created island in the Westman Islands off the southern coast of Iceland. Named for a fire giant in Norse mythology, Surtsey announced its presence with undersea turmoil on November 14, 1963. An island appeared the next day and grew rapidly, with every step in its progress being closely observed. The main vent became inactive for a time (beginning May 7, 1965), but two associated ones then spewed up smaller independent islands (Syrtlingur and Jolnir) of their own; both subsequently disappeared. All activity ended on June 5, 1967, yet Surtsey continues to be of scientific interest as ecologists note its increasingly diverse complement of plant and animal life.

Heimay ("home island") is the largest and most permanent of the Westman Islands, the same group to which Surtsey belongs. It is approximately five thousand years old but had never been known to erupt. On January 23, 1977, however, a 2-kilometer fissure opened at the edge of the major town (five thousand inhabitants), which was then evacuated. The new vent, Eldfell, quickly developed a cone and poured lavas down into the deserted village, eventually destroying more than three hundred homes. Accumulations of heavier-than-air noxious gases made salvage operations hazardous and endangered farm animals (many of which had to be destroyed). While various attempts to impede the encroaching lava achieved only limited success, the volcano stopped of its own accord on June 28.

After erupting nine times between 1831 and 1857, Mount St. Helens, in the state of Washington, was dormant until March 27, 1980, when significant activity began. The major eruption, of May 18, blew off the top of the mountain and was said to be the most forceful in the contiguous forty-eight states during historic times. Mudslides triggered by the eruption then destroyed bridges, logging camps, homes, and thousands of trees. About sixty people perished in one way or another. Further eruptions followed on May 25, June 12, July 22, and August 7, with occasional activity thereafter. These volcanic eruptions were the first in the contiguous United States since those of Mount Lassen. They attracted immense popular interest and were closely monitored by scientists.

In a now-familiar pattern, El Chichón Volcano, in northern Chiapas, Mexico, awoke from a prolonged dormancy (it had been considered "solfataric") to erupt without warning on March 28, 1982. At least ten persons who fled into a local church died when earthquakes demolished their haven; others were killed by falling debris and molten rock. A long ash cloud rained cinders across southern Mexico. Later eruptions in April left more people dead and homeless; survivors were then evacuated. Little had previously been known about this hitherto inactive killer.

Although the eruption episodes listed above are among those most likely to appear in discussions of modern volcanism—either because of their historical significance or because of the thoroughness with which they were observed—it would be extremely misleading to suggest that they include most, or even the most typical, examples. Thousands of eruptions take place every year, most of which scientists know nothing about because the volcanoes remain hidden within the depths of the oceans. There are still remote areas on earth, moreover, in which even a moderate land-based eruption might go unnoticed and therefore unrecorded.

Among the volcano records that do exist, all of the following have erupted twenty times or more since 1800: the three classic Italian volcanoes, Vesuvius, Stromboli, and Etna; Nyamuragia in Africa; Karthala (Grand Comoro) and Piton de la Fournaise (Reunion) in the Indian Ocean; White Island, Tarawera, Tongariro, Ngauruhoe, and Ruapehu in New Zealand; Bam and Manam, New Guinea; Ambrym and Lopevi, New Hebrides; Marapi and Dembo, Sumatra; Anak ("son of") Krakatoa, between Sumatra and Java; Gedeh, Gunter, Slamet, Merapi, Semeru, Tengger, Lamogan, and Ruang, Java; Batur, Bali; Soputan and Lokon-Empung, Sulawesi; Apu Siau, Sangihe Islands; Gamalama, Halmahera; Mayon and Taal, the Philippines; Sakurajima, Kirishima, and especially Aso, Kyushu; Yake-Dake, Asama, Zao, and Iwake, Honshu; Oshima, Izu Islands; Komaga-Take and Taramai, Hokkaido; Karymsky and Kliuchevskoi, Kamchatka; Akatan and Shishadlin, Aleutian Islands; Pavlof, Alaskan peninsula; Kilauea and Mauna Loa, Hawaii; Colima, Mexico; Fuego, Guatemala; Izalco and San Miguel, El Salvador; Cerro Negro and Masaya, Nicaragua; Poas and Irazu, Costa Rica; Purace, Colombia; Reventador, Cotopaxi, and Sangay, Ecuador; Fernandina, Galápagos; and Rupungatito, Llaima, and Villarrica, Chile. It will be noted that a number of the most famous volcanoes do not appear; they have erupted either less frequently or not at all since 1800.

Methods of Study

Broadly speaking, data relevant to any given volcanic eruption fall into three separate but related categories, depending upon whether they were collected prior to the eruption, during it, or afterward. Data collected prior to an eruption include the previous eruptive history of the volcano: the types, the frequency, and the magnitude of its known eruptions. Illuminating as this information can be, however, it has not been shown to have reliable predictive value. Yet, a volcano on the verge of eruption will often signal its upcoming behavior in a variety of ways. These include increased fumarolic activity, changes in the chemistry of associated waters, seismic disturbances, and tumescence (swelling). Certain changes in electrical and magnetic phenomena have been noted also. Volcanologists, therefore, monitor all of these indicators carefully.

To ensure objectivity, precision, and thoroughness, much of this monitoring is accomplished through instruments. Pyrometers and thermometers, for example, measure the heat of volcanic rocks (the temperature of the cone rises before an eruption) or of associated waters, as in a crater lake. As rocks adjacent to the vent heat up, they become less powerfully magnetic, as surface or aerial surveillance with a magnetometer will establish. In some cases, rising temperature can also be detected through a series of aerial infrared photographs. Unfortunately, none of these methods predicts eruptions definitively. The two most commonly used instruments (and, on the whole, the most reliable) are the seismometer and the tiltmeter. The seismometer measures and records vibrations within the earth. When strong enough to be detected by humans, these vibrations constitute earthquakes. It has been known for centuries that seismic activity often precedes volcanic eruptions. Since the invention of the seismometer in the latter nineteenth century, however, it has also been possible to detect fainter, more deep-seated vibrations; the subterranean rise of molten lava within the volcanic vent. Typically, such vibrations become stronger and more frequent as the time of eruption nears. Seismometry, therefore, is one of the most useful observational techniques available to the volcanologist.

As molten lava rises within the vent of an already existing volcano, the cone of that volcano subtly changes shape. Although the changes are generally too small to be obvious, they can be ascertained and verified through the use of a tiltmeter. This instrument can be in any of several forms (a pendulum, for example) but most commonly depends upon the displacement of a liquid, as in a carpenter's level. Laser beams have been used to measure tilt with great precision. On the other hand, some tumescence takes place so rapidly and conspicuously (generally along a seacoast) that detection by the naked eye is possible. Subsidence usually follows in any case, regardless of whether there has actually been an eruption.

Actual volcanic eruptions can last either seconds or days; a period of activity, characterized by frequent eruptions, can continue for years. During all or part of the times involved, it may well not be possible to inspect the volcano closely. First at Vesuvius, then at Hawaii and elsewhere, a number of the most active (or the most dangerous) volcanoes are monitored constantly, by instruments and by humans, at volcano observatories. During a very violent eruption, instruments, observers, and even the observatory itself may be lost. In any case, volcanologists have always relied upon whatever eyewitness data they could gather to supplement their understanding of a volcanic event. Any eruption is a contribution to history. In this regard, the most significant new technique has been the use of photography.

Although photography was invented in 1839, early techniques did not permit the recording of subjects in motion. Thus, no photographs of any eruption exist before 1872, when on April 26, memorable views of Vesuvius were obtained. The first eruption to be photographed in the Western Hemisphere was that of Izalco (El Salvador) in 1894. There were then striking images of nuées ardents at Mount Pelée by Tempest Anderson and Alfred Lacroix in 1902. Anderson's *Volcanic Studies in Many Lands* (London, 1903) became the first picture book of volcanic phenomena. Eruptions of Mount Lassen in 1914-1915 were the first within the continental United

States to be photographed. B. F. Loomis' *Pictorial History of the Lassen Volcano* (Anderson, California, 1926) is perhaps the earliest example of its kind.

Even in more modern times, it is possible for a major eruption to take place without being observed by humans; Katmai, 1912, is a well-known example. The only recourse then available to volcanologists is that of hypothetical reconstruction, based upon the physical evidence. This evidence generally consists of changes in the volcano's cone, adjacent geological effects, and the type and pattern of its discharges, insofar as they can be ascertained.

Context

In the early days of geology, it was believed by some that volcanoes were only local phenomena, deriving their power from the ignition of coal beds. If so, then they were of little consequence in the history of the earth and hardly more than a curiosity within it. Since the latter eighteenth century, when this view was promulgated (primarily by Abraham G. Werner), scientists have come to realize how mistaken it is.

In his *Theory of the Earth* (1788; 1795 in book form), James Hutton (1726-1797) first emphasized that the earth's internal heat was the fundamental driving force underlying its surface permutations. Volcanoes for him were safety valves intended to prevent his heat-engine earth from blowing up. Unlike certain predecessors, however, he did not emphasize their major constructive role in creating the surface of the earth. The first theorist to hold essentially modern views regarding volcanoes was George Julius Poulett Scrope (1797-1876) in *Considerations on Volcanoes* (1825; revised editions, 1862, 1872). As more volcanoes were studied, three facts strikingly emerged. First, volcanoes are not randomly placed on earth but are instead grouped into narrow belts—one surrounding the Pacific, in particular. Second, volcanoes are of different types, differing in origin, appearance, and behavior. Third, eruptions are also of different types, though individual volcanoes often produce more than one type, even during the same eruptive episode. By the beginning of the twentieth century, therefore, the understanding of volcanism and its role in the history of the earth had become much more sophisticated.

Two of the most important twentieth century discoveries about volcanoes are that they are even more numerous in ocean basins than they are on land and that those on land are almost always associated with established or incipient plate boundaries. With very few exceptions, continental volcanoes occur as a by-product of subduction, the process through which one plate slides underneath another and is melted. Many theorists believe that the earth is less thermal than previously—that its geological activity in this respect is slowing down. Others insist that the most powerful episodes of volcanicity with which humans are familiar have taken place in relatively recent geological times. However this debate is eventually to be resolved, scientists regard volcanoes as the most visible indicators of the earth's still-active interior processes.

Bibliography

Bullard, Fred M. *Volcanoes: In History, in Theory, in Eruption.* Austin: University of Texas Press, 1962, 1973. A college-level text, dated in some particulars. Includes several nineteenth and twentieth century eruptions, classified by type.

Francis, Peter. *Volcanoes.* Harmondsworth, England: Penguin, 1976. Good narrations for Krakatoa and Pelée. One of the best introductions available.

Jagger, Thomas A. *Volcanoes Declare War: Logistics and Strategy of Pacific Volcano Science.* Honolulu: Paradise of the Pacific, 1945. Well-illustrated accounts of volcanic eruptions around the Pacific rim, by a very distinguished volcanologist.

Macdonald, Gordon A. *Volcanoes.* Englewood Cliffs, N.J.: Prentice-Hall, 1972. Similar to Bullard, but with a larger number of eruption narratives, this remains a very useful book, lucidly written and richly informative.

Sigurdsson, Haraldur, and Steven Carey. "The Far Reach of Tambora." *Natural History* 97 (June, 1988): 67-73. A nontechnical, well-illustrated account of the 1815 eruption and its worldwide effects.

Simkin, Tom, et al. *Volcanoes of the World: A Regional Directory, Gazetteer, and Chronology of Volcanism During the Last Ten Thousand Years.* Stroudsburg, Pa.: Hutchinson Ross, 1981. A book to be consulted rather than read, this one lists all known eruptions for all known volcanoes.

Wilcoxson, Kent H. *Chains of Fire: The Story of Volcanoes.* New York: Chilton, 1966. Responsibly summarizes eruption narratives from all over the world, with footnotes. A fine introduction to the topic.

Dennis R. Dean

Cross-References

Lava Flow, 367; Volcanoes: Types of Eruption, 685.

Volcanoes: Types of Eruption

No two volcanic eruptions are identical. There are, however, sufficient similarities between some eruptions that they serve as models to describe the activity of the remainder. Quantification of these descriptive models has permitted a scientific basis for more accurate eruption prediction, for interpreting past unwitnessed eruptions, and for assessing the nature of volcanic activity on extraterrestrial bodies.

Field of study: Volcanology

Principal terms

CRYSTAL: a solid with a regular atomic arrangement

LITHIC: having to do with rock

PHREATIC ERUPTION: an eruption in which water plays a major role; also called hydrovolcanic

PHREATOPLINIAN ERUPTION: a Plinian-type eruption in a wet environment

PUMICE: pyroclastic rock full of vesicles (spherical voids originally occupied by gas)

PYROCLASTIC ROCK: fragmented rock produced during a volcanic eruption (the term includes essentially all volcanic products except lava flows)

Summary

There are many ways to classify volcanic eruptions, whether by the shape of the vent (linear versus point source), the location (submarine or subaerial), the composition of the erupted material, the form of the erupted material (lava, gas, or pyroclastics), or the style of the eruption. All these categories are interrelated to a certain degree. Volcanologists classify eruptions according to style in order to have tools for readily describing the particular type of activity of any given volcano.

By the 1920's, Alfred Lacroix had defined four distinct eruption styles, which he termed Hawaiian, Strombolian, Vulcan, and Peléan; these terms remain in common usage. The names make reference to volcanoes where that style of eruption is most common or had been most accurately described. Style terms that have been adopted since that time follow this system. (An exception is the term "Plinian," which makes reference to Pliny the Elder, who died in A.D. 79 during the eruption of Vesuvius that destroyed the town of Pompeii, or to Pliny the Younger, who described the eruption.) For example, the eruption on May 12, 1980, of Mount St. Helens led to the introduction of the term "directed blast eruption" to the volcanological literature (although a similar style had been previously exhibited by the Soviet volcano Bezymianny). The eruption of Surtsey, off the coast of Iceland, in the 1960's led to the definition of the Surtseyan eruption style.

No attempts to quantify eruption style were made until the pioneering work of George Walker in the early 1970's, with subsequent refinements by Walker and by graduate students who worked with him. In reality, no volcano has a single eruption style. For example, at each of its three vents, the volcano Stromboli might be erupting in a different style at the same moment. Thus, it is more accurate to group together those volcanoes where a single eruption style predominates, those that exhibit a repeated pattern or sequence of eruption styles, and those that have several eruptive styles but no pattern of their occurrence. (The figure illustrates the general relationships between different eruption styles.)

Walker's attempt to quantify eruption style is based on five parameters: magnitude, intensity, dispersive power, violence, and destructive potential. Magnitude has to do with either the quantity of material or the amount of energy released by an eruption. Because the purpose of quantifying style is to be able to apply measurable parameters to unwitnessed eruptions, the total volume emitted during a single event is a more useful parameter than is energy release. To facilitate comparison between volcanoes, the actual volume is converted to a dense rock equivalent (DRE), which makes allowance for the voids in the rock. Measured volumes range from less than 0.001 cubic kilometer to 1,000 cubic kilometers. Intensity is a measure of the discharge rate—that is, the volume emitted during a specified period of time. Measured intensities range from less than 0.1 to more than 10,000 cubic meters per second DRE. Intensity usually varies throughout the course of an eruption, so a more applicable measurement might be peak intensity. Although there is a strong correlation between intensity and magnitude, a wide range of magnitudes can exist for a given intensity. Dispersive power has to do with the area covered by the deposits of a single event. This parameter can be strongly affected by winds. Areas covered can range from less than one square kilometer to about 10^6 square kilometers. Violence is a measure of the momentum applied to particles in an eruption. Destructive potential refers to the areal extent of the damage to property and vegetation. Neither violence nor destructive potential is particularly applicable to past unwitnessed eruptions, but both are significant in volcanic hazard assessment.

Christopher Newhall and Steve Self have developed a volcanic explosivity index (VEI), which attempts to integrate quantitative data with more subjective style descriptions. As its name implies, the index is most applicable to

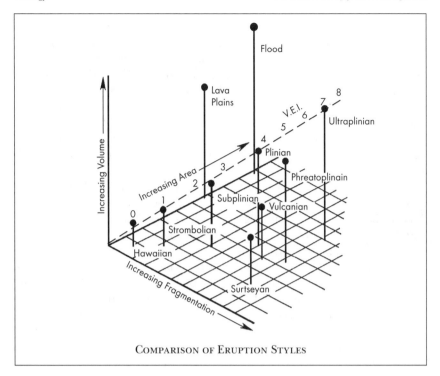

COMPARISON OF ERUPTION STYLES

pyroclastic eruptions and is unsuitable for large-scale lava eruptions. Furthermore, the VEI is based partly on eruption column height, which is a function of intensity, and intensity cannot be correlated directly with magnitude. In spite of the fact that calling an eruption a VEI 5 provides little in the way of a mental image of the volcanic activity, it does provide an extremely useful means of computer-based comparisons of the thousands of dated eruptions.

To graph distinctions between the different explosive eruption styles, one might plot fragmentation index (F) versus dispersal (D). D is calculated by determining the area (enclosed by an outline) where the deposit thickness is more than one-hundredth of the maximum deposit thickness. F represents the percentage of the deposit that has a grain size finer than 1 millimeter at a point where the deposit is one-tenth of the maximum thickness. In the case of terrestrial eruptions, the deposit is inevitably spread into an elliptical shape by the prevailing wind; the long axis of the ellipse is known as the axis of dispersion. The F measurement is made on this axis, downwind from the volcano. This graphic method has an advantage over the VEI because it permits a distinction to be drawn between those eruptions in which water plays a major role (phreatic) and "dry" eruptions. Water causes the erupted fragments to be extremely fine-grained; therefore phreatic eruptions have a high F number.

A refined version of the area-to-size graph, introduced by David Pyle, employs an exponential rather than a logarithmic scale. This graph employs as its axes b_t (called the thickness half-distance—the horizontal distance over which the deposit halves its thickness) and b_c/b_t where b_c is the clast half-distance (that is, the horizontal distance over which the maximum clast, or rock fragment, size is reduced by half). The morphology of the deposit is represented by b_t; low values result in a conical shape, and high values in a sheetlike form. Finally, b_c/b_t is a measure of the degree of fragmentation of the deposit; here, low values represent deposits dominated by fine material. Like the D/F graph, this graph system is applicable only to explosive eruptions.

Geoffrey Wadge proposed that lava eruptions could be subdivided into groups defined by magnitude and intensity: high effusion rate with low volume, low effusion rate with high volume, and, perhaps, high effusion rate with high volume and low effusion rate with low volume.

Attempts to quantify eruption style by direct measurement are restricted to subaerial (land) volcanoes on the earth. It has been estimated, however, that about 75 percent of the material erupted on earth is generated in the deep sea, where pressure and temperatures are vastly different from those on land. Only with the development of manned submersibles did the variety of volcanic products on the ocean floor become apparent. "Smokers"—small, chimneylike structures issuing hot, mineral-laden gases—have been filmed. On other bodies that exhibit volcanic activity, such as Io, there may be essentially zero pressure and much lower gravity. Even if eruption magnitude and intensity in such an environment were the same as those of a subaerial eruption, the style of eruption would be drastically different. Observation of active volcanism on Io, however, has been limited to two brief flybys by Voyager satellites.

Methods of Study

With temperatures on the order of a thousand degrees Celsius and velocities of some erupting materials of 100 meters per second, direct observation of active eruptions can be a hazardous undertaking. Fortunately, direct measurements to determine the magnitude and intensity of an eruption can be performed at a distance, using photographic or remote-sensing techniques. The height of the volcanic column can be measured directly, using ground-based or aerial photography, to provide an estimate of the intensity of the eruption. The volume of solid particles in the eruption column can be measured by determining the amount of light that can be passed through it. Velocities of entrained particles can be measured through viewing high-speed films of the eruption column.

Techniques pioneered by Japanese geologists and extended by George Walker involve measurement of the ejected material after an eruption has ceased. These techniques are little different from those applied to sedimentary rocks; they include the measurement of thickness, the maximum grain size, the grain-size distribution, the proportion of different components, the crystal content of pumice, and the density and porosity. Measurements of thickness of a deposit are used to construct an isopach map—that is, a map showing where the deposit thickness falls between a series of upper and lower limits. Thickness measurements are taken at numerous locations, the number being determined by the size of the deposit and the level of accuracy required. The resultant map provides an indication of the deposit volume, the location of the vent, and the way the deposit was dispersed.

Maximum grain-size measurements are useful for determining the energetics of the eruption column and the effects of wind velocities. If numerous grains are measured over the area of the deposit, an isopleth map can be constructed, indicating areas in which maximum particle sizes are equal. An isopleth map can sometimes be more useful than an isopach map, for erosion tends to remove fine material from a deposit, lessening its thickness, while larger particles remain. The grain-size distribution is obtained by simply sieving numerous samples. A certain amount of care has to be exercised, because pumice and glass are extremely fragile. The result of this analysis is a determination of the degree of sorting of the deposit. This information can be employed to distinguish between certain types of eruptions.

A deposit contains three components: pumice, lithics, and crystals. The varying proportions of each of these components throughout the deposit can be used to infer the conditions in the magma chamber prior to eruption, and the relative proportions at different sites can be employed in determining the eruption dynamics. The physical separation of these components can be an extremely time-consuming task, involving sorting by hand under

A spectacular lava dome emerging off the coast of Hawaii. *(Don Swanson/U.S. Geological Survey)*

a microscope. The crystal content of pumice is a further indicator of the pre-eruption conditions in the magma chamber. To determine the crystal content, pumice lumps can be crushed and the more durable crystals removed. Measurements of density and porosity are important for calculating the dense rock equivalent. In the case of pumice, which often has unconnected pore spaces, these measurements can be rather complicated. Simply, samples are variously measured dry in air, after soaking for several days in water, and under water.

Context

Determination of the type of a volcanic eruption is important for two major reasons: first, to permit theoretical reconstructions of unwitnessed eruptions and, second, so that scientists can predict the consequences of future eruptions at an active volcano. After the magnitudes and intensities have been established for the unwitnessed eruptions at a specific volcano, such as Mount St. Helens, by combining these data with age data for each deposit, it becomes possible to determine the eruptive history. The volcanologist is then armed with information regarding the largest magnitude eruption that has taken place in the past and the frequency at which volcanic events occur. It is then possible to forecast that a specific volcano is capable of producing an eruption of a given magnitude in the future and to ascertain the probability of such an event taking place within a given period of time.

If a volcano goes through a regular sequence of eruptive events prior to a paroxysmal eruption, recognition of this precursor activity may assist authorities in their decision to evacuate the local populace. The effects of failure to undertake such an analysis and to heed warning signs is well illustrated by the 1951 eruption of Mount Lamington, which had been regarded as an extinct volcano, in Papua New Guinea. A cloud of smoke was seen above the vent on January 15 of that year; five days later came the paroxysmal eruption, which caused six thousand deaths. In those areas with a high population density or rich agricultural land, insurance brokers are particularly interested in knowing details of possible eruptions in order to minimize their economic losses.

From the scientific viewpoint, types of volcanic activity provide windows into the interior of the planet and allow assessments of what is happening beneath plate margins, continental interiors, and the ocean floors—and beneath the surfaces of alien planets and satellites. Climatic changes and mass extinctions have been attributed to large-scale volcanic eruptions. Long-term changes in the magnitude and styles of eruption reflect important changes taking place in planetary interiors. The air we breathe, the water we drink, and perhaps life itself are the products of volcanic activity.

Bibliography

Cas, Ray A. F., and J. V. Wright. *Volcanic Successions: Modern and Ancient.* Winchester, Mass.: Allen & Unwin, 1987. George Walker's influence is readily evident in this book; one of its authors was his student. An excellent overview of volcanism from a largely sedimentological viewpoint. Aimed at college-level and research students.

Lipman, Peter W., and D. R. Mullineaux, eds. *The 1980 Eruptions of Mount St. Helens, Washington.* U.S. Geological Survey Professional Paper 1250. Reston, Va.: Department of the Interior, 1981. Detailed accounts of the most violent eruption witnessed in the contiguous United States at this writing. Aimed at a research-level audience, it can, however, be read by high school students. Of particular interest to anyone living in the western United States.

Macdonald, Gordon A. *Volcanoes.* Englewood Cliffs, N.J.: Prentice-Hall, 1972. A delightful, highly readable but scientifically informed account of volcanism. Suitable for the general reader as well as graduate students. Somewhat dated but full of useful information.

Simkin, Tom, et al. *Volcanoes of the World.* Stroudsburg, Pa.: Hutchinson Ross, 1981. The almanac of historic volcanic eruptions. One of the originators of the volcanic explosive index authored this book in part. Rather technical.

Williams, Howell, and A. R. McBirney. *Volcanology.* San Francisco: Freeman, Cooper, 1979. One of the few books on volcanoes suitable for use as an undergraduate textbook. The culmination of more than fifty years of work on volcanoes by the senior author.

Wood, C. A., and J. Kienle. *Volcanoes of North America: U.S.A. and Canada.* New York: Cambridge University Press, 1989. Descriptive accounts of recent and past activities of volcanoes in North America.

James L. Whitford-Stark

Cross-References

Magmas, 383; Plate Tectonics, 505; Volcanoes: Recent Eruptions, 675.

Water-Rock Interactions

Water-rock interactions occur as fluids circulate through rocks of the earth's crust. Isotopes of common elements are exchanged between a fluid and its host rock. As a result of these reactions, a rock may preserve a record of the fluids that have passed through its pore spaces. Studies of water-rock interactions provide information on the nature of fluid movement through rocks and on the origin of economic ore deposits.

Field of study: Geochemisrty

Principal terms

CONNATE FLUIDS: fluids that have been trapped in sedimentary pore spaces

DEHYDRATION: the release of water from pore spaces or from hydrous minerals as a result of increasing temperature

EXCHANGE REACTION: the exchange of isotopes of the same element between a rock and a liquid

FRACTIONATION: a physical or chemical process by which a particular isotope is concentrated in a solid or liquid

ISOTOPES: atoms of the same element with identical numbers of protons but different numbers of neutrons in their nuclei

JUVENILE WATER: water that originated in the upper mantle

MASS SPECTROMETER: a laboratory instrument that separates isotopes of a particular element according to their mass difference

METEORIC WATER: water that takes part in the surface hydrologic cycle

VOLATILES: dissolved elements and compounds that remain in solution under high-pressure conditions but would form a gas at lower pressures

Summary

Fluids that circulate within the crust of the earth take part in chemical reactions involving an exchange of elements between the fluid and the host rock. Such reactions, referred to as fluid-rock interactions, are an important mechanism in the concentration of minerals in economically valuable ore deposits. A useful means of studying fluid-rock interactions is by measurement of the stable isotopic composition of rocks that have undergone such an exchange history.

Rocks are exposed to fluids that are diverse in their origin and composition. The most important volatile constituents of natural fluids are water, carbon dioxide, carbon monoxide, hydrogen fluoride, sulfur compounds, and light hydrocarbons such as methane. In addition, fluids contain dissolved solids derived from the crustal rocks through which they pass. As the list of volatiles makes clear, the dominant elements present in fluids are oxygen, hydrogen, carbon, and sulfur.

Elements may be characterized by their atomic number and their atomic weight. Atomic number refers to the number of protons present in the nucleus of an atom and is a constant value for each element. Atomic weight is determined by adding together the number of protons and neutrons contained in an atom. Because the number of neutrons present often varies within a limited range, atoms of a particular element may have several different atomic weights. These atoms are referred to as isotopes. Oxygen, with an atomic number of 8, occurs most commonly with eight neutrons but may have nine or ten. Therefore, three isotopes of oxygen occur in nature: oxygen 16, oxygen 17, and oxygen 18. Similarly, carbon occurs as carbon 12 or carbon 13. (Carbon 14 is a radioactive isotope that forms in the earth's upper atmosphere and will not be considered here.) Hydrogen contains only one proton; however, a small fraction of hydrogen atoms also contain a neutron, which doubles the mass of the atom. These heavy hydrogen atoms are called deuterium. Sulfur has four stable isotopes: sulfur 32, 33, 34, and 36.

For any given mineral or fluid, the relative concentrations of the isotopes of a particular element may be expressed as a stable isotopic ratio, or the ratio of the second most abundant isotopic species over the most abundant isotope. Any physical process that results in the enrichment or depletion of the concentration of a heavy isotope is referred to as fractionation. A common fractionation process is evaporation of water. Water molecules containing the lighter isotope of oxygen (oxygen 16) will preferentially evaporate so that the remaining liquid will be enriched in the heavier oxygen isotope (oxygen 18). Fractionation also occurs during the growth of minerals in either a magma or a water-rich solution. Some minerals, because of the nature of their chemical bonds and their crystal structure, tend to concentrate a greater number of heavy isotopes than do other minerals. Quartz, dolomite, and calcite are common examples of minerals that contain high concentrations of heavy oxygen, while oxides such as ilmenite and magnetite have very little of the heavy isotope. The effectiveness of fractionation during mineral growth is dependent upon temperature. Low temperatures permit minerals to be more selective in choosing atoms for growing crystal sites, resulting in large differences in isotopic ratios between different minerals. At high temperatures, the selection of atoms is a more random process, and differences between isotopic ratios become progressively smaller.

Because the heavier isotopes of elements naturally occur in such small concentrations, isotopic ratios of oxygen and carbon, for example, have numerical values that are very small and difficult to measure accurately. For this reason, isotopic ratios for a particular sample are presented as a relative enrichment or depletion of the heavy isotope of an element as compared with a defined standard. The difference between the sample and the stan-

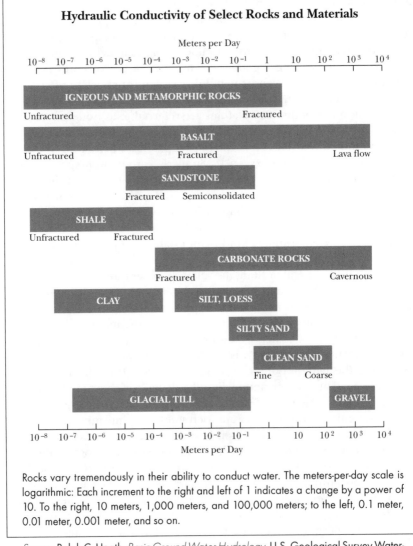

Hydraulic Conductivity of Select Rocks and Materials

Rocks vary tremendously in their ability to conduct water. The meters-per-day scale is logarithmic: Each increment to the right and left of 1 indicates a change by a power of 10. To the right, 10 meters, 1,000 meters, and 100,000 meters; to the left, 0.1 meter, 0.01 meter, 0.001 meter, and so on.

Source: Ralph C. Heath, *Basic Ground-Water Hydrology,* U.S. Geological Survey Water-Supply Paper 2220, 1983.

dard is measured in parts per thousand or per million and is expressed by the Greek letter delta (δ). For carbon isotopic values, for example, the accepted standard is called PDB and is obtained from a belemnite fossil of the Cretaceous-age Pee Dee formation of North Carolina.

As fluids migrate through rocks, reactions occur that involve the exchange of stable isotopes between the fluid and the solid. The exchange process may be pervasive, where fluid movement is diffusive and affects the entire rock mass, or localized along specific fluid channelways, such as fractures, where only the wall of the rock is altered along the route of water movement. The degree to which the isotopic composition of the rock is altered depends on the initial composition of the rock and the fluid, the temperature at which isotopic exchange is occurring, and the amount of fluid present. Typically, the oxygen and hydrogen isotopic values for crustal water are light compared to those for most rocks. Therefore, as isotopic exchange proceeds, the isotopic composition of the rock becomes progressively lighter, while the water becomes increasingly enriched in the heavier isotopes. If the fluid-rock interaction has occurred under constant temperature conditions, the final isotopic value of the rock is proportional to the volume of water that has passed through the rock.

There are four principal sources, or reservoirs, of crustal fluids. Each of these fluid reservoirs contains stable isotopic ratios that reflect the fractionation mechanisms at work and the result of chemical reactions between the fluids and their host rocks. "Meteoric water" is a term applied to fluids that take part in the surface hydrologic cycle. Water that undergoes evaporation, precipitation, and runoff to lakes and to the ocean is capable of penetrating the earth's crust to a depth of several kilometers. This penetration is usually accomplished by fluid migration along weaknesses in the crust, such as faults or fracture systems. Natural hot springs are an example of meteoric water that has been heated deep in the crust and then reemerges at the surface. The oxygen and hydrogen isotopic ratios associated with meteoric water are controlled principally by the distillation effect of evaporation and precipitation. As a result of the general transport of air masses from the equator toward the poles, meteoric water isotopic values vary in a systematic manner, with heavier ratios found near the earth's equator and progressively lighter isotopic values occurring toward the poles. Meteoric water that is trapped in the pore spaces of accumulating sediments is referred to as connate water. When loose sediments have lithified to form hard sedimentary rock, the enclosed pore fluids may become isolated for very long periods. Connate water reveals an isotopic trend similar to that of meteoric water, except that the oxygen values tend to be heavier because of the capacity of the lighter oxygen (oxygen 16) contained in the trapped pore fluids to be exchanged for some of the heavier oxygen (oxygen 18) of the host sedimentary rock. Isotopic exchange between connate water and the host sediments continues until equilibrium is achieved. As rocks undergo increases in temperature and pressure associated with metamorphism, water is frequently released in a process called dehydration. The escaping water may be from pore spaces in the rock or from hydrous minerals such as micas or amphiboles. Because of their origin, these dehydration fluids are also referred to as metamorphic fluids. Dehydration water has a very wide range of isotopic compositions, which reflect the diversity of the original sediments. Juvenile water, which originates in the upper mantle, escapes from ascending magma and represents the fourth important fluid source. Not all water derived from magma should be considered juvenile, as meteoric or connate water will frequently be present in sediments that undergo melting deep in the crust. True juvenile water has a very narrow range of isotopic composition. Because of mixing of crustal fluids and exchange reactions with igneous rocks, samples of unaltered juvenile water are found very rarely.

Isotopic compositions of rocks reflect the formation history of the particular rock type. Igneous rocks are controlled by the composition of their magmatic source area, which is usually in the lower crust or upper mantle. Other factors include the temperature of crystallization, the type of minerals, and the degree to which the magma remains isolated or mixes with other constituents. Unaltered igneous rocks typically have a narrow range of isotopic values as compared with natural fluids. Deep mantle rocks contain minerals that are low in oxygen 18 and therefore fall in a narrow range of isotopically light compositions. Crustal rocks, with a higher proportion of silicate minerals, which concentrate oxygen 18, are isotopically heavy. Sedimentary rocks have two distinct modes of formation. Clastic sedimentary rocks are made of transported particles of weathered material and therefore have isotopic ratios that reflect the individual components. Chemical sedimentary rocks precipitate directly and often involve biological activity. The oxygen isotopic values of these rocks are usually much heavier than those of fluids. Because of fractionation factors associated with organisms, carbon isotope ratios are highly variable. Rocks that have undergone metamorphism contain the widest range of isotopic compositions, as a result of chemical reactions in the presence of fluids that may be derived from any of the reservoirs previously described.

Methods of Study

The most important analytical tool in the study of fluid-rock interactions is the stable isotope mass spectrometer. This instrument separates isotopes according to their mass differences, as determined by the deflection of charged ions within a magnetic field. Elements of interest are extracted from minerals through appropriate chemical reactions and then converted to a gas, which is entered into the mass spectrometer. The sample gas is bombarded by a stream of electrons, converting the gas molecules to positively charged ions. The ions are accelerated along a tube, where a powerful magnet deflects the charged molecules into curved pathways. Lighter particles are deflected more than heavier ones, so several streams of ions result, each with a particular mass. The relative proportions of each isotope are measured by comparing the induced current produced by each ion stream at a collector.

Studies of oxygen and hydrogen isotopes have been particularly useful in research on water-rock interactions involving igneous rocks. Hydrous minerals, such as biotite and hornblende, are separated out of granitic rocks in order to extract hydrogen isotopic values. Oxygen isotopic compositions are usually measured from feldspars and quartz. Two areas of water-rock study have been of particular interest. The first concerns the origin and quantity of water responsible for the isotopic alteration of large igneous intrusions within the continental crust. The second area of investigation addresses the interaction of ocean water with ocean-crust basalt and the formation of associated sulfide ore deposits.

The initial isotopic composition of igneous rocks is predictable according to their mineral content and temperature of crystallization. Therefore, it is possible to recognize when igneous rocks have been influenced by exchange with a fluid. Many examples have been found of shallow igneous intrusions that have been depleted in oxygen 18 through interaction with large volumes of isotopically light meteoric water. Hydrogen isotope exchange is even more sensitive than is oxygen exchange, because igneous rocks have so little hydrogen relative to the amount in water. A very small quantity of water may produce a large isotopic shift in a rock's hydrogen value, while the oxygen is largely unaffected.

Isotopes associated with the ocean crust have been studied through cores obtained from seabed drilling and through the analysis of ophiolites, which represent portions of ocean crust exposed on land. In the vicinity of the mid-ocean ridge, where temperatures exceed 300 degrees Celsius, ocean water circulates to a depth of 3 or 4 kilometers and is responsible for depleting oxygen 18 by one or two parts per thousand.

Fluids associated with metamorphism have also been intensively studied by stable isotopic methods. Increasing metamorphic grade is associated with progressively lighter isotopic values for oxygen and hydrogen. Areas of regional metamorphism may show two end-member types of fluid behavior. Consistent depletion of oxygen and hydrogen isotopic values throughout a large terrain points to exchange with an external source of water that flows pervasively through the region. Alternatively, only the trapped connate water is involved in exchange reactions, leading to higher isotopic composition values and to enrichment of deuterium in hydrous minerals. Because of the large difference in isotopic composition between magma and sedimentary rocks, contact metamorphism is particularly appropriate for study. Samples from intrusions indicate that the margin of the igneous rock is enriched in oxygen 18 through exchange with the country rock, while the interior of the body remains unaltered. The early stages of fluid-rock interaction are dominated by magmatic water, while meteoric water becomes increasingly important as cooling proceeds.

Sedimentary rocks that form by precipitation, such as limestone and chert, have heavy isotopic compositions as a result of the large fractionation between water and either calcite or quartz at low temperature. Clastic sandstones, which contain transported quartz grains, are characterized by the lighter isotopic values of the component particles. Sedimentary rocks selected from a large sample area frequently show a progressive change in the amount of isotopic depletion that has resulted from water-rock exchange. As water continues the exchange process, the isotopic composition of the fluid also shifts. During the next increment of water-rock exchange, the potential amount of depletion of the rock will not be as great. By plotting isotopic values of sedimentary rocks, it is possible to determine directions of fluid motion.

Context

Water-rock interactions are important primarily for their role in the formation of economically vital ore deposits. These are localized regions where metals such as gold, silver, lead, copper, zinc, and tungsten occur in unusually high concentrations and can be extracted. Water, circulating within the crust of the earth, plays an important role in the formation of most ore deposits by leaching elements from rocks and concentrating them in zones of new mineral growth. During this process, the isotopic compositions of the host rock and fluid are progressively changed. Analysis of the resulting stable isotope ratios of the ore rocks allows geologists to understand the sources and quantity of the mineralizing fluids. Better understanding of the formation of ore deposits has led to greater success in the discovery of new mineral resources.

Stable isotope research into water-rock interaction is not limited to studying water-rich fluids associated with precious mineral deposits. Petroleum and natural gas flow from source rocks rich in organic material to porous reservoir rocks, where they may become trapped. Rocks through which hydrocarbons have migrated frequently show well-depleted carbon isotope values and thus preserve a record of fluid movement. Oil companies have used carbon data to track the migration history of hydrocarbons and to locate regions where petroleum leaks to the surface. The use of carbon isotopes has also proved successful in identifying source rocks associated with producing oil fields. This technology will become even more important as resources become increasingly scarce.

Another area of concern associated with water-rock interaction is the safe, long-term storage of toxic and nuclear waste. One of the most important criteria for the isolation of dangerous wastes is that the enclosing rocks be relatively dry and impermeable to water movement so that hazardous material is not transported into water supply aquifers. The record of water flow recorded by stable isotopes is sensitive to even very small fluid volumes and provides one of the means of assessing the risks associated with a toxic disposal site.

The source of fresh drinking water for almost half the population in the United States is subsurface groundwater. This crucial resource is jeopardized by contamination with common pollutants, such as pesticides, and by depletion through the withdrawing of water faster than it is replenished at recharge zones. Research involving stable isotope studies has become important in hydrology to track fluid movement within aquifers and identify sources of recharge water.

Bibliography

Faure, Gunter. *Principles of Isotope Geology*. New York: John Wiley & Sons, 1977. A college-level text that covers both radioactive and stable isotopes. The first five chapters are introductory in nature and include a good historical review of the development of isotope geology and mass spectrometry. The last four chapters cover stable isotopes and include figures reproduced from class research papers. Each chapter includes a detailed reference list.

Hoefs, Jochen. *Stable Isotope Geochemistry*. 3d ed. New York: Springer-Verlag, 1987. Suitable for an advanced college student who seeks a detailed discussion of isotope fractionation, sample preparation, and laboratory standards. The material is introduced in three sections. The first chapter provides theoretical principles, the second chapter is a systematic description of the most common stable isotopes, and the third summarizes the occurrence of stable isotopes in nature. An extensive list of references is included at the end of the book.

Krauskopf, Konrad B. *Introduction to Geochemistry*. 2d ed. New York: McGraw-Hill, 1979. A comprehensive advanced text that covers most aspects of the chemistry of natural fluids. Radioactive and stable isotopes are briefly treated in chapter 21, while chapter 17 contains a discussion of ore-forming solutions. This resource is particularly useful for students who seek detailed information on the chemistry and interaction of crustal water. Suggestions for further reading are provided at the end of each chapter.

O'Neil, J. R. "Stable Isotope Geochemistry of Rocks and Minerals." In *Lectures in Isotope Geology*, edited by Emilie Jaëger and Johannes C. Hunziker. New York: Springer-Verlag, 1979. This source provides a brief and clear introductory section on stable isotope nomenclature. The remainder of the chapter outlines major conclusions drawn from isotope analysis of igneous, metamorphic, sedimentary, and ore deposit rocks. Examples are provided from pioneering research studies. Although the text is oriented toward the college level, high school students interested in the results of isotope studies will find this chapter useful.

Smith, David G., ed. *The Cambridge Encyclopedia of Earth Sciences*. New York: Crown Publishers, 1981. Chapter 8, "Trace Element and Isotope Geochemistry," is a brief, well-illustrated summary of the occurrence of trace elements, stable isotopes, and radiogenic elements. The chapter emphasizes how trace element and isotope studies have enhanced understanding of processes such as the generation of magma and the occurrence of ore deposits. The discussion of water-rock interaction associated with the ocean crust would be accessible to advanced high school students. Few additional references are offered.

Grant R. Woodwell

Cross-References

Igneous Rock Bodies, 307; Magmas, 383.

Weathering and Erosion

The weathering process breaks down the rocks of the earth's surface into soluble material and into the particles that form the basis for soils and agriculture. These weathered products are then swept away by the various erosional agents, such as rivers and glaciers, that shape the planet's rocky surface.

Field of study: Sedimentology

Principal terms

ABRASION: the wearing away of rock by frictional contact with solid particles moved by gravity, water, ice, or wind

ACID RAIN: rain with higher levels of acidity than normal; the source of the high levels of acidity is polluted air

CHEMICAL WEATHERING: the chemical decomposition of solid rock by processes that change its original materials into new chemical combinations

EROSION: the general term for the various processes by which particles already loosened by weathering are removed

GRANITE: a coarse-grained, igneous rock composed primarily of the minerals quartz and feldspar

LIMESTONE: a sedimentary rock composed of calcium carbonate

MECHANICAL WEATHERING: the physical disintegration of rock into smaller particles having the same chemical composition as the parent material

MINERAL: a naturally occurring, inorganic, crystalline material with a unique chemical composition

SANDSTONE: sedimentary rock composed of grains of sand cemented together

WEATHERING: the general term for the group of processes that break down rocks at or near the earth's surface

Summary

Although the landscape appears not to change, constructive and destructive forces are at work within the earth, building the crust up and breaking the rocks down and carrying the resulting debris away. The destructive forces are known as weathering and erosion. "Weathering" refers merely to the mechanical disintegration and chemical decomposition of rocks and minerals at or near the earth's surface. No movement of these materials is implied. Exposure to weather causes rocks to change their character and either crumble into soil or become transformed into even smaller particles that are readily available for removal. "Erosion" refers to the processes by which particles already loosened by weathering are removed. This process involves two steps: First, the loose materials must be picked up, or entrained; second, the materials must be physically carried, or transported, to new locations. The major ways by which earth materials are eroded are by means of rivers, underground water, moving ice, waves, wind, and landslides.

Weathering is a near-surface phenomenon because it involves the response of earth materials to the elements of sunlight, rain, snow, and the like. It does not affect rocks that are deeply buried within the earth's crust. Only after these rocks are exposed at the earth's surface, after a long period of uplift and removal of overlying material, does weather begin to affect them. In this changed environment they are subject to the comparatively hostile action of acid rain, subfreezing temperatures, and high humidity. The resulting transformations that take place in the rock are what is called weathering.

Scientists recognize two different types of weathering: mechanical and chemical. Although the two types are generally discussed separately, it is important to keep in mind that in nature they generally work hand in hand. Mechanical weathering (also known as disintegration) involves the physical breakdown of the rock into smaller and smaller grains, usually because of the application of some kind of pressure, such as the expansion of water during freezing or the growth of plant roots in rock crevices. The chemical composition of the rock, however, remains unchanged. The end result of mechanical weathering is smaller pieces of rock that are identical in composition and appearance to the original larger rock mass.

Chemical weathering (also known as decomposition) involves a complex alteration in the materials that compose the original rock. These materials are chemically changed into new substances by the addition or removal of certain elements, usually through the action of water. The familiar rusting of iron is an example of chemical weathering; there is a total change in the composition and appearance of the original material. At first, there is a hard, silvery metal; afterward, all that remains is a soft, yellowish-brown powder.

Consider the effects of mechanical weathering and chemical weathering on a cube of rock that measures 6 centimeters on a side. Assume that by means of mechanical weathering, the cube is broken down into 216 cubes measuring 1 centimeter on a side. Now, much more surface area is available for chemical attack. For this reason, chemical weathering proceeds much faster when a rock is first broken into smaller pieces by mechanical weathering.

The effects of weathering are visible in the impressive isolated formations of Utah's Monument Valley. *(Robert McClenaghan)*

In nature, mechanical weathering can proceed in a variety of ways. The best-known example involves the action of freezing water. Because water increases about 9 percent in volume as it freezes, enormously large outward-directed pressures develop within a rock when water freezes in its pore spaces and cracks. The result is the angular rock fragments found scattered about most mountain tops and sides. Soil contains water, too, and horizontal lenses of ice may form within the soil when water freezes in it. These create bumps in lawns and the familiar "frost heaves" of mountain roads. When heavy trucks rumble over these heaved pavements during thaws, the pavement gives way to create potholes.

In deserts, soil water is drawn upward through the rock and evaporates at the hot upper surface, leaving its dissolved salts behind as crystals growing in the pore spaces of the rock. These growing salt crystals also exert powerful pressures within the rock, so that porous rocks, such as sandstone, undergo continuous grain-by-grain disintegration in desert climates. The pressures needed for mechanical weathering can also be produced when the extreme heat from a forest fire or a lightning strike causes flakes to chip off a rock or when a growing plant or tree extends its root system into cracks and splits a rock apart. Another type of rock splitting is known as exfoliation; it is caused by the spontaneous expansion of rock masses when they are freed from the confining pressures of overlying and surrounding rock. This process produces large, dome bedrock knobs with an onionlike structure. Stone Mountain, Georgia, and Half Dome, in Yosemite National Park, are examples.

Chemical weathering is a more complex process than mechanical weathering, because in this case, the original rock material is actually transformed into new substances. The rusting of iron has already been mentioned as an example of chemical weathering. Many common rocks and minerals contain iron. During chemical weathering, the iron in these substances combines with oxygen from the air to form various compounds known as iron oxides.

Another way in which chemical weathering attacks rocks is by dissolving them. Large areas of the earth's surface are underlain by a rock type known as limestone. Limestone is readily dissolved by water containing small quantities of acid. All rainfall is weakly acidic as a result of its dissolving carbon dioxide from the air to produce dilute carbonic acid. Rains that originate in areas of high air pollution are even more acidic, a condition known as acid rain. When rainfall that contains carbonic acid comes in contact with limestone bedrock, the acid reacts with the calcium carbonate in the rock to produce calcium bicarbonate, a soluble substance that is readily carried off in solution.

A final example of chemical weathering involves the weathering of granite, a hard igneous rock composed primarily of feldspar and quartz. When granite undergoes chemical weathering, each mineral is affected differently. The feldspar is gradually transformed into a new mineral, clay, which is soft and easily molded when wet. Clay offers very little resistance to erosion. Quartz, on the other hand, is highly resistant to chemical attack and is left behind when the clay is removed. Some of the quartz grains remain in the soil, but most will be carried off by rivers, becoming rounded as they tumble along. Eventually they form the sands of beaches and, in time, the sedimentary rock known as sandstone.

The term "erosion" refers to those processes by which the loose particles formed by weathering are picked up and carried to new locations. Erosion is a highly significant phenomenon at the earth's surface. Examples of it range from small gullies in a farmer's field to a catastrophic landslide in a high mountain valley. Nevertheless, the general principle involved in all types of erosion is the same: Weathered earth materials move downslope from their place of formation to a new location, with gravity as the driving force. The materials may simply slide downhill as a landslide, or they may be carried down the hill by an erosional agent, such as running water. Worldwide, running water in the form of streams and rivers is probably the single most important erosional agent. Locally, other erosional agents may be highly significant, including underground water flow, glaciers, waves, and wind action.

The downhill movement of weathered materials under the influence of gravity alone results in landslides if the downslope movement is rapid but in "creep" if the movement is imperceptibly slow. When large quantities of water are present in the weathered material, the downslope movement is called a mudslide. Running water can erode material from its channel banks in four different ways: Soluble material can be dissolved by weakly acidic river water, bedrock can be worn smooth as a result of abrasion by sand and gravel carried along the stream bed, unconsolidated bank and bed materials can be swept away by a strong current, resulting in bank caving, and upwardly directed turbulent eddies in the water may lift small particles from the bottom and entrain them in this fashion. Underground water erodes bedrock primarily by dissolving it, whereas glaciers act more like rivers, abrading the underlying rock by means of rock fragments frozen in the ice. Glaciers are also able to pluck rock masses from their channel walls when these rock masses have frozen to the main ice mass; the rocks are torn loose as the ice moves on.

Waves erode shorelines, wearing rock surfaces smooth by means of the sand and gravel they carry. Waves can also dislodge particles from a cliff face. Cracks quickly open in cliffs, seawalls, and breakwaters, and when water is forced into these cracks, the air in the cracks becomes highly compressed, exerting still further pressure on the rock. Wind erosion, on the other hand, relies on the abrasive action of sand grains transported by the wind and on the lifting power of eddies, which are able to entrain finer-grained soil particles.

Methods of Study

Scientists have analyzed the rate at which rocks weather and have found that the important factors are rock type, mineral content, amount of moisture present, temperature conditions, topographic conditions, and amount of plant and animal activity. A rock type may be highly resistant to weathering in one climate and quite unresistant in another. Limestone, for example, which forms El Capitan, the highest peak in the desert region of southwest Texas, underlies the lowest valleys in the humid climate of the Appalachian Mountains.

Using field observations and laboratory experiments, scientists have studied the rate at which different minerals are attacked by chemical decomposition. Among the minerals formed by igneous activity, quartz is least susceptible to chemical attack, whereas olivine, a greenish-colored mineral rich in iron and magnesium, is one of the most susceptible. The reason is that olivine forms at high temperatures and pressures when melted rock first begins to cool and is consequently unstable at the lower temperatures and pressures that prevail at the earth's surface. Quartz, on the other hand, forms late in the cooling process, when the temperatures and pressures are more similar to those encountered at the earth's surface; therefore, quartz more readily resists attack by chemical weathering. Scientists have concluded that the more the conditions under which a mineral forms are akin to those at the earth's surface, the more resistant to chemical attack the mineral will be.

Numerous observations have been made relating to the rapidity with which weathering takes place. The eruption of Mount St. Helens in Washington State on May 18, 1980, has provided a natural laboratory for such study. During the eruption, vast quantities of volcanic ash were hurled into the air and deposited to depths exceeding several meters in the vicinity of the volcano. Scientists are carefully analyzing the changes that are taking place in the ash because of mechanical and chemical weathering and the rate at which this ash is being converted into a productive soil for the growth of vegetation. Scientists also study the rate at which tombstones and historic monuments of known age are attacked by weathering. For marble tombstones in humid climates, the amount of weathering within a single lifetime may amount to several millimeters.

The rate at which earth materials are moved from place to place on the earth's surface by the various agents of erosion is also of interest. One way to approach this problem is to measure the quantity of sediment being carried by a river each year and then to calculate how much of a loss this amount represents for the entire area drained by the river. Data from various locations in the United States suggest that the overall rate of erosion amounts to approximately 6 centimeters per one thousand years. Corroborating evidence comes from another source: Photographs made in scenic areas compared with photographs made from the same vantage point one hundred years ago or more show surprisingly little erosional modification of the earth's surface. It is believed, however, that once human beings occupy an area intensively, erosion rates increase significantly.

Context

Weathering affects not only bedrock outcrops but also human-made structures. Unless they are continually repaired and restored, all structures become weather-beaten and, in time, weaken and fall into ruin. Beginning in the early 1970's, people also became aware of the harmful consequences of acid rain. Concrete, limestone building blocks, and the marble used for statuary are all susceptible to the dissolving action of acid rain. In fact, many statues adorning public buildings in Europe become unrecognizable because of this process.

The earth materials produced by weathering are of great significance. The larger grains are known as regolith and by the addition of partly decomposed organic matter are turned into soil, the basis for agriculture. Other grains are carried by rivers into the sea to become the raw materials from which beaches are made. Residues of weathered materials are sometimes left behind, such as clay and ores of iron, aluminum, and manganese, which may form valuable mineral deposits.

Erosional processes also affect human life. Gravity's influence may bury small villages in catastrophic landslides, or it may trigger the imperceptible downslope movements, known as creep, that cause structures on hillsides to collapse. When the Alaska pipeline was being built, construction of all kinds was hampered by problems caused by soil flowage when the permafrost in the ground thawed. The erosional activity of rivers shapes the landscape, cutting gorges and supplying sediment to alluvial fans, floodplains, levees, and deltas. Sometimes the erosional activity of a river gets out of hand, causing devastating floods. Even where no stream channel is present, farmlands can be seriously damaged by soil erosion.

The erosional power of moving ice sculptures some of the world's most spectacular mountains. Along coastlines, wave erosion creates cliffs or threatens structures such as lighthouses, seawalls, and breakwaters. Coastal currents may carry sand away, causing severe beach erosion. The wind contributes to erosion when it moves sand grains to create a sandstorm. This blinding cloud can sandblast the paint off a car or break a telephone pole in two. Even dust-sized material, when lifted from the ground in the form of a dust storm, can have a devastating effect. During the 1930's, an area known as the Dust Bowl developed in the Great Plains region of the United States. A prolonged drought and unwise agricultural practices resulted in severe dust storms that blew away valuable topsoil, lowering the ground level by nearly a meter in some places.

Bibliography

Bertin, Léon. *The Larousse Encyclopedia of the Earth.* New York: Prometheus Press, 1965. This reference book has well-written sections on the weathering processes of disintegration and decomposition. There are also lengthy sections on mass wasting, transportation by the wind, subsurface or groundwater, running water, wave erosion, and the work of glaciers. The text is copiously illustrated with excellent black-and-white and color plates. Suitable for general readers.

Cain, J. A., and E. J. Tynam. *Geology: A Synopsis.* Vol. 1, *Physical Geology.* Dubuque, Iowa: Kendall/Hunt, 1980. A condensed treatment of physical geology that gives the reader a quick overview of the various weathering and erosion processes. There are many helpful diagrams, tables, and photographs that illustrate the principles involved. An excellent introduction to geology for the non-scientist. Suitable for high school readers.

Jensen, M. L., and A. M. Bateman. *Economic Mineral Deposits.* 3d ed. New York: John Wiley & Sons, 1979. An outstanding economic geology text, containing detailed information on the processes of formation and the occurrence of residual deposits of iron, manganese, aluminum, and clay as a result of weathering. There are cross sections of individual deposits. Suitable for college-level readers and the interested layperson with some technical background.

Judson, S., M. E. Kauffman, and L. D. Leet. *Physical Geology.* 7th ed. Englewood Cliffs, N.J.: Prentice-Hall, 1987. Chapter 5, "Weathering and Soils," discusses the various types of weathering and the methods for studying the rates of weathering and erosion. The text is illustrated with photographs, diagrams, and tables that provide important data. Written at a level suitable for undergraduates.

Plummer, Charles C., and David McGeary. *Physical Geology*. 4th ed. Dubuque, Iowa: Wm. C. Brown, 1988. An unusually readable text. Extended discussions of weathering and the various erosion processes are supplemented by excellent photographs and line drawings. Each chapter has an extended list of supplementary readings, and there is an excellent glossary. Suitable for college-level readers and interested nonspecialists.

Shelton, J. S. *Geology Illustrated*. San Francisco: W. H. Freeman, 1966. This book contains some of the finest black-and-white photographs ever taken of geologic features, along with explanatory text. Part 3, "Sculpture," has numerous aerial views showing how the earth's surface has been modified by the erosion processes of landslides, streams and rivers, groundwater, glaciers, waves, and wind. Suitable for laypersons.

Tarbuck, E. J., and F. K. Lutgens. *Earth Science*. 5th ed. Columbus, Ohio: Merrill, 1988. This popular text offers an introduction to weathering and erosional processes. It is concisely written and generously illustrated with color photographs and color line drawings. Key terms are in boldface in the body of the text, and there is a helpful glossary. Suitable for high school readers.

Donald W. Lovejoy

Cross-References

Alluvial Systems, 1; Floodplains, 241; Glacial Landforms, 281.

Glossary

AA: A Hawaiian term (pronounced "ah-ah") that has been adopted for lava flows with rough, clinkery surfaces.

ABSOLUTE DATE OR AGE: The numerical timing (in years or millions of years) of a geologic event, as contrasted with relative (stratigraphic) timing.

ABYSSAL PLAINS: Flat-lying areas of the sea floor, located far from continents; they cover more than half the total surface area of the earth.

ACCRETION: The process by which small bodies called planetesimals are attracted by mutual gravitation to form larger bodies called protoplanets.

ACHONDRITE: A stony meteorite that contains mostly silicate minerals and a small amount of metal formed from the cooling of molten rock.

ACID RAIN: Rain with higher levels of acidity than normal; the source of the high levels of acidity is polluted air.

ALBEDO: The fraction of incident light on a surface that is reflected back in all directions.

ALLUVIUM: Sediment deposited by flowing water.

ALPHA PARTICLE: A helium nucleus emitted during the radioactive decay of uranium, thorium, or other unstable nuclei.

ALPINE GLACIER: A small, elongate, usually tongue-shaped glacier commonly occupying a preexisting valley in a mountain range.

ANDESITE: A volcanic igneous rock type intermediate in composition and density between granite and basalt.

ANGLE OF REPOSE: The maximum angle of steepness that a pile of loose material such as sand or rock can assume and remain stable; the angle varies with the size, shape, moisture, and angularity of the material.

ANION: An atom that has gained electrons to become a negatively charged ion.

ANORTHOSITE: A plutonic igneous rock, solidified from the molten state, consisting mostly of feldspar.

ANTICLINE: A folded structure created when rocks arch upward; the limbs of the fold dip in opposite directions, and the oldest rocks are exposed in the middle of the fold.

APHANITIC: A textural term that applies to an igneous rock composed of crystals that are microscopic in size.

AQUEOUS SOLUTION: A term synonymous with "hydrothermal fluid" and "intergranular fluid" for a fluid mixture that is hot and has a high solvent capacity, which permits it to dissolve and transport chemical constituents; it becomes saturated upon cooling and may precipitate hydrothermal minerals.

AQUIFER: A water-bearing bed of rock, sand, or gravel, capable of yielding substantial quantities of water to wells or springs.

ARCHEAN EON: The period of geologic time from about 4 billion to 2.5 billion years ago.

ARKOSE: A sandstone in which more than 10 percent of the grains are feldspar or feldspathic rock fragments; also called feldspathic arenite.

ASBESTOSIS: The deterioration of the lungs, caused by the inhalation of very fine particles of asbestos dust.

ASH: Fine-grained pyroclastic material less than 2 millimeters in diameter, ejected from an erupting volcano.

ASTEROID: A small, rocky body in orbit around the sun; a minor planet.

ASTEROID BELT: The region between the orbits of Mars and Jupiter, containing the majority of asteroids.

ASTHENOSPHERE: The partially molten portion of the outer mantle, ranging to a depth of 250 kilometers, that lies at the base of the lithosphere and exhibits low seismic velocities and strong attenuation of seismic wave strength.

ASTROBLEME: The remnant of a large impact crater on earth; erosion will have altered the superficial appearance, but confirmation can be made from deeper structural damage and the presence of characteristically shattered and shocked rock.

ASTRONOMICAL UNIT: The average distance between the earth and the sun: 150 million kilometers.

ATMOSPHERE: The five clearly defined regions composed of layers of gases and mixtures of gases, water vapor, and solid and liquid particles, extending up to 483 kilometers above the earth.

ATOLL: A tropical island on which a massive coral reef, often ringlike, generally rests on a volcanic base.

AUREOLE: A ring-shaped zone of metamorphic rock surrounding a magmatic intrusion.

AVALANCHE: Any large mass of snow, ice, rock, soil, or a mixture of these materials that falls, slides, or flows rapidly downslope; velocities may reach in excess of 500 kilometers per hour.

AXIS: A line parallel to the hinges of a fold; also called fold axis or hinge line.

BARCHAN DUNE: A crescent-shaped sand dune of deserts and shorelines that lies traverse to the prevailing wind direction.

BARREL: The standard unit of measure for oil and petroleum products, equal to 42 U.S. gallons or approximately 159 liters.

BARRIER ISLAND: A long, low sand island parallel to the coast and separated from the mainland by a salt marsh and lagoon; a common coastal feature on depositional coasts worldwide.

BASALT: A volcanic rock that results when lava rich in iron and magnesium and low in silica is cooled rapidly, resulting in a fine-grained, dark-colored appearance.

BASEMENT: The crystalline, usually Precambrian, igneous and metamorphic rocks that occur beneath the sedimentary rock on the continents.

BASIN: A regionally depressed structure in which sediments accumulate.

BATHOLITH: A large-mass igneous rock with more than 100-square-kilometer exposure, made up of multiple related smaller bodies called plutons.

BATHYMETRIC CONTOUR: A line on a map of the ocean floor that connects points of equal depth.

BAUXITE: The principal ore of aluminum; a mixture of aluminum compounds produced by prolonged weathering of bedrock in tropical or subtropical climates.

BEDLOAD: Sediment in motion in continuous or semicontinuous contact with sediment bed by sliding, rolling, or hopping.

BENIOFF ZONE: The dipping zone of earthquake foci found below island arcs; it is named for Hugo Benioff, the seismologist who first defined it.

BETA PARTICLE: An electron emitted from the nucleus of a radioactive element.

BIOABRASION: Physical and chemical erosion or removal of rock as a result of the activities of marine organisms.

BIOGENIC SEDIMENTS: The sediment particles formed from skeletons or shells of microscopic plants and animals living in seawater.

BIOGEOGRAPHIC PROVINCE: A geographic region distinguished by a unique set of endemic organisms that live in the region.

BIOMARKERS: Chemicals found in oil with a chemical structure that links their origin with specific organisms; also called geochemical fossils.

BIOSPHERE: The total living material—plants and animals and their environment—of a specific area of the earth.

BIOSTRATIGRAPHY: The discipline that defines rock layers on the basis of their fossil content.

BIREFRINGENCE: The difference between the maximum and minimum indices of refraction of a crystal.

BITUMEN: A generic term for a very thick, natural semisolid; asphalt and tar are classified as bitumens.

BLOCKING TEMPERATURE: The temperature at which a magnetic mineral becomes a permanent recorder of a magnetic field.

BODY WAVE: A seismic wave that propagates interior to a body; there are two kinds—P waves and S waves—that travel through the earth, reflecting and refracting off of the several layered boundaries within the earth.

BOLIDE: A meteorite or comet that explodes upon striking the earth.

BORE: A nearly vertical advancing wall of water that may be produced by tides, a tsunami, or a seiche.

BORT: A general term for diamonds that are suitable only for industrial purposes; these diamonds are black, dark gray, brown, or green in color and usually contain many inclusions of other minerals.

BOTTOM-WATER MASS: A body of water at the deepest part of the ocean, identified by similar patterns of salinity and temperature.

BOUGUER GRAVITY: A residual value for the gravity at a point, corrected for latitude and elevation effects and for the average density of the rocks above sea level.

BRACKISH WATER: Water with salt content between that of salt water and fresh water; it is common in arid areas on the surface, in coastal marshes, and in salt-contaminated groundwater.

Braided river: A relatively shallow river with many intertwined channels; its sediment is moved primarily as riverbed material.

BRECCIA: A coarse-grained clastic rock composed of angular broken fragments held together by a mineral cement or fine-grained matrix.

BRINE: Water with a higher content of dissolved salts than is normally found in seawater.

CALDERA: A large, flat-floored volcanic depression that is formed on top of a large, shallow magma chamber during the eruption or withdrawal of magma; calderas are usually tens of kilometers across and can be a kilometer or more in depth.

CALORIS BASIN: The largest known structure on Mercury; it is similar to the moon's Imbrium Basin.

CARBONACEOUS CHONDRITES: Stony meteorites that are slightly metamorphosed agglomerates containing chondrules, unmelted aggregates, and volatile-rich matrix materials.

CARBONATE COMPENSATION DEPTH (CCD): The depth in the oceans at which the rate of supply of calcium carbonate equals the rate of dissolution of calcium carbonate.

CARBONATE ROCK: A sedimentary rock composed of grains of calcite (calcium carbonate) or dolomite (calcium magnesium carbonate).

CASING: A tubular material, usually a metal, that is inserted into the raw borehole of a well for the purpose of preventing the collapse of geologic material outside the borehole.

CAST: A fossil that displays the form of the original organism in true relief.

CATACLASIS: The crushing and shearing process that breaks up rocks.

CATASTROPHISM: The theory that the large-scale features of the earth were created suddenly by catastrophes in the past; catastrophism is the opposite of uniformitarianism.

CATION: An element that has lost one or more electrons such that the atom carries a positive charge.

CENOZOIC ERA: The period of geologic time from about 65 million years ago to the present.

CERAMICS: Nonmetal compounds, such as silicates and clays, produced by firing the materials at high temperatures.

CHEMICAL BOND: The force holding two chemical elements together as part of a molecule.

CHEMICAL WEATHERING: The chemical decomposition of solid rock by processes that change its original materials into new forms that are chemically stable at the earth's surface.

CHERT: A hard, well-cemented sedimentary rock that is produced by recrystallization of siliceous marine sediments buried in the sea floor.

CHONDRITE: A stony meteoritic material containing glassy spherical inclusions called chondrules, which are usually composed of iron, aluminum, or magnesium silicates.

CINDER CONE: A small volcano composed of cinder or lumps of lava containing many gas bubbles, or vesicles; often the early stage of a stratovolcano.

CIRQUE: A steep-sided, gentle-floored, semicircular hollow produced by erosion at the head of a glacier high on mountain peaks.

CIRRUS: Trail or streak clouds, ranging from 5 to 13 kilometers above the ground, that are feathery or fibrous in nature.

CLADISTICS: A method of determining relationships in which shared advanced characteristics exhibited by the organism are used.

CLASTIC ROCK: A sedimentary rock, typically a sandstone, composed of broken fragments of minerals and rocks.

CLAY: A term with three meanings: a particle size (less than 2 microns), a mineral type (kaolin, illite), and a fine-grained soil that is puttylike when damp.

CLEAVAGE: The tendency for minerals to break in smooth, flat planes along zones of weaker bonds in their crystal structure; foliation that imparts a preferred direction of fracturing to a rock.

CLINKER: Irregular lumps of fused raw materials to which gypsum is added before grinding into finely powdered cement.

COAL: Dark brown to black rock formed by heat and compression from the accumulation of plant material in swampy environments.

COASTAL WETLANDS: Shallow, wet, or flooded shelves that extend back from the freshwater-saltwater interface; they may consist of marshes, bays, lagoons, tidal flats, or mangrove swamps.

COEFFICIENT OF THERMAL EXPANSION: The linear expansion ratio (per unit length) for any particular material as the temperature is increased.

COHESION: The strength of a rock or soil imparted by the degree to which the particles or crystals of the material are bound to one another.

COLD FRONT: The contact between two air masses when a bulge of cold, polar air surges southward into regions of warmer air.

COLD POLAR GLACIER: A glacier that is below the pressure-melting temperature of ice throughout.

COLOR INDEX: The percentage by volume of dark minerals in a rock; it is calculated for quick identification of rocks.

COLUMNAR JOINTING: The formation of columns, often with hexagonal cross sections, as joints grow inward from the outer surfaces of cooling igneous rock bodies.

COMA: An envelope of gas and dust surrounding the comet nucleus that has not yet been pushed into the comet tail by solar wind and by radiation pressure.

COMET: A solar system body, usually in an elongated and randomly oriented orbit, composed of rocky and icy materials that form a flowing head and extended tail when the body nears the sun.

COMET NUCLEUS: The central core of a comet, composed of frozen gases and dust; the source of all cometary activity.

COMPETENCE: A concept that expresses the size of the largest sediment particles that can be moved by a given fluid flow.

COMPLEX CRATER: An impact crater of large diameter and low depth-to-diameter ratio caused by the presence of a central uplift or ring structure.

COMPONENT: A chemical entity used to describe compositional variation within a phase.

COMPRESSIVE STRENGTH: The ability to withstand a pushing stress or pressure, usually given in pounds per square inch.

CONCORDANT AGE: A situation in which several naturally radioactive elements, such as uranium, thorium, strontium, and potassium, all give the same age for a rock sample.

CONDENSATION TEMPERATURE: The temperature at which a component of the primitive solar system condensed into solid particles.

CONDUCTION: The transfer of heat through the process of atomic or molecular collisions from a hotter to a colder body.

CONE: The hill or mountain, more or less conical, surrounding a volcanic vent and created by its ejecta; it is normally surmounted by a crater.

CONE OF DEPRESSION: The depression, in the shape of an inverted cone, of the groundwater surface that forms near a pumping well.

CONFINED AQUIFER: An aquifer that is completely filled with water and whose upper boundary is a confining bed; also called a water table aquifer.

CONFINING PRESSURE: Pressure acting in a direction perpendicular to the major applied stress in a rock deformation experiment.

CONGLOMERATE: Sedimentary rock composed of gravel in a sandy matrix.

CONNATE FLUIDS: Fluids that have been trapped in sedimentary pore spaces.

CONTACT METAMORPHISM: The change in mineralogy and/or composition of rock as a result of high temperatures around magmatic intrusion of limited extent.

CONTACT METASOMATISM: Metasomatism in proximity to a large body of intrusive igneous rock, or pluton; the intensity of metasomatism increases as igneous contact is approached.

CONTAMINANT: Any ion or chemical that is introduced into the environment, especially in concentrations greater than those normally present.

CONTINENTAL DRIFT: A hypothesis that attributes the present arrangement of continental shields and ocean floors to the breakup of the original supercontinent, Pangaea.

CONTINENTAL GLACIER OR ICE SHEET: A glacier of considerable thickness that completely covers a large part of a continent, obscuring the relief of the underlying surface.

CONTINENTAL MARGIN: That part of the earth's surface that separates the emergent continents from the deep sea floor; it generally consists of the continental shelf, continental slope, and continental rise.

CONTINENTAL RISE: The broad and gently sloping ramp that rises from the abyssal plain to the base of the continental slope; submarine fans are found here.

CONTINENTAL SHELF: The submerged offshore portion of a continent, ending where water depths increase rapidly from a few hundred to thousands of meters.

CONTINENTAL SLOPE: The steeply sloping region between the continental shelf and the continental rise.

CONVECTION: The transfer of heat by the movement or circulation of the heated parts of a liquid or gas as a result of density differences caused by heating.

CONVECTION CELL: A single closed path of rising warm material and sinking cold material.

CONVERGENT ORGANISMS: Evolutionarily unrelated organisms that come to resemble one another through adaptation to the same life habitat.

CONVERGENT PLATE MARGIN: The boundary between two plates that are moving toward each other, which may result in island arc development or volcanic arcs on land.

CORDILLERA: A long, elevated mountain chain marked by a valley-and-ridge structure.

CORE: The spherical, mostly liquid mass located 2,900 kilometers below the earth's surface; a central, solid part is known as the inner core.

CORE DRILLING: A method of extracting samples of the materials being penetrated in a deep-drilling project.

CORE RING OR CORE EDDY: A mass of water that is spun off an ocean current by the current's meandering motion.

CORIOLIS EFFECT: A phenomenon caused by the earth's rotation in which moving objects unattached to the earth move to the right of their path of motion in the Northern Hemisphere and to the left in the Southern Hemisphere.

CORONAE: Ring structures on Venus, consisting of alternating concentric ridges and valleys and higher than the external terrain; they are possibly volcanic in origin.

CORRELATION: Matching the sequence of events (distinctive layers, fossils, magnetic polarity intervals) between two stratigraphic sections.

COUNTRY ROCK: The rock into which magma is injected to form an intrusion.

COVALENT BONDING: A type of chemical bonding produced by overlapping orbitals of adjacent atoms sharing electrons; covalently bonded solids usually have low solubility in water.

CRANIAL CAPACITY: The internal volume (in cubic centimeters) of the braincase of a skull.

CRATER: The circular depression atop a volcanic cone or formed by meteoritic impact.

CRATON: The part of the continent that is covered with a variable thickness of sedimentary rock but that has not been affected by mountain building for a considerable length of time.

CREEP: The slow, more or less continuous downslope movement of earth material.

CROSS-BEDDING: Layers of rock or sand that lie at an angle to horizontal bedding, formed by currents flowing in the direction toward which the beds slope.

CRUST: The outermost layer of the earth; the continental crust, composed of dominantly silicon-rich igneous rocks, metamorphic rocks, and sedimentary rocks, is between 30 and 40 kilometers thick, while the oceanic crust, composed of magnesium- and iron-rich rocks such as basalt, is merely 5 kilometers thick.

CRYSTAL: A solid with an internally ordered arrangement of component atoms.

CRYSTAL STRUCTURE: The regular arrangement of atoms in a crystalline solid.

CRYSTALLIZATION: The formation and growth of a crystalline solid from a liquid or from a gas.

CUMULATE: An igneous rock composed chiefly of crystals that accumulated by sinking or floating from a magma.

CUMULONIMBUS: The tall, billowy variety of cloud with precipitation falling; lightning, hail, violent winds, and strong vertical wind drafts are common with this type of cloud.

CUMULUS: Clouds with vertical development, or heap clouds, ranging from ground level to 6 kilometers above the earth's surface.

CURIE TEMPERATURE: The temperature above which a permanently magnetized material loses its magnetization.

CURRENT: A sustained movement of seawater in the horizontal plane, usually wind-driven.

DAUGHTER PRODUCT: An isotope that results from the decay of a radioactive parent isotope.

DEBRIS FLOW: A flowing mass consisting of water and a high concentration of sediment with a wide range of size, from fine muds to coarse gravels.

DECLINATION: For a particular location on the earth's surface, the horizontal angle between true north and the compass needle direction.

DÉCOLLEMENT: The detachment surface beneath a fold belt, usually located in a weak layer of rock such as shale; the maximum depth to which the rocks are folded and faulted.

DEFLATION: The sorting out, lifting, and removal of loose, dry, silt- and clay-sized soil particles by a turbulent eddy action of the wind.

DELTA: A deposit of sediment, often triangular, formed at a river mouth where the wave action of the sea is low.

DENDROCHRONOLOGY: The study of tree rings; it provides a means of calibrating radiocarbon dates with absolute chronology.

DENSITY: The mass per unit of volume (grams per cubic centimeter) of a solid, liquid, or gas.

DEPOSITION: The physical or chemical process by which sedimentary grains come to rest after being eroded and transported.

DESALINIZATION: The process of removing salt and minerals from seawater or from saline water occurring in aquifers in order to render it fit for agriculture or other human use.

DESERT VARNISH: A thin, dark, hard, shiny or glazed iridescent film coating, stain, or polish in rocks that is composed largely of iron and manganese oxides and silica; it is formed by weathering of dust films and by microbial action.

DESERTIFICATION: A process by which deserts are created, expanded, or changed by clearing away peripheral forestry or brush, thus allowing the desert to occupy new spaces.

DETACHMENT FAULT: A horizontal or gently dipping, regionally extensive fault; the hanging wall usually contains numerous smaller, steeper normal faults that end at the detachment fault.

DETRITAL MINERALS: Minerals that have been eroded, transported, and deposited as sediments.

DETRITAL REMANENT MAGNETIZATION (DRM): Sedimentary rock magnetization acquired by magnetic sediment grains aligning with the magnetic field.

DIAGENESIS: The conversion of unconsolidated sediment into consolidated rock after burial by the processes of compaction, cementation, recrystallization, and replacement.

DIAPIR: An unstable mass of magma ascending in the earth's crust.

DIATREME: A volcanic vent or pipe formed as the explosive energy of gas-charged magmas break through crustal rocks.

DIFFERENTIATION: The process by which a primitive planet, when heated, separates into a high-density metallic core and one or more silicate outside layers; also, the process of developing more than one rock type from a common magma.

DIFFUSION: The movement of ions or molecules through a medium (solid, liquid, or gaseous) from a location of high concentration to one of a lower concentration.

DIKE: A tabular igneous rock body that cuts across the fabric of the solid rocks.

DIP: The angle between a structural feature (for example, a joint, fault, or bedding plane) and the horizontal, which is usually the earth's surface.

DIPOLE FIELD: The field shape produced by two electrically charged particles, such as a proton and an electron, or by two magnetic poles, such as north and south.

DISCHARGE: The total amount of water passing a point on a river per unit of time.

DISCORDANT: Having sides that are at a substantial angle to the layering in the country rock.

DISCORDANT AGE: A situation in which several radioactive elements do not give the same age because of gain or loss of decay products from a rock sample.

DISSOCIATION: The breaking up of a compound into simpler components, such as the separation of a molecule into its constituent elements.

DISTRIBUTARY CHANNEL: A river that is divided into several smaller channels, thus distributing its flow and sediment load.

DIVERGENT PLATE MARGIN: The boundary where two plates move apart, resulting in the upwelling of magma from the mantle.

DIVERSITY: The variety of life, usually described in terms of the number of species present.

DOLDRUMS: The equatorial zone, where winds are light and variable and there is a heavy thunderstorm rainfall.

DOWNBURST: A downward outflowing of air and the associated wind shear from a thunderstorm that is especially hazardous to aircraft.

DRAINAGE BASIN: The land area that contributes water to a particular stream; the edge of such a basin is a drainage divide.

DRUMLIN: A smooth, elongate, oval-shaped hill or ridge formed under a moving glacier.

DUCTILITY: The ability to change shape without breaking when subjected to stress.

DUNE: A large bed form oriented dominantly transverse to flow, produced at moderate to high flow speeds.

DUST DEVIL: A rotating column of rising air, made visible by the dust it contains; smaller and less destructive than a tornado, it has winds of less than 60 kilometers per hour.

DUST TAIL: Dust particles blown off the cometary nucleus by radiation pressure after sublimation; it appears yellow as a result of reflected sunlight.

EARTH TIDE: The change in slope of the globe in response to the gravitational attraction of the sun and moon.

EARTHQUAKE: A sudden release of strain energy in a fault zone as a result of violent motion of a part of the earth along the fault.

ECLIPTIC PLANE: The plane in which the earth's orbit lies as it revolves around the sun.

ECOLOGICAL SUCCESSION: The process of plant and animal changes from simple pioneers such as grasses to stable, mature species such as shrubs or trees.

ECOLOGY: The relationship between organisms and their environment.

EJECTA: The material ejected from the crater made by a meteoric impact; also, material thrown out of a volcano during eruption.

EL NIÑO: Part of a gigantic meteorological system called the Southern

Oscillation that links the ocean and atmosphere in the Pacific, causing periodic changes in climate.

ELASTIC: A substance that when compressed, bent, stretched, or deformed in any way undergoes an amount of deformation proportional to the applied force and returns back to its original shape as soon as the force is removed.

ELECTROLYSIS: A process whereby molecular water is broken down into oxygen and hydrogen.

ELECTROMAGNETIC RADIATION: Forms of energy, such as light and radio waves, that consist of electric and magnetic fields that move through space.

EOLIAN EROSION: A mechanism of erosion or crater degradation caused by wind.

EPEIRIC SEA: A shallow sea that temporarily (in geologic terms) covers a portion of a craton; also termed an epicontinental sea.

EPEIROGENY: Uplift or subsidence of the crust within a region, without the internal disturbances characteristic of orogeny.

EPHEMERAL STREAM: A river or stream that flows briefly in response to nearby rainfall; such streams are common in arid and semiarid regions.

EPICENTER: The point on the earth's surface directly above the focus of an earthquake.

EPOCH: A relative time unit and a subdivision of a period; in the Cenozoic era, the two periods are divided into seven epochs.

EQUILIBRIUM: A stable state of a system; no further change will occur unless pressure, temperature, or composition is changed.

EQUILIBRIUM LINE: The line or zone that divides a glacier into the upper zone of accumulation and the lower zone of wastage.

ERA: A large division of geologic time, composed of more than one geologic period.

EROSION: The removal of weathered rock and mineral fragments and grains from an area by the action of wind, ice, gravity, or running water.

ERUPTION: Volcanic activity of such force as to propel significant amounts of magmatic products over the rim of the crater.

ESKER: A sinuous ridge of stratified drift formed in a tunnel under the ice.

ESTUARY: A thin zone along a coastline where freshwater system(s) and river(s) meet and mix with an ocean.

EUKARYOTIC CELL: The cell type present in all animals, plants, fungi, and protists; it has a distinct nucleus and mitochondria, chloroplasts, and other subcellular structures that are absent in prokaryotic cells.

EUSTATIC SEA-LEVEL CHANGE: A change in sea level worldwide, observed on all coast-lines on all continents.

EVAPORATION: The process by which water is changed from a liquid or solid into a vapor.

EVAPORITE: A rock largely composed of minerals that have precipitated upon evaporation of seawater or lake water.

EVAPOTRANSPIRATION: The movement of water from the soil to the atmosphere in response to heat, combining transpiration in plants and evaporation.

EVOLUTION: Change in species through time, with accompanying change in the gene pool.

EVOLUTIONARY TRENDS: Statistical directions of evolutionary changes; major trends have been toward increased body size, brain size, and complexity.

EXCHANGE REACTION: The exchange of isotopes of the same element between a rock and a liquid.

EXFOLIATION: The splitting off of curving sheets from the outside of a body of rock; also called sheeting.

EXOSKELETON: A bony armor covering the outside of the animal.

EXOSPHERE: The outermost layer of the earth's atmosphere.

EXTENSION: The expansion or stretching apart of rocks.

EXTINCTION: The disappearance of a species or large group of animals or plants.

EXTRUSION: The emission of magma or lava and the rock so formed onto the earth's surface.

EXTRUSIVE ROCK: Igneous rock that has been erupted onto the surface of the earth.

EYE: The calm central region of a hurricane, composed of a tunnel with strong sides.

FABRIC: A penetrative geometrical property of a rock, such as compositional layering or preferred orientation of platy grains.

FAMILY: A grouping of types of organisms, ranking above the level of a genus.

FATHOM: A unit of length equal to 6 feet (or 1.8 meters), used principally in the measurement of marine depths.

FAULT: A fracture or system of fractures across which relative movement of rock bodies has occurred.

FAULT DRAG: The bending of rocks adjacent to a fault.

FAULT SLIP: The direction and amount of relative movement between the two blocks of rock separated by a fault.

FAUNAL SUCCESSION: The sequence of life forms, as represented by the fossils within a stratigraphic sequence.

FERROMAGNETIC: Relating to substances with high magnetic permeability, definite saturation point, and measurable residual magnetism.

FETCH: The area or length of the sea surface over which waves are generated by a wind having a constant direction and speed.

FINENESS: A measure of the purity of gold or silver, expressed as the weight proportion of these metals in an alloy; gold fineness considers merely the relative proportions of gold and silver present, whereas silver fineness considers the proportion of silver to all other metals present.

FIREBALL: A very large and bright meteor that often explodes with fragments falling to the ground as meteorites; sometimes called a bolide.

FISSION FRAGMENT: One of the lighter nuclei resulting from the fission of a heavier element.

FISSION TRACK: The damage along the path of a fission fragment traveling through an insulating solid material.

FIXED CARBON: The solid, burnable material remaining after water, ash, and volatiles have been removed from coal.

FLOODPLAIN: The relatively flat valley floor on either side of a river which may be partly or wholly occupied by water during a flood.

FLOW RATE: The amount of water that passes a reference point in a specific amount of time, measured in liters per second.

FLUID: A material capable of flowing and hence taking on the shape of its container; gases and liquids are both examples of fluids.

FLUID INCLUSION: A bubble, within a mineral, which is filled partly with liquid and partly with gas.

FLUORESCENCE: Light emitted as the result of the decay of an atom from an excited state back to its ground state.

FLUVIAL: Of or related to streams and their actions.

FOCUS: The region within the earth from which earthquake waves emanate; also called the hypocenter.

FOLDING: The process of bending initially horizontal layers of rock so that they dip; folds are the features produced by folding and can be as short as millimeters and as long as kilometers.

FOLIATION: A planar feature in metamorphic rocks.

FOOD CHAIN: A hierarchical arrangement of the organisms of an ecological community according to the order of predation, in which each uses the next, usually lower, member on the scale as a food source.

FOOTWALL: The rock body located below a nonvertical fault.

FOSSIL: The remains of a once-living organism that have been naturally preserved in the earth's crust.

FOSSIL FUEL: A general term used to refer to petroleum, natural gas, and coal.

FRACTIONAL CRYSTALLIZATION: The process by which minerals form in a magma and either float or settle from it, depending on whether they are lighter or heavier than the magma.

FRACTIONATION: A physical or chemical process by which a particular element or isotope is concentrated in a solid or liquid.

FRACTURE ZONES: Large, linear zones of the sea floor, characterized by steep cliffs, irregular topography, and faults; such zones commonly cross and displace oceanic ridges by faulting.

FREE-AIR GRAVITY: A residual value for the gravity at a point, corrected for latitude and elevation effects; this value allows the scientist to determine differences in the densities of subsurface rocks.

FRESH WATER: Water with less than 0.2 percent dissolved salts, such as is found in most streams, rivers, and lakes.

FUMAROLE: A vent that emits only gases.

GABBRO: A silica-poor intrusive igneous rock consisting mostly of calcium-rich feldspar and iron and magnesium silicates; its volcanic equivalent is basalt.

GALILEAN MOONS: The four largest moons of Jupiter, discovered by Galileo in the seventeenth century.

GAMMA RAYS: Short-wavelength electromagnetic radiation similar to X rays; they are emitted by many radioactive elements.

GENUS: A group of closely related species; for example, *Homo* is the genus of humans, and it includes the species *Homo sapiens* (modern humans) and *Homo erectus* (Peking Man, Java Man).

GEOARCHAEOLOGY: The technique of using ancient human habitation sites to determine the ages of landforms.

GEOCHEMICAL CYCLE: The path of earth materials as erosional products of rock are deposited and reformed again into rock; also called the rock cycle.

GEOCHRONOLOGY: The study of the absolute ages of geologic samples and events.

GEOLOGIC RECORD: The history of the earth and its life forms as recorded in successive layers of sediment and the fossil specimens they contain.

GEOMORPHOLOGY: The study of the origins of landforms and the processes of landform development.

GEOPHYSICS: The quantitative evaluation of rocks by electrical, gravitational, magnetic, radioactive, and elastic wave transmission and heat-flow techniques.

GEOSTATIONARY ORBIT: An orbit in which a satellite appears to hover over one spot on the planet's equator; this procedure requires that the orbit be high enough that its period matches that of the planet's rotation and has no inclination relative to the equator; for earth, that altitude is 35,900 kilometers.

GEOSTROPHIC: An upper-level wind that flows in response to a balance of pressure gradient and Coriolis acceleration.

GEOSYNCLINE: A major depression in the surface of the earth where sediments accumulate; geosynclines are long and narrow and lie parallel to the edges of continents.

GEOTHERMAL GRADIENT: The rate at which temperature increases with depth in the earth.

GEOTHERMOMETERS: Minerals whose components can be used to determine temperatures of mineral formation.

GEYSER: A type of hot spring that periodically erupts.

GLACIATION: A major formation of land ice; also, the period in which it occurs.

GLACIER: An accumulation of ice that flows viscously as a result of its own weight; a glacier forms when snowfall accumulates and recrystallizes into a granular snow (firn, or névé), which becomes compacted and converted into solid, interlocking glacial ice.

GLASS: A solid without a periodic ordered arrangement of atoms; it frequently forms when molten material is rapidly cooled.

GNEISS: A coarse-grained metamorphic rock that shows compositional banding and parallel alignment of minerals.

GONDWANALAND: An ancient, large continent in the Southern Hemisphere that included Africa, South America, India, Australia, and Antarctica.

GRABEN: A roughly symmetrical crustal depression formed by the lowering of a crustal block between two normal faults that slope toward each other.

GRADUAL EVOLUTION: The theory that evolution occurs throughout much of a species' existence, mostly at slow rates.

GRANITE: A light-colored igneous rock made up mainly of three minerals—two feldspars and quartz—with variable amounts of darker minerals.

GRAVIMETER: A device that measures the attraction of gravity.

GRAVITY ANOMALIES: Differences between observed gravity readings and expected values after known irregularities have been taken into account.

GRAYWACKE: A sandstone in which more than 10 percent of the grains are mica or micaceous rock fragments; also called lithic arenite.

GREENHOUSE EFFECT: The condition that develops in a planetary atmosphere whereby radiative gases, such as carbon dioxide, hold heat that enters the atmosphere by solar radiation; the warming effect is similar to that inside a glass enclosure such as a greenhouse.

GROUNDMASS: The fine-grained material between phenocrysts of a porphyritic igneous rock.

GROUNDWATER: Water that is located beneath the surface of the earth in interconnected pores.

GUYOT: An extinct oceanic volcano, presently submerged far below sea level, with a top that has been leveled flat by wave erosion; some of these tops are capped by coral reefs at or near their summits.

HADLEY CELL: A circulatory pattern of winds driven by heat energy, proposed by George Hadley in 1735 to explain the trade winds.

HALF-LIFE: The time for half the atoms in a radioactive sample to decay, having a different value for each radioactive material.

HARDNESS: Resistance to abrasion or surface deformation.

HARMONIC TREMOR: A movement or shaking of the ground accompanying volcanic eruptions.

HEAD WALL: The block of rock that lies directly above the plane of a fault; also known as a hanging wall.

HIGHLANDS: Densely cratered regions on the lunar surface, which, when seen with the naked eye, take on a pale yellow color; highlands are primarily anorthositic breccia.

HIGH-LEVEL WASTES: Wastes containing large amounts of dangerous radioactivity.

HORIZON: A layer of soil material approximately parallel to the surface of the land that differs from adjacent related layers in physical, chemical, and biological properties.

HORNFELS: The hard, splintery rocks formed by contact metamorphism of finegrained sediments and certain lavas.

HORSE LATITUDES: The belts of latitude located approximately 30 degrees north and 30 degrees south of the equator, where the winds are very light and the weather is hot and dry.

HOT SPOT OR MANTLE PLUME: A zone of hot, upwelling rock that is rooted in the earth's upper mantle; as plates of the earth's crust and lithosphere glide over a mantle plume, a trail of hot spot volcanoes is formed and the earth's surface bulges upward in a several-hundred-kilometer-wide by 1-kilometer-high dome.

HUMIC ACID: Organic matter extracted by alkalis from peat, coal, or decayed plant debris; it is black and acidic but unaffected by other acids or organic solvents.

HUMMOCKY: Refers to a topography characterized by slope composed of many irregular mounds (hummocks), which are produced during sliding or flowage movements of earth and rock.

HURRICANE: A severe tropical storm with winds exceeding 119 kilometers per hour that originates in tropical regions.

HYDROCARBONS: Chemicals composed chiefly of the elements hydrogen and carbon; the term is usually applied to crude oil, natural gas, and their by-products.

HYDROLOGIC CYCLE: The circulation of water as a liquid and vapor from the oceans to the atmosphere and back to the oceans.

HYDROLOGY: Broadly, the science of water; the term is often used in the more restricted sense of flow in channels.

HYDROSPHERE: The waters of the earth, including rivers, lakes, and oceans.

HYDROTHERMAL: Related to hot water, particularly involving the production or dissolution of minerals.

HYPOCENTER: The central underground location of an earth tremor; also called the focus.

ICE AGES: Periods in the earth's past when large areas of the present continents were glaciated.

IGNEOUS ROCK: Any rock that forms by the solidification of molten material.

IGNIMBRITE: An igneous rock deposited from a hot, mobile, ground-hugging cloud of ash and pumice.

IMMISCIBLE FLUIDS: Two fluids incapable of mixing to form a single homogeneous substance; oil and water are common examples.

IMPACT BASIN: A large cavity produced by a meteorite impact.

IMPACT BRECCIA: Angular, fragmental rock produced by meteorite impact.

IMPACT CRATER: A depression, usually circular, in a planetary surface, caused by the high-speed impact of rocky debris or comet nuclei.

INCLINATION: The angle in the vertical plane between the horizontal and the direction of magnetization of a rock.

INCLUSION: A foreign substance enclosed within a mineral; often, very small mineral grains and cavities filled with liquid or gas.

INDEX FOSSIL: A fossil that can be used to identify and determine the age of the stratum in which it is found.

INDEX MINERAL: An individual mineral that has formed under a limited or very distinct range of temperature and pressure conditions.

INDEX OF REFRACTION: The ratio of the speed of light in a vacuum to the speed of light in a particular transparent medium.

INFRARED RADIATION: Electromagnetic radiation extending from just below the sensitivity of the human eye to millimeter-wavelength radio waves; the band is divided into optical infrared and thermal infrared and is also known as heat radiation.

INNER EDGE POINT: The landward edge of a wave-cut terrace at the base of a sea cliff; the elevation of this point is the position of highest sea level during formation of the terrace.

INSOLATION: Incoming solar radiation; differences in global insolation at various places on the earth's surface create weather and climate patterns.

INTENSITY: An arbitrary measure of an earthquake's effect on people and buildings, based on the modified Mercalli scale.

INTERFERENCE: The combining of waves or vibrations from different sources so that they are either in step and reinforce each other or out of step and oppose each other.

INTERMEDIATE ROCK: An igneous rock that is transitional between a basic and a silicic rock, having a silica content between 54 and 64 percent.

INTERNAL DRAINAGE: The condition in which a river system has no outlet; instead, the river system drains into a saline lake or playa.

INTERSTELLAR MEDIUM: The material that lies between the stars; it consists mainly of grains of dust and gas, mostly hydrogen, along with heavier elements released by supernova explosions.

INTRUSIVE ROCKS: Igneous rocks formed from magmas that have cooled and crystallized underground.

ION: An atom that has lost or gained one or more electrons.

IONIC BOND: The strong electrical forces holding together positively and negatively charged ions (for example, sodium and chloride ions in common salt).

IONOSPHERE: The ionized layer of gases in the earth's atmosphere, occurring between the thermosphere and the exosphere; it starts at about 50-100 kilometers above the surface of the planet.

IRON FORMATION: A layered sedimentary deposit that consists mostly of chemically precipitated sedimentary rock with more than 15 percent iron.

ISLAND ARC: A curved chain of volcanic islands, generally located a few hundred kilometers from a trench where active subduction of one oceanic plate under another is occurring.

ISOBAR: A line of equal pressure.

Isobath: The contour lines of continental slopes.

Isochron: On a radioactive isotope (parent) versus radiogenic isotope (daughter) diagram, a line connecting points representing samples of equal age.

Isograd: A line on a geologic map that marks the first appearance of a single mineral or mineral assemblage in metamorphic rocks.

Isostasy: The concept of balance by which continental and oceanic crusts are "floating" on the denser substrata of the mantle.

Isostructural: Having the same structure but a different chemistry.

Isotope: A species of an element having the same number of protons but a different number of neutrons and, therefore, a different atomic weight.

Isotropic: Having properties that are the same in all directions; the opposite of anisotropic, or having properties that vary with direction.

I-type granitoid: Granitic rock formed from magma generated by partial melting of igneous rocks in the upper mantle or lowermost crust.

Jet stream: A narrow current of high-speed winds in the upper atmosphere.

Joint: A fracture in a rock across which there has been no substantial slip parallel to the fracture.

Juvenile water: Water that originated in the upper mantle.

Karat: A unit of measure (abbreviated "K") of the purity of gold; it is equal to 1/24 part gold in an alloy.

Karst: A type of topography developed above beds of limestone as a result of partial dissolution of the limestone by through-flowing water.

Kerogen: Fossilized organic material in sedimentary rocks that is insoluble and that generates oil and gas when heated; as a form of organic carbon, kerogen is one thousand times more abundant than coal and petroleum, combined, in reservoirs.

Kettle: A depression created by the melting of a chunk of ice.

Kimberlite: An unusual, fine-grained variety of peridotite that is believed to be congealed magma from the deep mantle; it contains not only olivine—which is commonly altered to serpentine—and pyroxenes but also calcite, mica, and trace amounts of diamond.

Kingdom: One of the five large subdivisions of life, differentiated on the basis of gross body plan (single-celled versus multicellular) and method of obtaining nutrients (produces own food or obtains it from other organisms).

La Niña: The part of the Southern Oscillation that brings cold water to the South American coasts, which makes easterly trade winds stronger, the waters of the Pacific off South America colder, and ocean temperatures in the western equatorial Pacific warmer than normal.

Lahar: A mudflow composed chiefly of volcanic debris on the flanks of a volcano.

LANDSLIDE: A general term that applies to any downslope movement of materials; landslides include avalanches, earthflows, mudflows, rockfalls, and slumps.

LAPILLI: Pyroclastic fragments between 2 and 64 millimeters in diameter.

LAPSE RATE: The rate of decrease in temperature with increasing height.

LATERITE: A deep red soil, rich in iron and aluminum oxides and formed by intense chemical weathering in a humid tropical climate.

LAURASIA: An ancient, large continent that once existed in the Northern Hemisphere.

LAVA: The fluid rock issued from a volcano or fissure and the solidified rock it forms when it cools.

LAVA TUBE: A cavern structure formed by the draining out of liquid lava in a pahoehoe flow.

LEACH: To dissolve from the soil or from a mineral grain.

LEACHATE: Water that has seeped down through the landfill refuse and has become polluted.

LEVEE: A dikelike structure, usually made of compacted earth and reinforced with other materials, that is designed to contain the stream flow in its natural channel.

LIGNIN: A family of compounds in plant cell walls, composed of an aromatic nucleus, a side chain with three carbon atoms, and hydroxyl and methoxyl groups.

LIMESTONE: A common sedimentary rock containing the mineral calcite; the calcite originated from fossil shells of marine plants and animals or by precipitation directly from seawater.

LINEATION: Any linear or curvilinear rock fabric.

LIQUEFACTION: The loss in cohesiveness of water-saturated soil as a result of ground shaking caused by an earthquake.

LITHIC FRAGMENT: A grain composed of a particle of another rock; in other words, a rock fragment.

LITHIFICATION: The hardening of sediment into rock through compaction, cementation, recrystallization, or other processes.

LITHOLOGY: A term used by many geologists to refer generally to the composition and texture of a rock.

LITHOSPHERE: The outer shell of the earth, including both the crust and the upper mantle, which behaves rigidly over time periods of thousands to millions of years.

LITHOSPHERIC PLATES: Segments of the lithosphere that are similar in size to continents; these plates form a mosaic that covers the earth's surface.

LOCAL SEA-LEVEL CHANGE: A change in sea level in one particular area of the world, usually by land rising or sinking in that specific region.

LONGITUDINAL BAR: A midchannel accumulation of sand and gravel with its long end oriented roughly parallel to the river flow.

LONGITUDINAL DUNE: An elongate sand dune parallel to the prevailing wind.

LONG-PERIOD COMET: A comet whose period of orbital revolution is greater than two hundred years.

LONGSHORE CURRENT: A slow-moving current between a beach and the ocean breakers, moving parallel to the beach; the current direction is determined by the wave refraction pattern.

LOW-LEVEL WASTES: Wastes that are much less radioactive and, thus, less likely to cause harm than other wastes.

LUMINESCENCE: The emission of light by a mineral.

LUSTER: The reflectivity of the mineral surface; there are two major categories of luster: metallic and nonmetallic.

MACRONUTRIENT: A substance that is needed by plants in large quantities; nitrogen, phosphorus, and potassium are macronutrients.

MAFIC: A dark-colored mineral, such as olivine, pyroxene, amphibole, or biotite, or a rock composed mainly of such minerals.

MAGMA: Molten silicate liquid plus any crystals, rock inclusions, or gases trapped therein.

MAGNETIC ANOMALIES: Patterns of reversed polarity in the ferromagnetic minerals present in the earth's crust.

MAGNETIC DOMAIN: Within a mineral, a region with a single direction of magnetization.

MAGNETIC POLARITY TIME SCALE: The geologic history of the changes in the earth's magnetic polarity.

MAGNETIC POLE: The location on the earth's surface where the earth's magnetic field is perpendicular to the surface.

MAGNETIC REMANENCE: The ability of the magnetic minerals in a rock to "lock in" the magnetic field of the earth prevailing at the time of their formation.

MAGNETIC REVERSAL: A change in the earth's magnetic field from the north-oriented magnetic pole to the south-oriented magnetic pole.

MAGNETOPAUSE: The outer boundary of the earth's magnetic field.

MAGNITUDE: A measure of the amount of energy released by an earthquake, based on the relation between the logarithm of ground motion at the detecting instrument and its distance from the epicenter.

MANGANESE NODULES: Rounded, concentrically laminated masses of iron and manganese oxide found on the deep sea floor.

MANTLE: The portion of the earth's interior extending from about 60 kilometers in depth to 2,900 kilometers; it is composed of relatively high-density minerals that consist primarily of silicates.

MARIA: Dark, mostly circularly shaped basaltic regions on the moon that fill the largest impact craters, called basins, or the low-lying regions around these basins.

MARINE: Referring to a seawater, ocean environment.

MASS SPECTROMETER: An apparatus that is used to separate the isotopes of an element and to measure their relative abundances.

MASS WASTING: The downslope movement of earth materials under the direct influence of gravity.

MATRIX: The fine-grained material of a meteorite that surrounds both chondrules and inclusions; it consists of hydrous silicate minerals, troilite, magnetite, and other lower-temperature phases.

MAUNDER MINIMUM: The period from 1645 to 1715, when sunspot activity was almost nonexistent.

MEAN SEA LEVEL: The average height of the sea surface over a multiyear time span, taking into account storms, tides, and seasons.

MEANDER: A large, sinuous curve or bend in a stream of equilibrium on a floodplain.

MESOSPHERE: The layer of the earth's atmosphere occurring above the stratosphere and below the thermosphere, about 40-85 kilometers above sea level.

MESOZOIC ERA: The middle of the three eras that constitute the Phanerozoic eon (the past 570 million years), which encompasses three geologic periods—the Triassic, the Jurassic, and the Cretaceous—and represents earth history between about 250 and 66 million years ago.

METAMORPHIC FACIES: An assemblage of minerals characteristic of a given range of pressure and temperature; the members of the assemblage depend on the composition of the protolith.

METAMORPHIC GRADE: The degree of metamorphic intensity as indicated by characteristic minerals in a rock or zone.

METAMORPHIC ROCK: A rock formed when another rock undergoes changes in mineralogy, chemistry, or structure resulting from changes in temperature, pressure, or chemical environment at depth within a planet.

METAMORPHIC ZONES: Areas of rock affected by the same limited range of temperature and pressure conditions, commonly identified by the presence of a key individual mineral or group of minerals.

METAMORPHISM: The alteration of the mineralogy and texture of rocks because of changes in pressure and temperature conditions or chemically active fluids.

METASOMATISM: Chemical changes in rock composition that accompany metamorphism; significant metasomatism almost invariably involves the presence of an aqueous fluid.

METASTABLE: Referring to an apparently stable state that can persist indefinitely outside the conditions of equilibrium; most rocks and minerals are metastable at the earth's surface.

METEOR: A bright streak of light in the sky, sometimes called a shooting star, produced by a meteoroid entering the earth's atmosphere at high speed and heating to incandescence.

METEOR SHOWER: A meteor display caused by comet dust particles burning up in the upper atmosphere during the annual passage of earth through a cometary wake or debris field.

METEORIC WATER: Surface water that infiltrates porous and fractured crustal rocks; the same as groundwater.

METEORITE: The remnant of an interplanetary body that survives a fall through the earth's atmosphere and reaches the ground.

METEOROLOGY: The study of weather.

MICROFOSSIL: The characteristic imprint left by a microscopic organism in a geological formation such as a stromatolite.

MID-OCEAN RIDGE: A roughly linear, submarine mountain range where new sea-floor lithosphere is created by sea-floor spreading.

MIGMATITE: A rock exhibiting both igneous and metamorphic characteristics, which forms when light-colored silicate minerals melt and then crystallize, while the dark silicate minerals remain solid.

MILLIBAR: One thousand dynes per square centimeter; a dyne is a small unit of force that will cause a body of 1 gram to accelerate at 1 centimeter per second per second.

MILLIGAL: The basic unit of the acceleration of gravity, used by geophysicists to measure gravity anomalies equal to 0.001 centimeter per second per second.

MINERAL: A naturally occurring, inorganic crystalline substance with a restricted chemical composition.

MINING WASTES: Soil and rock removed in the process of extracting minerals, primarily overburden and unusable earth.

MITOCHONDRION: The eukaryotic organelle in which energy is generated by aerobic respiration.

MIXED TIDES: Tides having the characteristics of diurnal and semidiurnal tidal oscillations; these tides are found on the west coast of the United States.

MOBILE BELT: A linear belt of igneous and deformed metamorphic rocks, produced by plate collision at a continental margin; relatively young mobile belts form major mountain ranges.

MODE: The type and amount of minerals actually observed in a rock.

MODEL: A simulation of a phenomenon that is difficult to observe or specify by direct means; models abstract from phenomena under study those qualities that the investigator perceives to be essential for understanding.

MODERATOR: A material used in a nuclear reactor to diminish the speed of neutrons in order to increase their probability of causing fission.

MOHOROVIČIĆ DISCONTINUITY (MOHO): The boundary between the crust and the upper mantle, which was first defined by a rapid change in seismic velocities; it separates low-density crust from the denser mantle.

MOHS HARDNESS SCALE: A series of ten minerals arranged and numbered in order of increasing hardness, with talc (1) as the softest mineral known and diamond (10) as the hardest.

MOISTURE CONTENT: The weight of water in the soil divided by the dry weight of the soil, expressed as a percentage.

MOLD: A fossil that displays the form of the original organism in negative relief.

MOLE: The amount of pure substance that contains as many elementary units as there are atoms in 12 grams of the isotope carbon 12.

MOLECULE: The smallest entity of an element or compound that retains chemical identity with the substance in mass.

MONERANS: Generally, organisms having a single, prokaryotic cell; they often grow in colonies.

MONSOON: A seasonal pattern of wind at boundaries between warm ocean bodies and landmasses.

MORAINE: An arcuate ridge consisting of till and/or stratified drift, deposited at the margin of a glacier.

MORPHOLOGIC EVOLUTION: Changes in the body or in the skeleton shape of organisms through time.

MORPHOLOGY: The appearance (shape and form) of an organism or a mineral grain.

MUDFLOW: Both the process and the landform characterized by very fluid movement of fine-grained material with a high (sometimes more than 50 percent) water content.

MUTATION: A spontaneous change in a gene; the ultimate source of variation on which natural selection acts.

NAPPE: A complex, large-scale recumbent fold where some rock beds have been overturned.

NAPHTHENE HYDROCARBONS: Hydrocarbon molecules that have a ring-shaped structure in which any number of carbon atoms are bonded to one another with single bonds.

NATURAL GAS: A flammable vapor found in sedimentary rocks, commonly associated with crude oil; also known simply as gas or methane.

NATURAL SELECTION: The main process of biological evolution; the production of the largest number of offspring by individuals with traits that are best adapted to their environments.

NEAP TIDE: A tide with the minimum range, that is, when the level of the high tide is at its lowest.

NEBULA: A cloud of interstellar gas and dust.

NICHE: In an ecological environment, a position particularly suited for its inhabitant.

NONRENEWABLE RESOURCE: An earth resource that is fixed in quantity and will not be renewed within a human lifetime.

NORM: An idealized, calculated set of minerals meant to represent the chemical analysis of a rock.

NORMAL FAULT: A steep fault that forms during extension, or stretching, of the earth's surface; the rock above the fault shifts below the underlying rock.

NORMAL POLARITY: Orientation of the earth's magnetic field so that a compass points toward the Northern Hemisphere.

NORMAL STRESS: That component of the stress on a plane that acts in a direction perpendicular to the plane.

NOTCH OR NIP: An erosional feature found at the base of a sea cliff as a result of undercutting by wave erosion, bioabrasion from marine organisms, and dissolution of rock by groundwater seepage.

NUCLEOSYNTHESIS: The building up of the chemical elements from preexisting hydrogen by nuclear reaction processes.

NUÉE ARDENTE: A hot cloud of rock fragments, ash, and gases that suddenly and explosively erupt from some volcanoes and flow rapidly down their slopes.

NUTRIENT: A substance required for optimal functioning of a plant or animal; foods, vitamins, and minerals essential for life processes are nutrients.

OCEANIC CRUST: The upper 5-10 kilometers of the earth below ocean basins; it is heavier and younger than continental crust.

OIL: A common term for petroleum, a diverse mixture of mostly liquid hydrocarbons (combinations of hydrogen and carbon) obtained from oil wells.

OIL SHALE: A sedimentary rock containing sufficient amounts of hydrocarbons that can be extracted by slow distillation to yield oil.

OLDOWAN: A cultural tradition that is distinguished by stone tools characterized by rounded pebbles crudely chipped to a sharp edge on one side.

100-YEAR-FLOOD: A hypothetical flood whose severity is such that it would occur on average only once in a period of one hundred years, which equates to a 1 percent probability each year.

OORT CLOUD: The source of comets; a spherical shell reservoir of comet nuclei containing a trillion comets nearly one light-year from the sun.

OPHIOLITE COMPLEX: An assemblage of metamorphosed basaltic and ultramafic igneous rocks intimately associated with unmetamorphosed marine sediment; the rocks originate and are metamorphosed at marine ridges and are subsequently emplaced in mobile belts by plate-collision tectonics.

OPPOSITION: A situation in which one planet or other astronomical body approaches another in their respective orbits.

ORDER: A group of closely related genera; in mammals, orders include the rodents, bats, and whales.

ORE: Any concentration of economically valuable minerals.

ORGANELLES: Subcellular membrane-bound units that perform specific functions within the eukaryotic cell.

ORGANIC MOLECULES: Molecules of carbon compounds produced in plants or animals, plus similar artificial compounds.

OROGENIC BELT: A belt of crust that has been severely compressed, deformed, and heated, probably by convergence of crustal plates; a mountain belt.

OROGENY: A mountain-building episode, or event, that extends over a period usually measured in tens of millions of years; also termed a revolution.

ORPHAN LANDS: Unreclaimed strip mines created prior to the passage of state or federal reclamation laws.

OSMOSIS: The passage of a liquid from a weak solution to a strong solution across a semipermeable membrane.

OUTER CORE: A zone in the body of the earth, located at depths of approximately 2,885-5,144 kilometers, that is in a liquid state and consists of iron sulfides and iron oxides.

OUTGASSING: A process whereby planets emit gases into their atmospheres from their interiors.

OVERBURDEN: The material overlying the ore in a surface mine.

OXBOW LAKE: A lake formed from an abandoned meander bend when a river cuts through the meander at its narrowest point during a flood.

OXIDATION: A very common chemical reaction in which elements are combined with oxygen—for example, the burning of petroleum, wood, and coal; the rusting of metallic iron; and the metabolic respiration of organisms.

OZONE: A gas containing three atoms of oxygen; it is highly concentrated in a zone of the stratosphere.

P WAVE: A type of seismic wave generated at the focus of an earthquake, traveling 6-8 kilometers per second, with a push-pull vibratory motion parallel to the direction of propagation; *P* stands for "primary," as P waves are the fastest and first to arrive at a seismic station.

PAHOEHOE: A Hawaiian term (pronounced "pa-hoy-hoy") that is used in reference to lava flows with smooth, ropy surfaces.

PALEOBIOGEOGRAPHY: The study of the geographic distribution of past life forms.

PALEOBIOLOGY: The study of the most ancient life forms, typically through the examination of microscopic fossils.

PALEOCEANOGRAPHY: The study of the history of the oceans of the earth, ancient sediment deposition patterns, and ocean current positions compared to ancient climates.

PALEOCLIMATE: A climate that existed in the geologic past, usually one that existed before adequate climatic records were kept, that is, most often a prehistoric climate.

PALEOMAGNETISM: The study of magnetism preserved in rocks, which provides evidence of the history of earth's magnetic field and the movements of continents.

PALEONTOLOGY: The science of ancient life forms and their evolution as studied through the analysis of fossils.

PALEOZOIC ERA: The era that began about 570 million years ago and ended about 245 million years ago; it includes six periods: the Cambrian, the Ordovician, the Silurian, the Devonian, the Carboniferous, and the Permian.

PANGAEA: The supercontinent containing all continental crust that existed at the beginning of the Mesozoic era, about 250 million years ago.

PARAFFIN HYDROCARBONS: Hydrocarbon compounds composed of carbon atoms connected with single bonds into straight chains; also known as n-alkanes.

PARTHENOGENIC: Organisms in which unfertilized females produce viable, fertile offspring without copulation with males of the species.

PEDESTAL CRATER: A crater that has assumed the shape of a pedestal as a result of unique shaping processes caused by wind.

PEGMATITE: A very coarse-grained igneous rock that forms late in the crystallization of a magma; its overall composition is usually granitic, but it is also enriched in many rare elements and gem minerals.

PELAGIC: Meaning "of the open sea," a term that refers to sediments that are finegrained and deposited very slowly at great distances from continents.

PELITIC ROCK: Rock whose protolith contained abundant clay or similar minerals.

PERIDOTITE: A dense, dark-green rock that is composed mainly of the silicate mineral olivine and other magnesium- and iron-rich silicates such as pyroxenes; the earth's mantle and the ultramafic nodules derived from it are composed of peridotite.

PERIHELION: The point in a planet's orbit when it is nearest the sun.

PERIOD: A unit of geologic time comprising part of an era and subdivided, in decreasing order, into epochs, ages, and chrons.

PERMAFROST: Permanently frozen soil that is laced with water ice.

PERMEABILITY: The ability of rock, soil, or sediment to transmit a fluid.

PERMINERALIZATION: The filling of pores and cells with minerals without changing the material surrounding the pores or cells.

PERTURB: To change the path of an orbiting body by a gravitational force.

PETROGRAPHY: The description and systematic classification of rocks.

PETROLEUM: A natural mixture of hydrocarbon compounds existing in three states: solid (asphalt), liquid (crude oil), and gas (natural gas).

pH: A logarithmic measure of the hydrogen ion concentrations of solutions: a pH of 7 indicates neutrality, and greater than 7 is alkaline solution, while less than 7 is acid.

PHANERITIC: A textural term that applies to an igneous rock composed of crystals that are macroscopic in size, ranging from about 2 millimeters to more than 5 millimeters in diameter.

PHANEROZOIC EON: The period of geologic time with an abundant fossil record, extending from about 570 million years ago to the present.

PHASE: A chemical entity that is generally homogeneous and distinct from other entities in the system under investigation.

PHENOCRYST: A large, conspicuous crystal in a porphyritic rock.

PHOTOCHEMICAL REACTION: A chemical reaction occurring in polluted air that synthesizes new gases through the action of sunlight upon pollutant gases.

PHOTODISSOCIATION: The splitting of molecules by light, generally in the ultraviolet spectrum.

PHOTOSYNTHESIS: In plants, the process of fixing atmospheric carbon in organic compounds, with production of free oxygen as a by-product.

PHREATIC ERUPTION: An eruption in which water plays a major role; also called hydrovolcanic.

PHYLOGENY: The study of the evolutionary relationships among organisms.

PHYLUM: A major grouping of organisms, distinguished on the basis of basic body plan, grade of anatomical complexity, and pattern of growth or development.

PILLAR: Ore, coal, rock, or waste left in place underground to support the wall or roof of a mine opening.

PILLOW LAVA: A type of bulbous, glassy-skinned lava that forms only when basaltic lava flows erupt under water.

PLACER DEPOSIT: A mass of sand, gravel, or soil resulting from the weathering of mineralized rocks that contains grains of gold, tin, platinum, or other valuable minerals derived from the original rock.

PLANETARY WINDS: The large, relatively constant prevailing wind systems that result from the rotation of a planet.

PLANETESIMALS: The first small bodies to condense from the solar nebula, from which the planets are thought to have formed.

PLANKTON: Microscopic marine plants and animals that live in the surface waters of the oceans; these floating organisms precipitate the particles that sink to form biogenic marine sediments.

PLASMA TAIL: A cometary structure composed of molecules originating within the nucleus that are ionized by sunlight and pushed away from the comet by the magnetic field of the solar wind.

PLASTIC DEFORMATION: Nonrecoverable deformation, which does not disappear when the deforming stress is removed.

PLATE MARGIN: A region where the earth's crustal plates meet, as a converging (subduction) zone, a diverging zone (mid-ocean ridge), a transform fault, or a collisional interaction.

PLATE TECTONICS: The theory that the outer surface of the earth consists of large moving plates that interact to produce seismic, volcanic, and orogenic activity.

PLEISTOCENE EPOCH: The time from about 3 million years ago to about ten thousand years ago, during which large continental glaciers covered much of northern North America, Europe, and other parts of the world.

PLINIAN ERUPTION: A very violent, explosive type of volcanic eruption named either for Pliny the Elder, a Roman naturalist who died while observing the A.D. 79 eruption of Mount Vesuvius, or for Pliny the Younger, his nephew, who chronicled the eruption and whose detailed description possibly gave rise to the term.

PLUNGE: The inclination and direction of inclination of the fold axis, measured in degrees from the horizontal.

PLUTON: A generic term for an igneous body that solidifies well below the earth's surface; plutonic rocks are coarse-grained because they cool slowly.

POINT BAR: An accumulation of sand and gravel that develops on the inside of a meander bend.

POLAR JET: A stream of air blowing from a westerly direction at about 12 kilometers over the middle latitudes.

POLARIZATION: Filtration of light so that only rays vibrating in a specific plane are passed.

POLLUTION: A condition in which air, soil, or water contains substances that make it hazardous for human use.

POLYMORPHS: Minerals having the same chemical composition but different crystal structures.

PORE FLUIDS: Fluids, such as water (usually carrying dissolved minerals), gases, and hydrocarbons, found in the pore spaces in a rock.

POROSITY: The ratio of the volume of void space in a given geologic material to the total volume of that material.

PORPHYRITIC: A texture characteristic of an igneous rock in which large crystals are embedded in a groundmass of fine grain size.

PRECAMBRIAN: The interval of geologic time from the formation of the earth (about 4.6 billion years ago) to the beginning of the Cambrian period (about 570 million years ago).

PRECESSION: A wobbling of a planet's rotational axis, caused by gravitational forces on its nonspherical shape; the period of the earth's precession is about twenty-six thousand years.

PRECIPITATE: To condense from a solution.

PREFERRED ORIENTATION: A systematic bias or regularity in the orientation of mineral grains in a rock.

PRESSURE-MELTING TEMPERATURE: The temperature at which ice will melt under a specified pressure; under pressure, water can exist even at temperatures below freezing.

PRESSURE-TEMPERATURE REGIME: A sequence of metamorphic facies distinguished by the ratio of pressure to temperature, generally characteristic of a given geologic environment.

PRIMARY MINERALS: The minerals formed when magma crystallizes.

PRODELTA: A sedimentary layer composed of silt and clay deposited under water; the foundation upon which a delta is deposited.

PROGRADE METAMORPHISM: Recrystallization of solid rock masses, induced by rising temperature; it differs from metasomatism in that bulk rock composition is unchanged except for expelled fluids.

PROKARYOTIC CELL: The cell type found in the kingdom Monera, characterized by a number of criteria, including the absence of a cell nucleus, mitochondria, and chloroplasts.

PROSPECT: A limited geographic area identified as having all the characteristics of an oil or gas field but without a history of production.

PROTEROZOIC EON: The late Precambrian eon, from about 2.5 billion to 600 million years ago, before the proliferation of macroscopic life.

PROTOLITH: The original igneous or sedimentary rock later affected by metamorphism.

PUMICE: A vesicular glassy rock commonly having the composition of rhyolite; a common constituent of silica-rich explosive volcanic eruptions.

PUNCTUATED EQUILIBRIA: An alternative model of how evolution works, that is, by rapid speciation events that involve major changes in morphology.

PYCNOCLINE: A layer within a water body characterized by a rapid rate of change in density.

PYROCLASTIC FALL: The settling of debris under the influence of gravity from an explosively generated plume of material.

PYROCLASTIC FLOW: A highly heated mixture of volcanic gases and ash that travels down the flanks of a volcano; the relative concentration of particles is high.

PYROCLASTIC ROCKS: Rocks formed in the process of volcanic ejection and composed of fragments of ash, rock, and glass.

RADIATION: The transfer of heat energy through a transparent medium; the process by which the sun warms the earth.

RADIOACTIVE DECAY: A natural process by which an unstable, or radioactive, isotope transforms into a stable, or radiogenic, isotope.

RADIOACTIVE ISOTOPE: An isotope of an element that naturally decays into another isotope.

RADIOACTIVITY: The spontaneous emission from unstable atomic nuclei of alpha particles (helium nuclei), beta particles (electrons), and gamma rays (electromagnetic radiation).

RADIOGENIC ISOTOPE: An isotope resulting from radioactive decay.

RADIOMETRIC DATING: The use of radioactive elements that decay at a known rate to determine the ages of the rocks in which they occur.

RADIOSONDE: A balloon-borne instrument package for the simultaneous measurement and transmission of weather data.

RAMP: That portion of a thrust fault in which the fault cuts across a layer of relatively stiff rock at a higher angle than does the rest of the fault.

RAMPART CRATER: A crater shape found most often on Mars and produced by a subsurface mechanism that forms smooth rampartlike walls.

RAWINSONDE: A radiosonde tracked by radar in order to collect wind data in addition to temperature, pressure, and humidity.

REAL TIME: The term used for images that are transmitted immediately, rather than being stored for subsequent processing or viewing.

RECLAMATION: The array of human efforts—mainly slope reshaping, revegetation, and erosion control—meant to improve conditions produced by mining wastes.

RECRYSTALLIZATION: A solid-state chemical reaction that eliminates unstable minerals in a rock and forms new stable minerals; the major process contributing to rock metamorphism.

RECURRENCE INTERVAL: The average time interval, expressed in number of years, between occurrences of a flood of a given or greater magnitude than others in a measured series of floods.

REEF: A wave-resistant structure composed of organisms that precipitate calcium carbonate; also, a provincial term referring to a metalliferous mineral deposit, commonly of gold or platinum, which is usually in the form of a layer.

REFRACTORY: Having a high condensation temperature—a measure of resistance to evaporation in a vacuum; also, a mineral with a sufficiently high melting temperature that is unaffected by anatexis and remains in the solid residue.

REGIONAL GEOLOGY: A study of the geologic characteristics of a geographic area.

REGIONAL METAMORPHISM: Metamorphism characterized by strong compression along one direction, usually affecting rocks over an extensive region or belt.

REGOLITH: The layer of soil and rock fragments just above the solid planetary crust.

REGRESSION: The retreat of the sea from the land; it allows land erosion to occur on material formerly below the sea surface.

RELATIVE DATE: A date that places an artifact as older or younger than another object, without specifically giving an age for it.

RELATIVE HUMIDITY: The ratio, expressed as a percentage, of the actual amount of water vapor in a given volume of air to the amount that would be present if the air were saturated at the same temperature.

REMOTE SENSING: Any of a number of techniques, such as aerial photography, that collects information by gathering energy reflected or emitted from a distant source.

RESERVE: That part of the mineral resource base that can be extracted profitably with existing technology and under current economic conditions.

RESERVOIR: A body of porous and permeable rock; petroleum reservoirs contain pools of oil or gas.

RESOLUTION: The ability of an imaging system to discriminate the fine detail of the surface being imaged.

RESOURCE: A naturally occurring substance in such form that it can be currently or potentially extracted economically.

RESPONSE TIME: The time it takes a glacier to respond to changes in its mass balance, which it will generally do by advance or retreat of its terminus (tip).

RESURGENT DOME: A broad, oval area of uplift within a volcanic caldera that is marked by the upwarping and fracturing of caldera-filling deposits.

RETROGRADE METAMORPHISM: The reversal of prograde mineral reactions, caused by the reintroduction of water and/or carbon dioxide during the period of declining temperature following metamorphic culmination.

RETROGRADE ROTATION: The rotation of a planet in a direction opposite to that of its revolution.

REVERSE FAULT: A fault with a steep to moderate incline, in which the overlying block of rock has moved upward over the underlying block.

REVERSE POLARITY: Orientation of the earth's magnetic field so that a compass needle points to the Southern Hemisphere.

REVOLUTION: The yearly orbit of the earth around the sun.

RHYOLITE: A viscous, gas-rich type of magma that contains greater than about 70 percent silica (silicon and oxygen); rhyolite erupts nonexplosively as thick, slow-moving lava flows and explosively as widespread sheets of frothy pumice.

RICHTER SCALE: The scale, devised by C.F. Richter, used for measuring the magnitude of earthquakes.

RIFT VALLEY: A region of extensional deformation in which the central block has dropped down in relation to the two adjacent blocks.

RIFTING: The process whereby lithospheric plates break apart by tensional forces.

RIGHT-LATERAL STRIKE-SLIP: Sideways motion along a steep fault in which the block of the earth's crust across the fault from the observer appears to be displaced to the right; left-slip faults are displaced to the left.

ROCHE LIMIT: The inner orbital limit of a natural satellite, or moon, where the gravitational-tidal forces of the planet will shatter the moon.

ROCK: A naturally occurring, consolidated material of one or more minerals.

ROCKFALL: A relatively free-falling movement of rock material from a cliff or steep slope.

ROLL-TYPE DEPOSIT: A uranium deposit in sandstone that is characterized by oxidized (brightly colored) sandstone on one side of the deposit and chemically reduced (dull-colored) sandstone on the other.

ROTARY DRILLING: A fluid-circulating, rotating process that is the chief method of drilling oil and gas wells.

ROUGH: Gem-mineral material of suitable quality to be used for fashioning gemstones.

RUNOFF: The total amount of water flowing in a stream, including overland flow, return flow, interflow, and base flow.

S WAVE: The secondary seismic wave, traveling more slowly than the P wave and consisting of elastic vibrations transverse to the direction of travel; S waves cannot propagate in a liquid medium.

SABKHA: A supratidal environment of sedimentation formed under arid or semiarid conditions, characterized by evaporite salts, tidal flood, and windblown deposits.

SALINE LAKE: A lake with elevated levels of dissolved solids, primarily the result of evaporative concentration of salts; saline lakes lack an outlet to the sea.

SALINITY: A measure of the quantity of dissolved solids in ocean water.

SALT DOME: An underground structure in the shape of a circular plug, resulting from the upward movement of salt.

SALT WATER: Water with a salt content of 3.5 percent, such as is found in normal ocean water.

SALTATION: A mode of sediment movement in air wherein particles are lifted from the bed at a substantial angle to the horizontal and follow an arching trajectory downwind, little affected by the turbulence in the air.

SALTWATER INTRUSION: Aquifer contamination by salty waters that have migrated from deeper aquifers or from the sea.

SANDSTONE: Sedimentary rock composed of grains of sand cemented together.

SAPPING: A natural process of erosion at the bases of hill slopes or cliffs whereby support is removed by undercutting, thereby allowing overlying layers to collapse; spring sapping is the facilitation of this process by concentrated groundwater flow, generally at the heads of valleys.

SCABLAND: A region characterized by rocky, elevated tracts of land with little soil cover and by postglacial dry stream channels.

SCALE: The relationship between a distance on a map or diagram and the same distance on the earth.

SCARP: A steep cliff or slope created by rapid movement along a fault.

SCHIST: Metamorphic rock with subparallel orientation of micaceous minerals that dominate its composition.

SCHISTOSITY: A foliation defined by preferred orientation of elongate or platy mineral grains visible to the naked eye.

SEA-FLOOR SPREADING: The concept that new ocean floor is created at the ocean ridges and moves toward the volcanic island arcs, where it descends into the mantle.

SEAL: A rock unit or bed that is impermeable and inhibits upward movement of oil or gas from the reservoir.

SEAMOUNT: A submarine mountain, often a volcanic cone, 1,000 meters or higher.

SECONDARY CRATER: A crater resulting from impact of material thrown out of a primary impact crater.

SECONDARY MINERAL: A mineral formed later than the enclosing rock, either by metamorphism or by weathering and transport; placers are examples.

SECULAR GEOMAGNETIC VARIATION: A relatively slow, continuous change in the geomagnetic field, which at one time required decades to be observed.

SEDIMENT: Solid matter, either organic or inorganic in origin, that settles on a surface; it may be transported by wind, water, or glaciers.

SEDIMENT DISCHARGE: The rate of transport of sediment past a planar section normal to the flow direction, expressed as volume, mass, or weight per unit time; also called sediment transport rate.

SEDIMENTARY ROCK: A rock resulting from the consolidation of loose sediment that has accumulated in flat-lying layers on the earth's surface.

SEICHE: An oscillation in a partially enclosed body of water such as a bay or estuary.

SEISMIC: Pertaining to an earthquake.

SEISMIC BELT: A region of relatively high seismicity, globally distributed; seismic belts mark regions of plate interactions.

SEISMIC REFLECTION: Study of the layered sediments in ocean basins by directing sound waves into the sea floor so that they bounce off the different rock layers.

SEISMIC WAVES: Elastic oscillatory disturbances spreading outward from an earthquake or man-made explosion; they provide the most important data about the earth's interior.

SEISMICITY: The occurrence of earthquakes, which is expressed as a function of location and time.

SEISMOGRAM: An image of earthquake wave vibrations recorded on paper, photographic film, or a video screen.

SEISMOGRAPH: An instrument used for recording the motions of the earth's surface, caused by seismic waves, as a function of time.

SEISMOLOGY: The application of the physics of elastic wave transmission and reflection to subsurface rock geometry.

SEMIDIURNAL: Having two high tides and two low tides each lunar day.

SEMIMETALS: Elements that have some properties of metals but are distinct because they lack malleability or ductility.

SERIES: A time-rock unit representing rock deposition during a geologic epoch.

SESTON: A general term that encompasses all types of suspended lake sediment, including minerals, mineraloids, plankton, and organic detritus.

SHALE/MUDSTONE: Sedimentary rock composed of fine-grained products derived from the physical breakdown of preexisting rock; shales break along distinct planes and mudstones do not.

SHALLOW-FOCUS EARTHQUAKE: An earthquake having a focus less than 60 kilometers below the surface.

SHEAR: A stress that forces two contiguous parts of an object apart in a direction parallel to their plane of contact, as opposed to a stretching, compressing, or twisting force; also called shear stress.

SHEAR STRENGTH: The ability to withstand a lateral or tangential stress.

SHELF DAMS: Geologic formations that hold back sediments on the continental shelf.

SHEPHERD SATELLITES: The satellites that influence a planet's ring structure by removing debris from certain orbits.

SHIELD: A large region of stable, ancient rocks within a continent.

SHIELD VOLCANO: A volcano in the shape of a flattened dome, broad and low, built by flows of very fluid basaltic lava.

SHOCK WAVE: A compressional wave formed when a body undergoes a hypervelocity impact; it produces abrupt changes in pressure, temperature, density, and velocity in the target material as it passes through.

SHORT-PERIOD COMET: A comet whose orbital revolution period is less than two hundred years; an example is Comet Halley, at seventy-six years.

SILICATE MINERALS: Minerals composed of silicon, oxygen, and other metals, such as iron, magnesium, potassium, and sodium.

SILICEOUS OOZE: Sediment in which more than 30 percent of the particles are the remains of plants and animals whose skeletons are composed of silica.

SILL: A concordant sheet intrusion.

SIMPLE CRATER: A small impact crater with a simple bowl shape.

SINKHOLE: A hole or depression in the landscape, produced by dissolving bedrock; sinkholes can range in size from a few meters across and deep to kilometers wide and hundreds of meters deep.

SINUOUS RILLE: A riverlike channel produced by lava flowing across the lunar surface.

SLATE: A metamorphic rock that has a unique ability to be split into thin sheets; some slates are resistant to weathering and are thus good for exterior use.

SLICKENSIDES: Fine lines or grooves along a faulted body that usually indicate the direction of latest movement.

SLIP-FACE: The downwind or steep leeward front of a sand dune that continually stabilizes itself to the angle of repose of sand grains.

SLUMP: A term that applies to the rotational slippage of material and the mass of material actually moved; the mass has component parts called scarp, failure plane, head, foot, toe, and blocks; the toe may grade downslope in a flow.

SMOKERS: Undersea vents on the active rift areas, emitting large amounts of superheated water and dissolved minerals from deep inside the earth.

SOIL: All material that has been substantially altered at the earth's surface by interaction with the atmosphere, living things, or both, and which has not been laterally displaced subsequent to that alteration.

SOIL PROFILE: A vertical section of a soil, extending through its horizons into the unweathered parent material.

SOLAR NEBULA: The cloud of gas and dust that collapsed to form the solar system.

SOLAR SYSTEM: The sun and all the bodies that orbit it, including the nine planets and their satellites, plus numerous comets, asteroids, and meteoroids.

SOLAR WIND: Gases from the sun's atmosphere, expanding at high speeds as streams of charged particles.

SOLID: A substance that does not flow and has a definite shape.

SOLID SOLUTION: A solid whose composition shows a continuous variation in which two or more elements substitute for each other on the same position in the crystal structure.

Solubility: The tendency for a solid to dissolve.

SONAR: A subsonic sound system for measuring ocean depth and mapping the sea floor; an acronym for "sound navigation and ranging."

Sonde: The basic tool used in well logging; a long, slender instrument that is lowered into the borehole on an electrified cable and slowly withdrawn as it measures certain designated rock characteristics.

Speciation: The evolutionary process of species formation; the process through which new species arise.

Species: A group of individuals that can successfully interbreed only among themselves.

Specific gravity: The ratio of the mass of any volume of a substance to the mass of an equal volume of water.

Specific heat: The number of calories of heat required to raise the temperature of 1 gram of a substance 1 degree Celsius.

Spectroscopy: The study of the chemical and physical state of matter by splitting of a stream of radiation into the spectrum and observing the relative intensity at different wavelengths; this technique can be applied to radiation from an object (emission spectroscopy) or that passing through an object (absorption spectroscopy).

Spillway: A broad reinforced channel near the top of the dam, designed to allow rising waters to escape the reservoir without overtopping the dam.

Spokes: Dark radial features, similar to the spokes on a bicycle wheel, noted on the Saturnian ring system.

Spontaneous fission: Uninduced splitting of an unstable atomic nucleus into two smaller nuclei; an energetic form of radioactive decay.

Spore: A reproductive body with a tough, acid- and desiccation-resistant proteinaceous coat in land plants lacking seeds.

Spring: A place where groundwater reappears on the earth's surface; in karst topography, a spring represents the discharge point of a cave.

Spring sapping: A process in which water flows out of subsurface springs to surface level, forming a stream bed as it flows downslope.

Spring tide: A tide of maximum range, occurring when lunar and solar tides reinforce each other, a few days after the full and new moons.

Squall line: A line of vigorous thunderstorms created by a cold downdraft with rain, which spreads out ahead of a fast-moving cold front.

Stable isotope: An isotope of an element that does not change into another isotope.

Stage: A time-rock unit representing rock deposition during a geologic age.

Standard: A material of known isotopic composition; all enrichment and depletion is measured relative to the standard value.

Steinkern: An internal mold, preserving the interior form of an organism in negative relief.

Stope: An excavation underground to remove ore; the outlines of a stope are determined either by the limits of the ore body or by raises and levels.

Storm surge: A general rise above normal water level, resulting from a hurricane or other severe coastal storm.

STRAIN: The distortion of a rock or other material in response to stress.

STRAIN RATE: The rate at which deformation occurs, expressed as percent strain per unit time.

STRANDLINE: The position or elevation of the portion of the shoreline between high and low tide (at sea level); usually synonymous with "beach" and "shoreline."

STRATA: Layers of sedimentary rock.

STRATIFIED: Formed or lying in beds, layers, or strata.

STRATIFIED DRIFT: A sorted, layered sediment derived from glacier ice but subsequently reworked and resedimented by meltwater.

STRATIGRAPHIC SEQUENCE: A set of rock units that reflect the geologic history of a region.

STRATIGRAPHIC TIME SCALE: The history of the evolution of life on earth, broken down into time periods based on changes in fossil life in the sequence of rock layers; the time periods were named for the localities in which they were studied or from their characteristics.

STRATIGRAPHIC UNITS: Any rock layer or layers that can be easily recognized by specific characteristics, such as color, composition, or grain size.

STRATIGRAPHY: The study of sedimentary strata, which includes the concept of time, the possible correlation of the rock units, and the characteristics of the rocks themselves.

STRATOSPHERE: The layer of the atmosphere that lies immediately above the troposphere.

STRATOVOLCANO: A volcano constructed of layers of lava and pyroclastic rock; also called a composite volcano.

STRATUS: A sheet or layer cloud form ranging from 2 to 6 kilometers above ground.

STREAM EQUILIBRIUM: A state in which a stream's erosive energy is balanced by its sediment load such that it is neither eroding nor building up its channel.

STRESS: The force per unit area acting at any point within a solid body such as rock, calculated from a knowledge of force and area.

STRIATION: Parallel scratches cut in bedrock over which a glacier has passed.

STRIKE: The orientation of a horizontal line in a bed or a fault plane, measured relative to north.

STRIKE-SLIP FAULT: A fault across which the relative movement is mainly lateral.

STRUCTURAL GEOLOGY: The study of the form, pattern, and evolution of large-scale units of the earth's crust, such as basins, geosynclines, and mountain chains.

S-TYPE GRANITOID: Granitic rock formed from magma generated by partial melting of sedimentary rocks within the crust.

SUBDUCTION ZONE: A region where a plate, generally oceanic lithosphere, sinks beneath another plate into the mantle.

SUBLIMATE: Solid, crystalline material that is deposited directly from the vapor state; crystals of native sulfur around fumarole mouths are an example.

SUBMARINE CANYON: A submerged V-shaped canyon cut into the continental shelf and continental slope, through which turbidity currents funnel into the deeper parts of the oceans.

SUBSIDENCE: The sinking of the earth's surface because of the weight of a load such as unusually thick piles of sediments.

SUBSURFACE: Referring to features or characteristics—such as minerals, oil and gas, and structural features—that are not visible or apparent and that lie beneath the land surface.

SUPERNOVA: The final stage of a massive star's lifetime, when it has used all the nuclear fuel available to it and is thus unable to hold up against the force of gravity; its outer layers collapse, generating an intense amount of heat, which ignites further nuclear reactions and causes the star to explode.

SUPRATIDAL: Referring to the shore area marginal to shallow oceans that are just above high-tide level.

SURFACE WATER: Relatively warm seawater between the ocean surface and that depth marked by a rapid reduction in temperature.

SURFACE WAVE: A seismic wave that propagates parallel to a free surface and whose amplitudes disappear at depth; there are two kinds, Rayleigh waves and Love waves, that travel at the surface around the earth.

SUSPENDED LOAD: Sediment in motion well above a river's sediment bed, supported by the vertical motions of turbulent eddies.

SWELLS: Ocean waves that have traveled out of their wind-generating area.

SYENITE: A coarse-grained igneous rock that resembles granite and is rich in aluminum, sodium, and potassium.

SYMBIOSIS: A relationship or living together in close association, especially when mutually beneficial.

SYNCLINE: A folded structure created when rocks are bent downward; the limbs of the fold dip toward one another, and the youngest rocks are exposed in the middle of the fold.

SYSTEM: A time-rock unit representing rock deposition within a geologic period; also, any part of the universe, for example, a crystal, a given volume of rock, or an entire lithospheric plate.

TAPHONOMY: The study of all the processes that take place between the death of an organism and its discovery as a fossil.

TAR SAND: A natural deposit that contains significant amounts of bitumen; also called oil sand.

TECTONICS: The study of the processes that formed the structural features of the earth's crust; it usually addresses the creation and movement of immense crustal plates.

TENACITY: The resistance of a mineral to breakage, bending crushing, or tearing.

TEPHRA: All pyroclastic materials blown out of a volcanic vent, from dust to large chunks.

TERMINAL MORAINE: A ridge of unsorted glacial till deposited at the farthest edge or advance of a glacier.

TERRANE: A structurally distinct block of crust added to a continent by plate tectonic processes.

TERRESTRIAL PLANET: Any of the solid, rocky-surfaced bodies of the inner solar system, including the planets Mercury, Venus, Earth, Mars, and sometimes Earth's satellite, the Moon.

TERRIGENOUS: Originating from the weathering and erosion of mountains and other land formations.

TEST: An internal skeleton or shell precipitated by a one-celled planktonic plant or animal.

TETHYS: An ancient seaway embayed into Pangaea between the southeast corner of Asia, the western end of the Mediterranean, and the southeast end of Pangaea.

TEXTURE: The size, shape, and arrangement of minerals or particles in a rock.

THERMAL CONDUCTIVITY: The ability to permit heat to flow by induction.

THERMAL INVERSION: A region in the atmosphere of a planet in which a change from decreasing to increasing temperature with increasing altitude occurs.

THERMAL REMANENT MAGNETIZATION (TRM): The magnetization in igneous rock that results as magnetic minerals in a magma cool below their Curie temperatures.

THERMOCLINE: A layer within a water body, characterized by a rapid change in temperature.

THERMODYNAMICS: The area of science that deals with the transformation of energy and the laws that govern such changes; equilibrium thermodynamics is especially concerned with the reversible conversion of heat into other forms of energy.

THERMOHALINE CIRCULATION: Any circulation of ocean waters that is caused by variations in the density of seawater resulting from differences in the temperature and/or salinity of the water.

THERMOLUMINESCENCE: The process by which some minerals trap electrons in their crystal structure at a fixed rate and release them when heated.

THERMOSPHERE: The highest layer of the earth's atmosphere except for the exosphere; it begins at about 85 kilometers above sea level.

THRESHOLD OF MOVEMENT: The conditions for which a flow is just strong enough to move sediment particles at the surface of a given sediment bed.

THRUST BELT: A linear belt of rocks that have been deformed by thrust faults.

THRUST FAULT: A fault, usually dipping less than 30 degrees, across which the hanging wall has moved upward relative to the footwall.

TIDAL DISRUPTION: A gravitational mechanism that keeps ring particles revolving about a planet from accreting into larger bodies by tearing them apart.

TIDAL HEATING: The heating of a moon as tides are raised in its crust.

TILL: An unsorted, unconsolidated sediment of clay- to boulder-sized particles deposited directly by glacier ice without subsequent reworking by meltwater.

TOPOGRAPHIC MAP: A line-and-symbol representation of natural and selected manmade features of a part of the earth's surface, plotted to a definite scale.

TOPOGRAPHY: The collective physical features of a region or area, such as hills, valleys, streams, cliffs, and plains.

TOPSOIL: In reclamation, all soil that will support plant growth; normally, the 20-30 centimeters of the organically rich top layer.

TORNADO: A violent rotating column of air extending downward from a thunderhead cloud and having the appearance of a funnel, rope, or column.

TOTAL DISSOLVED SOLIDS (TDS): A quantity of solids, expressed in weight percent, determined from the weight of dry residue left after evaporation of a known weight of water.

TOUGHNESS: The degree of resistance to fragmentation or to plastic deformation.

TRACE FOSSIL: Indirect evidence of an organism's presence through tracks, trails, and burrows.

TRADE WINDS: Winds in the tropics that blow from the subtropical highs to the equatorial low.

TRANSCURRENT FAULT: A fault in which relative motion is parallel to the strike of the fault (that is, horizontal); also known as a strike-slip fault.

TRANSFORM FAULT: A fault connecting offset segments of an ocean ridge along which two plates slide past each other.

TRANSGRESSION: The flooding of a large land area by the sea, either by a regional downwarping of continental surface or by a global rise in sea level.

TRANSPIRATION: The process by which plants give off water vapor through their leaves.

TRANSVERSE BAR: A flat-topped body of sand or gravel oriented transversely to the river flow.

TRANSVERSE VALLEY: A river-cut valley or gorge that runs perpendicular to the main strike direction of a mountain chain.

TRAP: A structure in the rocks that will allow petroleum or gas to accumulate rather than flow through the area.

TRENCH: A long and narrow deep trough on the sea floor that forms where the ocean floor is pulled downward because of plate subduction.

TRIPLE JUNCTION: A point where three plate boundaries meet.

TROPICAL RAIN FOREST: A land area with monthly temperatures that average greater than 18 degrees Celsius and monthly precipitation that averages more than 6 centimeters.

TROPICAL STORM: A severe storm with winds ranging from 45 to 120 kilometers.

TROPOPAUSE: The boundary layer between the troposphere and the stratosphere.

TROPOSPHERE: The lowest region of the earth's atmosphere, which extends upward from the surface to about 8 kilometers; the troposphere contains about 85 percent of the total mass of the atmosphere, almost all the water vapor in the atmosphere, and most of the greenhouse gases.

TSUNAMI: A seismic sea wave created by an undersea earthquake, a violent volcanic eruption, or a landslide at sea.

TUFF: A general term for all consolidated pyroclastic rocks.

TURBIDITY CURRENT: A turbid, relatively dense mixture of seawater and sediment that flows downslope under the influence of gravity through less dense water.

TURBULENT FLOW: A high-velocity sediment flow in which individual sediment particles move in very chaotic or nonstreamline directions.

TWINNING: A phenomenon in which two or more intergrown crystal parts, though they have the same atomic arrangement and properties, differ in orientation by rotation or reflection.

ULTRAMAFIC: A term used to describe certain igneous rocks and most meteorites that contain less than 45 percent silica; they contain virtually no quartz or feldspar and consist mainly of ferromagnesian silicates, metallic oxides and sulfides, and native metals.

UNCONFINED AQUIFER: An aquifer whose upper boundary is the water table; also called a water table aquifer.

UNCONFORMITY: A significant break in a stratigraphic sequence because of nondeposition or erosion of the missing rock layers; much younger rocks are positioned above older rocks.

UNIFORMITARIANISM: The theory that processes currently operating in nature have been operating through the earth's geologic history; it suggests that the large-scale features of the earth were developed very slowly over vast periods of time.

UNREINFORCED MASONRY (URM): Materials not constructed with reinforced steel, for example, bricks, hollow clay tiles, adobe, concrete blocks, and stone.

UNSTABLE AIR: A condition caused by unusual coolness of the air above rising air so that the rising air, being warmer, will accelerate upward.

UPLIFT: An elevated block of the earth's lithosphere; coastal areas undergoing uplift tend to emerge above sea level.

UPPER MANTLE: The region of earth immediately below the Mohorovi/ci/c discontinuity, believed to be composed largely of peridotite (olivine and pyroxene rock), which is thought to melt to form basaltic liquids.

UPWELLING: A phenomenon in which warm surface waters are pushed away from the coasts by the rotation of the earth; cold waters that carry an abundance of nutrients are brought up from depth to replace the warm waters.

U-SHAPED VALLEY: The classic shape of a glaciated mountain valley.

VAN DER WAALS FORCE: A weak electrostatic attraction that arises because certain atoms and molecules are distorted from a spherical shape so that one side carries more of the charge than does the other.

VARVE: A pair of contrasting layers of sediment deposited over a year's time; the summer layer is light, and the winter layer is dark.

VECTOR: A term used in waste disposal when referring to rats, flies, mosquitoes, and other disease-carrying insects and animals that infest dumps; also, a mineral-filled fault or fracture in rock.

VEIN: A mineral deposit that fills a crack; veins form by precipitation of minerals from fluids.

VENT: A break or tear on the side of a mountain through which magma and pressure can escape.

VENTIFACT: Any stone or pebble that is shaped, worn, faceted, cut, or polished by the abrasive action of windblown sand, generally under desert conditions.

VESICULAR: Blistered, or containing bubblelike cavities.

VISCOSITY: A substance's ability to flow; the lower the viscosity, the greater the ability to flow.

VOLATILES: Chemical elements and compounds that become gaseous at fairly low temperatures, allowing them to be released from solids easily.

VOLCANIC EARTHQUAKES: Small-magnitude earthquakes that occur at relatively shallow depths beneath active or potentially active volcanoes.

VOLCANIC ROCKS: Igneous rocks formed at the surface of the earth.

VOLCANIC TREMOR: A long, continuous vibration, detected only at active volcanoes.

VOLCANO: A vent at the earth's surface in which gases, rocks, and magma erupt at the surface and build a more or less cone-shaped mountain.

VORTEX: A flow of fluid about a central axis that draws material into its center.

WARM TEMPERATE GLACIER: A glacier that is at the pressure melting temperature throughout.

WATER TABLE: The upper surface of the zone of saturation; above the water table, the pores in the soil and rock have air and water.

WATERSPOUT: A tornado over water; less violent and smaller waterspouts form in fair weather just as dust devils do over dry land.

WAVE BASE: The depth to which water particles of an oscillatory wave have an orbital motion; generally, the wave base is equal to one-half the distance of the length of the wave.

WAVE REFRACTION: The process by which a wave crest is bent as it moves toward shore.

WAVE-CUT BENCH: A gently seaward-sloping platform cut into the bedrock of a coast by wave erosion and landsliding; wave-cut benches are proof of sea-level variations and tectonic uplift and subsidence of coastal areas.

WEATHERING: The mechanical disintegration and chemical decomposition of rocks and sediments.

WELL LOG: A stripchart with depth along a well borehole plotted on the long axis and a variety of responses plotted along the short axis; there are many varieties, including borehole logs, geophysical logs, electric logs, and wireline logs, from which information may be obtained about lithology, formation fluids, sedimentary structures, and geologic structures.

WET-CLIMATE GEOMORPHOLOGY: The study of landforms in regions having average annual precipitation levels greater than 100 centimeters.

X RAY: A photon with a much higher energy and shorter wavelength than that of visible light; its wavelength is of the same order of magnitude as the spaces between atoms in a crystal.

XENOBIOTIC: Referring to a chemical that is foreign to the natural environment; a man-made chemical that may not be biodegradable.

XENOCRYSTS: Minerals found either as crystals or fragments in some volcanic rocks; they are foreign to the body of the rock in which they occur.

XENOLITHS: Various rock fragments that are foreign to the igneous body in which they are present.

ZONE OF SATURATION: That zone beneath the land surface where all the pores in the soil or rock are filled with water rather than with air.

Categorized List of Essays

GEOCHEMISTRY
Earth's Composition
Elemental Distribution
Geologic and Topographic Maps
Water-Rock Interactions

**GEOCHRONOLOGY AND
PALEONTOLOGY**
Archaeological Geology
Catastrophism
Cretaceous-Tertiary Boundary
Earth's Age
Earth's Oldest Rocks
Earth's Origin
Fossil Record
Geologic Time Scale

GEOMORPHOLOGY
Caves and Caverns
Karst Topography
Marine Terraces
Permafrost
River Valleys

GEOPHYSICS
Earth's Core
Earth's Crust
Earth's Differentiation
Earth's Mantle
Earth's Shape
Earth's Structure
Earthquakes
Lithosphere
Rock Magnetism

GLACIAL GEOLOGY
Alpine Glaciers

Continental Glaciers
Glacial Landforms

**MINERALOGY AND
CRYSTALLOGRAPHY**
Minerals: Physical Properties
Minerals: Structure

**PETROLEUM GEOLOGY AND
ENGINEERING**
Oil and Gas: Distribution
Oil and Gas: Origins

PETROLOGY
Andesitic Rocks
Basaltic Rocks
Granitic Rocks
Igneous Rock Bodies
Igneous Rocks
Magmas
Metamorphic Rocks
Rocks: Physical Properties
Sedimentary Rocks
Siliciclastic Rocks
Ultramafic Rocks

SEDIMENTOLOGY
Alluvial Systems
Coastal Processes and Beaches
Deltas
Drainage Basins
Floodplains
Lakes
Reefs
River Flow
Sand Dunes
Weathering and Erosion

STRATIGRAPHY
Biostratigraphy
Stratigraphic Correlation
Transgression and Regression
Uniformitarianism

STRUCTURAL GEOLOGY
Discontinuities
Folds
Joints
Meteorite and Comet Impacts
Normal Faults
Stress and Strain
Thrust Faults
Transform Faults

TECTONICS
Continental Crust
Continental Drift
Continental Growth
Continents and Subcontinents

Hot Spots and Volcanic Island
  Chains
Island Arcs
Mountain Belts
Ocean Basins
Ocean Ridge System
Oceanic Crust
Plate Tectonics
Subduction and Orogeny
Supercontinent Cycle

VOLCANOLOGY
Calderas
Eruptions
Intraplate Volcanism
Lava Flows
Pyroclastic Rocks
Recent Eruptions
Shield Volcanoes
Stratovolcanoes

Geology

Subject Index

A

Aa, 367, 709
Ablation, 8, 93
Abrasion, 701
Absolute dating, 23, 24, 202, 709
Abyssal plains, 709
Accessory minerals, 586
Accretion, 113, 162, 210, 709
Achondrite, 709
Acid rain, 701, 709
Acme zones, 38
African Rift Valley System, 327, 626
Aggradation, 534
Albedo, 709
Alkaline rocks, 318
Allogenic sediment, 357
Alluvial fans, 2
Alluvial systems, 1-7
Alluvium, 241, 709
Alpha particle, 709
Alpine glaciers, 8-15, 709
Alteration, 401
Altitude, 141
Alvarez, Luis, 120
Alvarez, Walter, 120
Anastomosing rivers, 4
Andesite, 179, 232, 332, 505, 599, 709
Andesite Model, 83
Andesitic magmas, 368, 385
Andesitic rocks, 16-22
Angle of repose, 709
Anion, 709
Anorthosite, 709
Anticline, 248, 709
Aphanitic, 290, 709
Aqueous solution, 709
Aquifer, 710
Archaeological geology, 23-29
Archaeology, 23-29

Archaeomagnetism, 24, 26
Archaeometry, 23
Archean eon, 162, 710
Arête, 281
Arkose, 568, 710
Asaro, Frank, 120
Asbestosis, 710
Ash, volcanic, 307, 360, 512, 599, 675, 710
Assemblage zones, 38
Assimilation, 383
Asteroid, 53, 407, 710
Asteroid belt, 710
Asthenosphere, 78, 224-225, 297, 324, 375, 383, 464, 710
Aston, Francis William, 204, 278
Astrobleme, 710
Astronomical Unit, 710
Atmosphere, 710
Atoll, 710
Atomic absorption spectrometry, 321
Augite, 30
Aureole, 710
Autoclastic breccia, 367
Avalanche, 710
Axial plane, 248
Axis, 248, 710

B

Barchan dune, 559, 710
Barlow, William, 431
Barrel, 710
Barrier island, 389, 710
Basalt, 18, 179, 225, 232, 307, 332, 375, 456, 464, 505, 542, 710
Basalt flows, 18
Basalt lava, 368
Basaltic magmas, 385

Basaltic rocks, 30-36
Base level, 534
Basement, 711
Basement rocks, 461
Basin and Range province, 225
Basin order, 141
Basins, 141-147, 456-463, 481, 711
Batholiths, 291, 309, 711
Bathymetric charts, 476
Bathymetric contour, 711
Bauxite, 711
Beaches, 71-77
Becquerel, Antoine-Henri, 204, 276
Bedload, 711
Bed-load transport, 527
Bedrock, 348
Benioff zone, 332, 711
Bermuda Rise, 475
Beta particle, 711
Bezymianny, 679
Bioabrasion, 389, 711
Biochronology, 39
Biogenic sediment, 357, 456, 711
Biogeographic province, 711
Biomagnetism, 545
Biomarkers, 711
Biosphere, 711
Biostratigraphy, 37-42, 593, 711
Birefringence, 711
Bitumen, 711
Black smokers, 629
Block lava, 367
Blocking temperature, 542, 711
Bluff, 241
Body fossil, 257
Body wave, 711
Bolide, 118, 711
Boltwood, Bertram, 204, 276
Bore, 711
Bort, 711
Bottom-water mass, 711
Bouguer gravity, 712
Bowen's reaction principle, 16
Brachiopod, 272
Brackish water, 712
Bragg, William Henry, 432
Bragg, William Lawrence, 432
Braided river, 1
Bravais, Auguste, 431

Breccia, 367, 712
Brine, 712
Brongniart, Alexandre, 38, 595
Bryozoan, 272
Buckland, William, 668
Buffon, Comte de, 54, 274
Bulk modulus, 551
Burnet, Thomas, 273
Bushveld complex, 660

C

Calc-alkali rocks, 318
Calcareous algae, 518
Calcite, 62, 429
Calderas, 43-52, 599, 712
California, 642
Caloris Basin, 712
Canadian Shield, 81, 103, 482
Canyons, 536
Carangeot, Arnould, 434
Carbonaceous chondrites, 712
Carbonate compensation depth
   (CCD), 712
Carbonate rocks, 518, 568, 712
Carbonic acid, 348
Carlsbad Cavern, 63
Casing, 712
Caspian Sea, 142
Cast, 712
Cataclasis, 712
Catastrophism, 53-61, 155, 665, 712
Cation, 712
Caves and caverns, 62-70, 142, 344, 348
Cenozoic era, 118, 712
Centrifugal force, 218
Ceramics, 712
Channel, 141
Chatham Rise, 475
Chemical bond, 712
Chemical rocks, 571
Chemical weathering, 340, 701, 704,
   712
Chert, 713
Chondrites, 162, 713
Cinder cone, 599, 713
Cirque, 8, 281, 713
Cirrus, 713

Cladistics, 713
Clastic rock, 568, 584, 713
Clastic sediments, 357
Clay, 357, 713
Clay minerals, 357
Claystone, 568
Cleavage, 415, 428, 713
Clinker, 713
Coal, 713
Coastal processes and beaches, 71-77
Coastal wetlands, 713
Coastlines, 389-397
Coefficient of thermal expansion, 550, 713
Cohesion, 713
Cold front, 713
Cold polar glacier, 93, 713
Color index, 315, 713
Columnar jointing, 340-341, 713
Coma, 713
Comet, 407, 714
Comet nucleus, 714
Competence, 714
Complex crater, 407, 714
Component, 714
Composite volcanoes, 599
Compressive strength, 550-551, 714
Concentric folds, 250
Concordant, 307
Concordant age, 714
Condensation temperature, 714
Conduction, 714
Cone, 714
Cone of depression, 714
Confined aquifer, 714
Confining pressure, 714
Conglomerate, 714
Conjugate shear sets, 340
Connate fluids, 692, 714
Contact goniometer, 434
Contact metamorphism, 398, 714
Contact metasomatism, 714
Contaminant, 714
Contamination, 383
Continental crust, 78-86, 182, 191, 233, 377
Continental drift, 86-92, 174, 464, 623, 714
Continental glacier or ice sheet, 714

Continental glaciers, 93-101
Continental growth, 102-109
Continental margin, 615, 715
Continental rift, 505
Continental rise, 715
Continental shelf, 438, 646, 715
Continental slope, 715
Continents and subcontinents, 86-92, 110-117
Continents and supercontinents, 623-630
Contour lines, 267
Convection, 171, 715
Convection cells, 86, 324, 715
Convergence, 623; and boundaries, 297, 324, 505, 617, 715; and plates, 464
Convergent organisms, 715
Copernicus, Nicolaus, 156
Coralline algae, 518
Corals, 518-525
Cordillera, 715
Cordilleran geosyncline, 106
Core, Earth's, 171, 224, 715
Core drilling, 715
Core-mantle boundary, 171
Core ring or core eddy, 715
Coriolis effect, 715
Coronae, 577, 715
Correlation, stratigraphic, 37, 593, 715
Cotton, Charles, 578
Country rock, 307, 310, 715
Covalent bonding, 715
Cranial capacity, 715
Crater Lake, 46, 601
Craters, 43-53, 407-414, 716
Cratons, 110, 182, 716
Creep, 716
Crestal plane, 248
Cretaceous period, 118
Cretaceous-Tertiary boundary, 59, 118-125
Crevasses, 10, 126
Cross-bedding, 559, 716
Crust, 30, 53, 78, 110, 133, 148, 224, 375, 615, 655, 716
Crust, oceanic, 471
Crustal differentiation, 188
Crystals, 290, 415, 428, 685, 716

Crystal structure, 716
Crystallization, 290, 383, 716
Cumulate, 716
Cumulonimbus, 716
Cumulus, 716
Curie, Marie, 204, 276
Curie, Pierre, 276
Curie temperature, 23, 542, 638, 716
Currents, 73, 716
Cuvier, Georges, 38, 53, 56, 595, 667
Cylindrical folds, 250

D

Dams, 146
Dana, James Dwight, 40, 103, 166, 440
Darwin, Charles, 258, 276, 670
Daughter product, 542, 716
Dead Sea, 142
Debris flow, 716
Debye, Peter, 434
Declination, 716
Décollement, 716
Deep-ocean trench, 297
Deep Sea Drilling Project, 107, 468
Deflation, 716
Deformation, 148, 218, 249, 607
Dehydration, 692
Deltas, 126-132, 716
Dempster, Arthur Jeffrey, 204
Dendrochronology, 717
Density, 110, 179, 224, 415, 550, 717
Deposition, 456, 717
Depositional systems, 1
Desalinization, 717
Desert varnish, 717
Desertification, 717
Deserts, 560
Detachment fault, 717
Detrital minerals, 717
Detrital remanent magnetization
   (DRM), 542, 545, 717
Detrital rock, 584
Devils Postpile, 341
d'Halloy, Omalius, 275
Diagenesis, 584, 717
Diagenetic minerals, 363
Diamonds, 196

Diapir, 717
Diatreme, 717
Dietz, Robert, 440
Differentiation, Earth's, 162, 188-195,
   213, 383, 717
Diffusion, 717
Dike, 307, 717
Dilatation, 607
Diluvial theory, 665
Dinosaurs, 37-42, 118-125
Dip, 448, 631, 717
Dip slips, 449
Dipole field, 717
Discharge, 241, 526, 718
Discontinuities, 133-140, 192, 195, 196,
   225
Discordant, 307, 718
Discordant age, 718
Disharmonic folds, 251
Dissociation, 718
Distortion, 607
Distributary channel, 126, 718
Divergence, 623
Divergent plate boundaries or
   margins, 297, 324, 464, 505, 617,
   638, 718. *See also* Plate margins
Diversity, 718
Doctrine of final causes, 665
Doldrums, 718
Dolomite, 62
Dolostones, 235
Downburst, 718
Drainage basins, 141-147, 718
Drift, 97, 284
DRM, 545
Drumlin, 96, 281, 718
Ductility, 718
Dunes, 559-567, 718
Dust devil, 718
Dust tail, 718

E

Earth tide, 218, 718
Earth's age, 155-161
Earth's composition, 162-170
Earth's core, 171-178, 224, 715
Earth's crust, 179-187

Earth's differentiation, 162, 188-195, 213, 383, 717
Earth's mantle, 195-201, 655
Earth's oldest rocks, 202-209
Earth's origin, 210-217, 412
Earth's rotation, 221
Earth's shape, 218-223
Earth's structure, 224-231, 375-382
Earthquake focus, 332, 505
Earthquake waves, 133
Earthquakes, 84, 86, 133, 378, 478, 603, 620, 625, 642, 718
East Pacific Rise, 467, 473, 506
Ecliptic plane, 718
Eclogite, 196
Ecological succession, 718
Ecology, 718
Ecosystem, 559
Ejecta, 718
El Chichón, 680
El Niño, 719
Elastic, 719
Elastic rebound, 148-149
Elastic waves, 183
Elasticity, 550
Electrolysis, 719
Electromagnetic radiation, 719
Electron microprobe analysis, 321
Elemental distribution, 232-233, 235-240
En echelon folds, 251, 310
Endogenic sediment, 357
Eolian erosion, 719
Epeiric sea, 719
Epeirogeny, 438, 719
Ephemeral stream, 1, 719
Epicenter, 719
Epoch, 719
Equilibrium, 719
Equilibrium line, 8, 93, 719
Era, 118, 719
Erosion, 141, 536, 584, 701, 719
Erratics, 12
Eruptions, 599, 675-691, 719
Esker, 281, 719
Estuaries, 127, 719
Eugeosynclines, 440
Eukaryotic cell, 719
Eustacy, 646

Eustatic sea-level change, 719
Evaporation, 719
Evaporites, 569, 720
Evapotranspiration, 526, 720
Evolution, 262, 720
Evolutionary trends, 720
Exchange reaction, 692, 720
Exfoliation, 340, 703, 720
Exoskeleton, 720
Exosphere, 720
Extension, 340, 720
Extinctions, 38, 53-61, 118, 413, 720
Extrusion, 383, 720
Extrusive rock, 16, 315, 720
Eye, 720

F

Fabric, 720
Facies, 402
Family, 118, 720
Fans, alluvial, 2
Fathom, 720
Fault drag, 448, 720
Fault slip, 720
Faulting, 615
Faults, 86, 102, 448-455, 631-645, 720
Faunal succession, 272, 720
Federov, Evgraf, 431
Feldspar minerals, 585
Felsic, 315
Ferromagnetic, 638, 720
Ferromagnetic material, 542
Fetch, 720
Field relations, 203
Fineness, 720
Fireball, 721
Firn, 9
Fission fragment, 721
Fission track, 721
Fixed carbon, 721
Fjords, 282
Flanks, 248
Flexure folding, 252
Floodplains, 1, 241-247, 721
Floods, 5, 14, 241-247, 529
Flow rate, 721
Fluid, 721

Fluid inclusion, 721
Fluorescence, 721
Fluvial, 241, 534, 721
Flyggberg, 97
Focus, 195, 721
Folding, 615, 721
Folds, 102, 248-256, 341
Foliation, 398, 721
Food chain, 721
Footwall, 448, 631, 721
Fossil fuel, 489, 721
Fossil record, 122, 257-266
Fossils, 37, 53, 118-125, 257, 389-397, 459, 721
Fractional crystallization, 721
Fractionation, 692, 721
Fractography, 340
Fracture, 419
Fracture zones, 471, 638, 721
Free-air gravity, 721
Fresh water, 722
Frontier, 481
Frost, 498-504
Fumaroles, 213, 722

G

Gabbro, 31, 235, 438, 473, 722
Galilean moons, 722
Gamma rays, 722
Ganges, 143
Genus, 722
Geoarchaeology, 23, 126, 722
Geochemical cycle, 722
Geochronology, 155-161, 202, 204, 722
Geographic Information Systems, 269
Geologic and topographic maps, 267-272, 345, 485, 538, 635, 747
Geologic record, 722
Geologic time scale, 37-42, 272-276, 278-280
Geological column, 53
Geomorphology, 241, 559, 722
Geophysics, 481, 722
Geostationary orbit, 722

Geostrophic, 722
Geosynclines, 102, 615, 722
Geothermal energy, 43-52
Geothermal gradient, 722
Geothermometers, 722
Geyser, 722
Glacial deposits, 283
Glacial landforms, 281-289
Glacial valleys, 282
Glaciation, 722
Glaciers, 93-101, 281-289, 358, 499, 647, 723; alpine, 8-15
Glass, 723
Glomar Challenger, 107, 468
Gneiss, 405, 723
Gondwanaland, 86, 723
Goniometer, 434
Gorges, 536
Graben, 448, 723
Gradual evolution, 723
Gradualism, 257
Granite, 102, 179, 224-225, 307, 332, 375, 438, 542, 701, 723
Granitic magmas, 386
Granitic rocks, 232, 290-296
Granitization, 290
Granulite, 80
Gravimeter, 723
Gravitational differentiation, 210
Gravity anomalies, 332, 723
Gravity faults, 448
Graywacke, 568, 723
Great Barrier Reef, 519
Great Lakes, 285
Great Salt Lake, 142
Greenhouse effect, 723
Groundmass, 16, 723
Groundwater, 62, 141, 723
Gulf of Aden, 327
Gutenberg Discontinuity, 136
Guyot, 389, 723
Gypsum, 62

H

Hadley cell, 723
Half-life, 202, 723
Hall, James, 102, 439

Hanging wall, 448
Hardness, 417, 550, 723
Harmonic tremor, 723
Harriot, Thomas, 429
Haüy, René-Just, 429
Hawaiian Island-Emperor Seamont
  Chain, 326
Head wall, 631, 723
Hekla, 679
Hematite, 544
Herodotus, 127
Hess, Harry, 510, 641
High-level wastes, 724
Highlands, 723
Hinge, 248
Holmes, Arthur, 278, 509, 578
Holotype, 272
Homogeneous accretion theory,
  188
Hooke, Robert, 273
Horizon, 724
Hornfels, 724
Horse latitudes, 724
Horst, 448
Hot spots, 34, 43, 297-306, 508, 577,
  724
Hull, Albert W., 434
Humboldt, Alexander von, 275
Humic acid, 724
Hummocky, 724
Hurricane, 724
Hutton, James, 56, 156, 274, 573,
  666
Huygens, Christiaan, 429
Hydraulic geometry, 526
Hydrocarbon reservoirs, 518
Hydrocarbon traps, 492
Hydrocarbons, 481, 489, 724
Hydrograph, 526
Hydrologic cycle, 526, 724
Hydrological, 141
Hydrology, 62-70, 526, 724
Hydrosphere, 724
Hydrostatic pressure, 249
Hydrothermal, 724
Hydrothermal vents, 471, 629
Hypabyssal rocks, 316
Hypervelocity impact, 407
Hypocenter, 638, 724

I

I-type granitoid, 726
Ice ages, 724
Ice sheets, 96, 99
Ice shelves, 94
Igneous, 30-36, 438
Igneous rock bodies, 307-314
Igneous rocks, 180, 195, 232, 290,
  315-318, 320-323, 724
Ignimbrite, 43, 512, 577, 724
Immiscible fluids, 724
Impact basin, 724
Impact breccia, 724
Impact craters, 407-414, 581, 725
Inclination, 725
Inclusions, 293, 725
Index fossil, 37, 593, 725
Index mineral, 725
Index of refraction, 725
Infrared radiation, 725
Inhomogeneous accretion theory,
  188
Inner core, 133
Inner edge point, 389, 725
Insolation, 725
Instrumental neutron activation
  analysis, 321
Intensity, 725
Interdistributary bay, 126
Interference, 725
Interfluve, 534
Intermediate rock, 16, 725
Internal drainage, 725
International Geophysical Year, 88
Interstellar medium, 725
Intraplate volcanism, 324-331
Intrusion, 383, 615
Intrusive rocks, 315, 725
Ion, 428, 725
Ionic bond, 428, 725
Ionosphere, 725
Iron formation, 725
Island arcs, 297, 332-339, 726
Islands, 31
Isobar, 726
Isobath, 726
Isochron, 202, 726
Isograd, 726

Isostasy, 179, 726
Isostatic adjustment, 93
Isostatic readjustment, 389
Isostructural, 726
Isotopes, 155, 162, 202, 290, 383, 692-700, 726
Isotropic, 726

J

Jet stream, 726
JOIDES Resolution, 107
Joints, 340-347, 726
Juvenile water, 692, 726

K

Kame, 281
Kame terraces, 11
Karat, 726
Karst, 62, 348, 726
Karst topography, 348-356
Kelvin, Lord, 276
Kepler, Johannes, 156, 429
Kerogen, 489, 726
Kettles, 281, 284, 358, 726
Kimberlite, 196, 726
Kingdom, 726
Komatiites, 167
Krakatoa, 599, 676
K/T boundary. *See* Cretaceous-Tertiary boundary
Kulp, J. L., 278

L

L waves, 150
La Niña, 726
Laccolith, 307
Lacroix, Alfred, 685
Lagerstatte, 257, 259
Lahar, 727
Lake Bonneville, 142
Lake Tanganyika, 626
Lake Toba, 47
Lakes, 357-366

Laminar flow, 526
Lamprophyre, 307
Landfills, 288
Landforms, glacial, 281
Landslides, 358, 727
Lapilli, 512, 727
Lapse rate, 727
Lapworth, Charles, 275
Lateral accretion, 102
Laterite, 727
Latitude and longitude, 218
Laue, Max von, 432
Laurasia, 86, 727
Lava, 17, 62, 307, 316, 675, 727
Lava domes, 19
Lava flow, 32, 83, 311, 367-374
Lava tube, 727
Leach, 727
Leachate, 727
Lehmann, Inge, 174
Leopold, Luna Bergere, 528
Levees, 245, 727
Lightfoot, Joseph Barber, 155
Lignin, 727
Limestone, 62, 141, 232, 235, 348, 350, 568, 570, 701, 727
Lineation, 727
Liquefaction, 727
Lithic, 685
Lithic fragments, 584, 587, 727
Lithification, 727
Lithology, 272, 593, 727
Lithosphere, 78, 86, 148, 224, 324, 332, 375-383, 389, 456, 727
Lithospheric crust, 218
Lithospheric plates, 30, 457, 727
Lithostatic pressure, 249
Little Ice Age, 8
Local base level, 241
Local sea-level change, 727
Logan, Sir William, 254
Long-period comet, 728
Longitudinal bar, 1, 728
Longitudinal dune, 559, 728
Longshore current, 71, 728
Longshore drift, 71
Low-level wastes, 728
Luminescence, 415, 420, 728

Luminophors, 74
Luster, 415, 417, 728
Lyell, Charles, 57, 156, 666, 676

M

Macrofossil, 257
Macronutrient, 728
Maddock, Thomas, Jr., 528
Mafic rocks, 110, 162, 315, 728
Magma, 17, 30-31, 238, 290, 297, 307,
  315, 367-374, 383-388, 601, 615, 728
Magmatic evolution, 383
Magnetic anomalies, 456, 638, 728
Magnetic domain, 728
Magnetic field, Earth's, 171, 172
Magnetic polarity time scale, 728
Magnetic pole, 728
Magnetic remanence, 728
Magnetic reversals, 230, 413, 506,
  542-549, 642, 728
Magnetism, 419, 542-549
Magnetite, 542, 544
Magnetopause, 728
Magnitude, 728
Maillet, Benoit de, 273
Mammoth-Flint Ridge Cave System, 63
Manganese nodules, 728
Mantle, 30, 78, 102, 110, 133, 148, 224,
  298, 324, 375, 655, 728
Mantle convection, 325
Mantle plume, 324
Map scale, 267
Marble, 62
Maria, 728
Marine, 729
Marine terraces, 389-397
Mass balance, 8, 93
Mass extinctions, 38, 413
Mass spectrometry, 202, 278, 692, 729
Mass wasting, 729
Matrix, 729
Matterhorn, 11
Matthews, Drummond, 642
Maunder minimum, 729
Mean sea level, 729
Meandering river, 1
Meanders, 241, 243, 729

Measured section, 646
Mechanical weathering, 701
Mediterranean Sea, 142
Mercalli scale, 151
Mesosphere, 729
Mesozoic era, 118, 729
Metallogenesis, 464
Metamorphic facies, 398, 729
Metamorphic grade, 398, 729
Metamorphic rocks, 236, 398-406, 729
Metamorphic zones, 729
Metamorphism, 290, 398, 438-447, 729
Metasomatism, 729
Metastable, 729
Meteor, 729
Meteor shower, 729
Meteoric water, 692, 730
Meteorite, 53, 407, 730
Meteorite and comet impacts, 407-414
Meteorites, 159, 163, 202-209, 214
Meteoritics, 210
Meteorology, 730
Methane, 489
Methodological uniformitarianism,
  665
Mica minerals, 586
Michel, Helen, 120
Michigan Basin, 81
Micro-earthquakes, 603
Microbreccias, 634
Microfossil, 730
Mid-Atlantic Ridge, 88, 90, 467, 478,
  506, 617
Mid-ocean ridges, 456, 638, 730
Migmatite, 290, 730
Miller, William H., 430
Millibar, 730
Milligal, 730
Mineral, 315, 357, 415, 428, 701, 730
Mineraloid, 357
Minerals, physical properties, 415-427;
  structure, 428-437
Minerals, magnetic, 543
Mining wastes, 730
Miogeosynclines, 440
Mississippi River, 128, 143, 242, 535
Mitochondrion, 730
Mitscherlich, Eilhardt, 430
Mixed tides, 730

Mobile belt, 730
Modal analysis, 295
Mode, 315, 730
Model, 730
Moderator, 730
"Mohole" project, 137
Mohorovičić, Andrija, 196
Mohorovičić Discontinuity, 78, 133, 135, 137, 181, 196, 215, 225, 375, 467, 472, 730
Mohs, Friedrich, 430, 551
Mohs hardness scale, 415, 730
Moisture content, 731
Mold, 731
Mole, 731
Molecule, 731
Monchiquite, 307
Monerans, 731
Monoclines, 250
Monoclinic folds, 250
Monogenetic, 577
Monsoon, 731
Moraine, 8, 281, 731
Morphologic evolution, 731
Morphology, 257, 731
Mount Etna, 601
Mount Fuji, 599
Mount Helka, 600
Mount Katmai, 602, 677
Mount Lamington, 679
Mount Mazama, 47
Mount St. Helens, 599, 601, 680, 686
Mount Vesuvius, 599, 676
Mountain belts, 438-447, 508, 616
Mountain ranges, 281
Mountains, undersea, 464
Mudflow, 731
Mudrocks, 235
Mudstone. *See* Shale
Murchison, Roderick Impey, 275
Mutation, 731
Mylonites, 634

N

Naphthene hydrocarbons, 731
Nappe, 731
National Mapping Program, 269

Natural bridge, 348
Natural gas, 489, 731
Natural levee, 126
Natural selection, 731
Neap tide, 731
Nebula, 731
Nebular hypothesis, 210
Neptunists, 54, 665
Névé, 9
Newton, Sir Isaac, 156
Niche, 731
Nier, Alfred Otto Carl, 204, 278
Nile delta, 129
Noncylindrical folds, 250
Nonrenewable resource, 731
Norm, 315, 731
Normal faults, 448-455, 731
Normal polarity, 732
Normal stress, 607, 732
Notch or nip, 389, 732
Nucleosynthesis, 732
Nuclide, 202
Nuée ardente, 675, 732
Nutrient, 732

O

Oblique slips, 449
Ocean basins, 456-463, 646, 648
Ocean currents, 222
Ocean Drilling Project, 107
Ocean ridge system, 464-470
Ocean trenches, 471
Oceanic basins, 506
Oceanic crust, 78, 86, 181, 191, 233, 471-480, 509, 732
Oceanic ridges, 30, 471, 640
Oceanic rise, 505
Oceanic trenches, 475
Ogives, 10
Oil, 732
Oil and gas, distribution, 481-488; origins, 489-497
Oil shale, 732
Oldham, Richard D., 174
Oldowan, 732
Olivine, 30, 164, 226, 235, 655
Olympus, 329

On the Origin of Species (Darwin), 258
100-year-flood, 732
Oort Cloud, 732
Ophiolite complex, 732
Ophiolite suite, 623
Opposition, 732
Order, 732
Ore, 732
Organelles, 732
Organic molecules, 732
Orogenesis, 78
Orogenic belts, 110, 438-447, 616, 733
Orogeny, 438, 615-622, 733
Orphan lands, 733
Orthoquartzite, 568
Orthorhombic folds, 250
Oscillatory wave, 71
Osmosis, 733
Outer core, 133, 733
Outgassing, 210, 733
Outwash plains, 284
Overburden, 733
Owen, Richard, 40
Oxbow lake, 1, 733
Oxbows, 241, 244
Oxidation, 733
Ozark Uplift, 81
Ozone, 733

P

P waves, 134, 149, 171, 174, 179, 184,
    192, 195, 196, 224, 227, 232, 376,
    733
Pahoehoe, 367, 733
Paleobiogeography, 733
Paleobiology, 733
Paleoceanography, 733
Paleoclimate, 733
Paleomagnetism, 86, 733
Paleontology, 257, 272, 733
Paleozoic era, 734
Pangaea, 86, 624, 626, 734
Paraffin hydrocarbons, 734
Parallel folds, 250
Parícutin, 679
Parthenogenic, 734
Partial melting, 188

Paterae, 577
Pauling, Linus, 432
Pedestal crater, 734
Pegmatite, 292, 316, 734
Pelagic, 734
Pelitic rock, 398, 734
Penetration funnels, 408
Peralkaline rocks, 318
Perfect spheroid, 218
Peridotite, 30, 196, 224-225, 232, 375,
    655, 734
Periglacial, 498
Perihelion, 734
Period, 734
Permafrost, 498-504, 734
Permeability, 481, 734
Permineralization, 734
Perovskite, 226
Perturb, 734
Petrographic microscopes, 589
Petrography, 734
Petroleum, 481-497, 734
Petrology, 25
pH, 734
Phaneritic, 290, 734
Phanerozoic eon, 734
Phase, 734
Phenocryst, 16, 735
Phillips, John, 119
Photochemical reaction, 735
Photodissociation, 735
Photosynthesis, 735
Phreatic, 62
Phreatic eruption, 685, 735
Phreatoplinian eruption, 685
Phylogeny, 735
Phylum, 735
Pillar, 735
Pillow lava, 367, 735
Pingo, 498
Placer deposit, 735
Plagioclase, 30, 235
Planetary winds, 735
Planetesimals, 412, 735
Plankton, 735
Plasma tail, 735
Plastic deformation, 735
Plate, 86
Plate margins, 318, 506, 735

Plate tectonics, 43, 78, 86, 87, 102, 110, 174, 179, 224, 232, 248, 298, 324, 390, 438, 464, 471, 505-511, 623, 735, 746
Playfair, John, 666
Pleistocene epoch, 389, 498, 735
Pleistocene ice age, 93, 96
Plinian eruption, 735
Plume, 298
Plunge, 248, 736
Pluton, 78, 290, 736
Plutonic rocks, 16, 315
Plutonists, 272, 665
Point bar, 1, 736
Polar axis, 221
Polar jet, 736
Polarization, 736
Pollution, 736
Polygenetic, 577
Polymorphism, 195
Polymorphs, 736
Pore fluids, 736
Porosity, 481, 736
Porphyritic, 290, 736
Porphyritic lava, 317
Porphyry, 16
Potassium-argon dating, 24
Precambrian, 736
Precambrian shields, 81
Precession, 736
Precipitate, 736
Preferred orientation, 736
Pressure-melting temperature, 93, 736
Pressure-temperature regime, 398, 736
Price, Neville, 341
Primary minerals, 736
Primary waves. *See* P waves
Principles of Geology, 669
Prodelta, 126, 736
Prograde metamorphism, 736
Prokaryotic cell, 737
Prospect, 737
Proterozoic eon, 737
Protolith, 398, 737
Protoplanet, 210
Pseudotachylites, 634
Pumice, 512, 599, 685, 737

Punctuated equilibrium, 257, 262, 737
Pycnocline, 737
Pyrite, 363
Pyroclastic fall, 512, 737
Pyroclastic flow, 512, 737
Pyroclastic rocks, 43, 316, 512-517, 577, 599, 685, 737
Pyroclastic surge, 512
Pyroxene, 164, 235, 655

R

Radiation, 737
Radioactive dating, 159
Radioactive decay, 202, 210, 737
Radioactive isotope, 737
Radioactive minerals, 421
Radioactivity, 155, 542, 737
Radiocarbon dating, 23-29
Radiogenic isotope, 202, 737
Radiometric dating, 737
Radiosonde, 737
Ramp, 737
Rampart crater, 737
Ramsay, William, 276
Range zones, 38
Rating curve, 526
Rawinsonde, 737
Rayleigh, Lord, 276
Real time, 737
Reclamation, 738
Recrystallization, 589, 738
Recurrence interval, 738
Red Sea, 327, 626
Reefs, 518-525, 738
Reflected wave, 375
Reflecting goniometer, 434
Reflection, 179, 224
Refraction, 179, 224, 375
Refractory, 738
Refractory (siderophile) elements, 162
Regional geology, 738
Regional metamorphism, 398, 738
Regolith, 738
Regression, 646-654, 738
Relative date, 23, 738
Relative humidity, 738

Remote sensing, 23-24, 254, 738
Reserve, 738
Reservoir, 489, 738
Reservoir rocks, 492, 524
Resistivity, of rocks, 553
Resolution, 738
Resource, 738
Response time, 93, 738
Resurgent dome, 739
Retrograde metamorphism, 739
Retrograde rotation, 739
Reverse fault, 631, 739
Reverse polarity, 739
Revolution, 739
Rhyolite, 368, 599, 739
Richter scale, 151, 739
Rift valley, 638, 739
Rift zones, 83, 324
Rifting, 456, 739
Right-lateral strike-slip, 739
Rigidity, 552
Ring of Fire, 18, 138
Rio Grande Rise, 328
Rittmann, Alfred, 578
River flow, 245, 526-533
River valleys, 243, 534-541
Rivers, 1-7, 241-247; anastomosing, 4
Roche limit, 739
Roche moutonnée, 97
Rock cycle, 665
Rock deformation, 607
Rock magnetism, 542-549
Rock mechanics, 249
Rock reservoirs, 635
Rockfall, 739
Rocks, 30-36, 166, 202-209, 290-296,
    307-315, 340-347, 398-406, 415,
    512-517, 568-576, 584-592, 607-614,
    655-664, 692-708, 739; andesitic,
    16-22; physical properties, 550-558.
    See also specific types
Rodinia, 628
Roll-type deposit, 739
Rotary drilling, 739
Rotation, 607
Rough, 739
Rugose corals, 518
Runoff, 739
Rutherford, Ernest, 204

S

S waves, 134, 149, 174, 192, 195, 196,
    227, 376, 739
S-type granitoid, 745
Sabkha, 739
Saline lake, 740
Salinity, 740
Salt dome, 740
Salt water, 740
Saltation, 559, 740
Saltwater intrusion, 740
San Andreas Fault, 150, 185, 467, 640,
    642
Sand dunes, 559-567
Sandstone, 232, 236, 568, 570,
    584-592, 701, 740
Sapping, 740
Scabland, 740
Scale, 740
Scandinavian Shield, 485
Scarp, 631, 740
Scherrer, Paul, 434
Schist, 404, 740
Schistosity, 740
Scleractinian corals, 518
Sea-floor spreading, 456, 458, 464,
    471, 617, 623, 638, 740
Sea level, 75, 646-654
Sea-level cycle, 646
Seal, 740
Seamounts, 471, 474, 740
Secondary crater, 740
Secondary mineral, 740
Secondary waves. See S waves
Secular geomagnetic variation, 740
Sedgwick, Adam, 275, 596
Sediment, 1-7, 126, 740
Sediment discharge, 741
Sedimentary rocks, 5, 37, 155, 232,
    181, 235, 438, 440, 481, 482, 545,
    568-576, 593, 741
Sediments, 110, 126, 464
Seiche, 741
Seismic, 741
Seismic activity, 471
Seismic belt, 741
Seismic reflection, 456, 741
Seismic refraction, 456

Seismic tomography, 199
Seismic waves, 171, 174, 188, 192, 196, 199, 227, 233, 254, 376, 384, 556, 741. *See also* P waves; S waves
Seismicity, 741
Seismogram, 741
Seismographs, 133, 151, 741
Seismology, 741
Semidiurnal, 741
Semimetals, 741
Sequence, 646
Series, 741
Seston, 741
Shale, 232, 235, 584, 741
Shallow-focus earthquake, 741
Shear, 741
Shear folding, 252
Shear strength, 550, 552, 741
Shear stress, 607
Sheet intrusion, 307
Shelf dams, 741
Shepherd satellites, 741
Shield volcanoes, 43, 577-583, 742
Shields, 102, 105, 404, 482, 578, 742
Shock waves, 407, 742
Shoreline, 75
Short-period comet, 742
Sial, 78
Silica, 315
Silicate minerals, 742
Silicates, 318
Siliceous ooze, 742
Siliciclastic rocks, 584-592
Sill, 307, 742
Sima, 78
Simple crater, 407, 742
Sinkhole, 348, 742
Sinking stream, 348
Sinuous rille, 742
Slate, 405, 742
Slickensides, 448, 742
Slip, 631
Slip-face, 559, 742
Slips, 449
Slump, 742
Smith, William, 38, 264, 275, 595
Smokers, 466, 688, 742
Snell's law, 179
Soddy, Frederick, 204, 278

Sohncke, Leonhard, 431
Soil, 742
Soil profile, 742
Soil resistivity, 25
Solar nebula, 210, 742
Solar system, 742
Solar wind, 210, 742
Solid, 743
Solid solution, 743
Solubility, 743
SONAR, 743
Sonde, 743
Speciation, 257, 743
Species, 257, 743
Specific gravity, 743
Specific heat, 743
Spectroscopy, 743
Speleology, 62
Speleothem, 62
Spillway, 743
Spinel, 226
Spokes, 743
Spontaneous fission, 743
Spore, 743
Spring, 348, 743
Spring sapping, 743
Spring tide, 743
Squall line, 743
Stable isotope, 743
Stage, 743
Standard, 743
Star dune, 559
Steinkern, 744
Steno, Nicolaus, 273, 429, 434, 595
Stensen, Neils, 273
Stocks, 309
Stone polygon, 498
Stope, 744
Storm surge, 744
Strain, 148, 607-614, 744
Strain rate, 744
Strandline, 389, 646, 744
Strata, 37, 118, 593, 744
Stratified, 744
Stratified drift, 281, 744
Stratigraphic correlation, 41, 593-598
Stratigraphic sequence, 272, 744
Stratigraphic time scale, 744
Stratigraphic units, 744

Stratigraphy, 23-24, 37, 53, 203, 593, 744
Stratosphere, 744
Stratovolcanoes, 16, 43, 512, 514, 599-606, 744
Stratus, 744
Streak, 417
Stream equilibrium, 534, 744
Streams, 534-541
Stress, 148, 448-449, 607, 744
Stress and strain, 607-614
Striation, 744
Strike, 631, 744
Strike-slip fault, 744
Strike slips, 449
Stromatolites, 518
Stromatoporoids, 518
Structural geology, 248, 744
Structure, 481
Subalkaline rocks, 318
Subcontinents, 110-117
Subduction, 102, 104, 332, 438-447, 472, 508
Subduction and orogeny, 615-622
Subduction zones, 16, 30, 298, 324, 505, 745
Sublimate, 745
Submarine canyon, 745
Submarine eruptions, 680
Subsidence, 390, 498, 745
Substantive uniformitarianism, 665
Subsurface, 745
Supercontinent cycle, 623-630
Supercontinents, 86-92, 110-117, 623
Supernova, 745
Supersaturated solution, 62
Supratidal, 745
Surface water, 745
Surface wave, 745
Surficial gravels, 665
Suspended load, 745
Suture zones, 82
Swells, 745
Syenite, 745
Symbiosis, 745
Syncline, 248, 745
System, 745

T

Tabulate "corals," 518
Talik, 498
Taphonomy, 257, 745
Tar sand, 745
Target material, 407
Tectonics. *See* Plate tectonics
Tenacity, 415, 419, 746
Tephra, 512, 599, 746
Terminal moraine, 746
Terraces, marine, 389-397
Terrane, 746
Terrestrial planet, 746
Terrigenous, 746
Tertiary period, 118
Test, 746
Tethys, 746
Texture, 398, 746
Thermal conductivity, 746
Thermal cracking, 491
Thermal inversion, 746
Thermal remanent magnetization (TRM), 542, 746
Thermocline, 746
Thermodynamics, 746
Thermohaline circulation, 746
Thermokarst, 498
Thermoluminescence, 25, 27, 746
Thermoluminescence dating, 24
Thermosphere, 746
Thin-section analysis, 20
Thin-sections, 294, 321
Tholeiitic rocks, 318
Tholus, 577
Threshold of movement, 746
Throw, 448
Thrust belt, 747
Thrust faults, 631-637, 747
Tidal disruption, 747
Tidal heating, 747
Till, 8, 281, 283, 747
Time scale, geologic. *See* Geologic time scale
TOC, 493
Tongue stones, 272

Topographic maps. *See* Geologic and topographic maps
Topography, 267-271, 348-356, 747
Topsoil, 747
Tornado, 747
Total dissolved solids (TDS), 747
Toughness, 550, 747
Trace elements, 34
Trace fossil, 257, 747
Trade winds, 747
Trans-Alaskan Pipeline, 503
Transcurrent fault, 638, 747
Transform faults, 505, 506, 638-645, 747
Transgression, 166, 747; and regression, 646-654
Transition metals, 320
Translation, 607
Transpiration, 747
Transverse bar, 1, 747
Transverse valley, 747
Traprock, 30
Traps, 492, 631, 748
Trenches, 475, 748
Triclinic folds, 250
Trilobite, 272
Triple junction, 748
Tropical rain forest, 748
Tropical storm, 748
Tropopause, 748
Troposphere, 748
Trough, 248
Tsunami, 71, 332, 748
Tuffs, 512, 514, 748
Turbidity current, 748
Turbulent flow, 526, 748
Twinning, 748

U

U-shaped valley, 281, 749
Ultramafic, 748
Ultramafic crust, 191
Ultramafic rocks, 655-664
Unconfined aquifer, 748
Unconformity, 272, 646, 748
Underfit stream, 534

Uniformitarianism, 53, 56, 155, 272, 670, 748
Uniformity of nature, 665
Unreinforced masonry (URM), 748
Unstable air, 748
Uplift, 390, 749
Upper mantle, 749
Upwelling, 749
Ural Mountains, 105

V

Vadose, 62
Valley glaciers, 8
Valleys, 534-541
Van der Waals force, 749
Varves, 158, 363, 749
Vector, 749
Vein, 749
Vent, 749
Ventifact, 749
Vesicular, 749
Vine, F. J., 642
Viscosity, 16, 599, 749
Volatile elements, 162
Volatiles, 512, 692, 749
Volcanic earthquakes, 749
Volcanic explosivity index, 686
Volcanic island chains, 297-306
Volcanic rocks, 16-22, 749
Volcanic tremor, 749
Volcanism, 53
Volcanoes, 43-52, 84, 168, 182, 235, 308, 324-331, 367-374, 478, 514, 577-583, 599-606, 749; recent eruptions, 675-684; types of eruption, 685-691. *See also* Shield volcanoes; Stratovolcanoes
Vortex, 749
Vulcanists, 54

W

Walker, George, 686, 688
Walvis Ridge, 328
Warm temperate glacier, 93, 749
Water table, 749

Water-rock interactions, 141-147, 348-356, 692-700
Waterspout, 750
Wave base, 71, 750
Wave energy, 126
Wave refraction, 71, 750
Wave-cut bench, 390, 750
Waves, 73
Weathering, 344, 701, 750
Weathering and erosion, 584, 701-708
Wegener, Alfred L., 87, 624, 625, 641
Well log, 750
Werner, Abraham, 595
Wet-climate geomorphology, 750
Whitford-Stark, James, 578
Wilson, J. Tuzo, 626, 642

Wilson cycle, 623
Wollaston, William Hyde, 434

X

X ray, 428, 750
X-ray diffraction, 27, 428
X-ray fluorescence, 321
Xenobiotic, 750
Xenocrysts, 750
Xenoliths, 750

Z

Zircons, 162
Zone of saturation, 750